Sustainable Machining and Green Manufacturing

Scrivener Publishing
100 Cummings Center, Suite 541J
Beverly, MA 01915-6106

Publishers at Scrivener
Martin Scrivener (martin@scrivenerpublishing.com)
Phillip Carmical (pcarmical@scrivenerpublishing.com)

Sustainable Machining and Green Manufacturing

Edited by

S. Thirumalai Kumaran

Department of Mechanical Engineering, PSG Institute of Technology and Applied Research, India

and

Tae Jo Ko

Department of Mechanical Engineering, Yeungnam University, South Korea

Scrivener Publishing

WILEY

This edition first published 2024 by John Wiley & Sons, Inc., 111 River Street, Hoboken, NJ 07030, USA and Scrivener Publishing LLC, 100 Cummings Center, Suite 541J, Beverly, MA 01915, USA

For more information about Scrivener publications please visit www.scrivenerpublishing.com.

Wiley Global Headquarters
111 River Street, Hoboken, NJ 07030, USA

For details of our global editorial offices, customer services, and more information about Wiley products visit us at www.wiley.com.

Library of Congress Cataloging-in-Publication Data

ISBN 978-1-394-19783-5

Cover image: Pixabay.Com
Cover design by Russell Richardson

Set in size of 11pt and Minion Pro by Manila Typesetting Company, Makati, Philippines

Printed in the USA

10 9 8 7 6 5 4 3 2 1

Contents

Preface

In an era defined by rapid technological advancements and an increasing awareness of environmental sustainability, the intersection of science and industry takes on a new dimension. From within this context, we delve into the diverse and compelling world of composite materials and sustainable manufacturing. In the following chapters, we embark on a journey through the realms of innovation, exploring the pivotal role of science and technology in reshaping our industries and fostering a more sustainable future.

The chapters assembled herein are a testament to the dedication and ingenuity of researchers and professionals alike, who have tirelessly pursued groundbreaking discoveries and practical solutions to some of the most pressing challenges facing the manufacturing sector. These chapters are thoughtfully organized to highlight the profound impact of various materials, techniques, and processes on both the performance of products and the preservation of our planet.

Chapter 1 investigates the intriguing synergy between natural fibers and epoxy composites, shedding light on how filler materials can enhance mechanical properties. Chapter 2 explores the potential of sustainable reinforcements in polymer composites, demonstrating the adaptability of these materials for diverse applications. Chapter 3 uncovers the critical role that manufacturing methods play in determining the mechanical prowess of bio fiber-reinforced composites.

Chapter 4 takes us into the realm of biomedical manufacturing, where advanced composite materials are reshaping the future of medical devices. Chapter 5 underscores the importance of environmentally conscious manufacturing processes, proving that green practices can coexist with industrial production. Chapter 6 delves into the innovative world of additive manufacturing and highlights the intricacies of composite filament production.

Chapter 7 explores the delicate balance between material selection and joining techniques, with an emphasis on sustainability in the manufacturing process. Chapter 8 introduces us to the ingenious use of natural materials in addressing environmental challenges, highlighting the significance of sustainable wastewater treatment. Chapter 9 underscores the role of welding in sustainable manufacturing practices, bridging the gap between tradition and innovation.

Chapter 10 offers a glimpse into the future of robotics, where sustainability plays a central role in engineering design. Chapter 11 provides an insightful overview of green manufacturing practices in the automotive industry, a sector undergoing profound transformation. Chapter 12 takes us on a journey toward waste reduction, demonstrating how green principles can optimize manufacturing processes.

Chapter 13 explores the synergy between design and sustainability in additive manufacturing, illustrating the potential for minimizing waste and energy consumption. Chapter 14 delves into the intricacies of process optimization in additive manufacturing, emphasizing efficiency and precision. Chapter 15 brings us to the world of precision machining, where cutting-edge technologies are transforming the way we work with materials.

It is our hope that this collection will inspire you to join the ranks of those committed to a more sustainable, efficient, and environmentally conscious future in manufacturing. Each chapter stands as a testament to the transformative power of science, technology, and the human spirit.

In closing, we extend our heartfelt appreciation to each author whose research and insights have made this book possible. Thank you for sharing your knowledge and expertise with us and with the broader community. Finally, we offer our sincere thanks to the Scrivener and Wiley publishing teams for their help with this book.

S. Thirumalai Kumaran
Tae Jo Ko
December 2023

1

Effect of Granite Filler on Mechanical Properties and Free Damping of Silk-Sisal–Reinforced Epoxy Composites

K. Sripriyan[1]* and S. Karthick[2]

[1]School of Mechanical Engineering, Vellore Institute of Technology, Bhopal, Madhya Pradesh, India
[2]School of Electrical and Electronics Engineering, VIT Bhopal University, Bhopal, Madhya Pradesh, India

Abstract
This paper discusses the effect of granite filler on the mechanical properties (flexural strength and impact strength), biodegradable characteristics, and damping behavior of silk-sisal composites. The hand layup method is used to fabricate the three layers of plain weave mat with varying weight percentages (2, 4, and 6 wt.%) of granite filler hybrid structure. According to the experimental results, the flexural strength value of the developed silk-sisal hybrid composites increases as the weight percentage of the granite filler content increases. In addition, adding 6 wt.% of granite filler content, which is approximately 75.8% higher damping than that of pure silk-sisal. The biodegradable behavior of silk-sisal hybrid composite has also been analyzed and well discussed in this paper.

Keywords: Silk-sisal fiber, granite filler, properties, damping characteristic, biodegradable

1.1 Introduction

Over the past 30 years, the development of materials to enhance the mechanical properties of low-carbon steels has been dominated by plastics, ceramics, and composites [1]. The quantity and variety of commercial

**Corresponding author*: sripriyan.k@vitbhopal.ac.in

S. Thirumalai Kumaran and Tae Jo Ko (eds.) Sustainable Machining and Green Manufacturing, (1–12) © 2024 Scrivener Publishing LLC

applications for composite materials have grown consistently as the substance has aggressively entered and dominated recent markets. Modern hybrid composite materials make up a sizable portion of the market for engineered materials, and they are used in everything from straightforward everyday objects to intricate specialized applications. Even though composites have already shown their value as materials that reduce weight, the current challenge is to make them cost-effective [2]. The composite sector is currently utilizing a number of cutting-edge production methods created to produce economically appealing composite components [3]. Improvements in manufacturing technology alone cannot overcome the cost barrier, particularly for composites. Composites must go through a number of integrated steps, including design, material, process, tooling, quality assurance, manufacturing, and even program management, in order to compete with metals [4, 5].

Natural fibers are resources that are reusable, affordable, and friendly to the environment and have gained popularity recently [6]. They have a history of being used as fundamental fillers in automobile parts. Natural fibers are increasingly being used in composite construction in place of other conventional materials because of environmental concerns [7, 8]. Due to their excellent mechanical properties, among other things, fibers like jute, silk, flax, banana, kenaf, sisal, and bagasse are frequently used in the fabrication of natural fiber–reinforced composites [9]. This proposed study aids in the creation of a hybrid composite made of silk and sisal to swap out conventional bumper materials.

Banana-coir-bagasse fibers were studied by Yan *et al.* [10] in hybrid composites. Among other things, they examined the amount of moisture, amount of ash and carbon, tensile strength, elemental analysis, and chemical analysis. Banana fiber had the highest levels of ash, carbon, and cellulose per the test results. Sanjeevi [11] studied the various research projects that researchers had been working on involving banana fiber composites. The majority of papers discussed the composition, mechanics, and metallurgy of reinforced fiber and banana composites.

Recent researchers were involved with the development of various hybrid composite structures, *viz.*, silk-sisal fibers, banana-sisal, glass-bamboo, and glass-sisal. All of these composites pose better mechanical, chemical, and metallurgical properties [12, 13]. Still, there may be a gap to further enhance the damping prosperity in the silk-sisal composite.

In this present work, an experimental study has been conducted to fabricate the 3-mm composite structure. Silk and sisal fiber are selected as a constant length of 25 mm and weight percentages of 60:40. Furthermore, to enhance the mechanical and damping behavior, granite nanofillers were added at different weight percentages (2%, 4%, and 6 wt.%). The novelty of the work is the fabrication of 3-mm silk-sisal added with granite nanofiller by hand layered structure. Further, the paper addresses the fundamental findings of hardness, flexural, impact properties, and damping behavior.

1.2 Material and Preparation

1.2.1 Materials Involved

The epoxy-based resin (LY556 grade) and hardener (HY951 grade) used in the study's matrix (10:1 blending ratio) are reinforced with silk-sisal. Silk is a very strong fiber, and it has a higher tensile strength of about 75%–85%. Its extensibility is about 15% when at the break [14]. Sisal is a plant fiber, and it has high strength and the ability to stretch, and it has an elongation of about 3.2% when at break [15]. Table 1.1 [16, 17] lists the characteristics of the hybrid composite made of silk, banana, and sisal fibers. The outcomes of the proposed silk-sisal hybrid composite are contrasted with those of the sisal-banana composition that has previously been studied [18].

Epoxy resin (LY556) has several desirable characteristics, including high strength, minimal shrinkage, strong adhesion to a variety of substrates, efficient electrical insulation, chemical insolvency resistance, low cost, and low toxicity. High compression strength and long-lasting durability are among the resin's attributes [1]. The hardener (HY 961) is an epoxy or fiberglass curing agent. The component that hardens the adhesive when combined with resin is known as a hardener, sometimes known as a catalyst, and epoxy resin must begin curing. The characteristics include excellent electrical qualities, better mechanical strength, and resistance to atmospheric and chemical deterioration [19].

Table 1.1 Silk, banana, and sisal fiber properties.

S. no.	Name of the composite	References	Tensile strength (MPa)	Flexural strength (MPa)	Impact strength (J)	Damping
1	Banana	4, 7	15.60	68.80	1.60	0.038
2	Sisal	6, 23	24.50	80.45	1.76	0.034
3	Sisal and banana	3, 5, 28	18.50	57.47	1.71	0.035
4	Silk	2, 23, 29	19.20	51.96	1.25	0.33

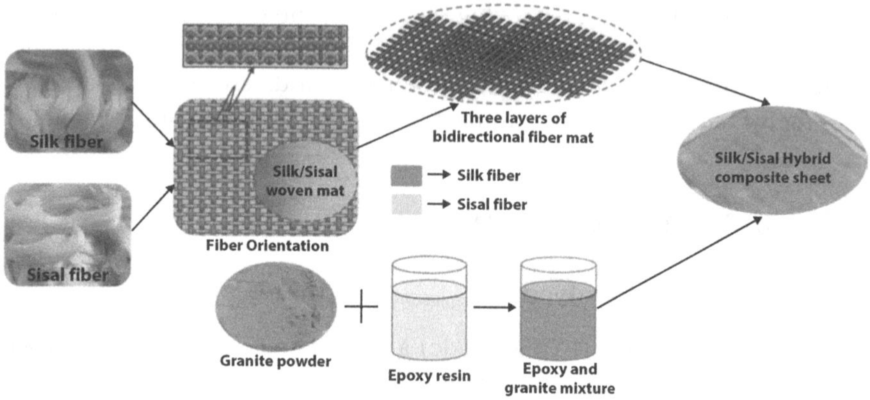

Figure 1.1 Steps involved in the preparation of the silk-sisal hybrid composite structure.

1.2.2 Composite Structure Preparation

The fibers are initially cleaned several times with regular water, then left to air-dry for a day. The fibers are soaked in the NaOH solution for a day in order to remove the excess debris. After washing with clean water, the clothes are dried in a hot oven for 3–4 hours at 70°C. For the purpose of molding the composites, the fibers are divided into 300-mm lengths. The composite (three-layer bidirectional fiber mat) structure is prepared using the manual hand layup open molding technique. Following that, weft yarns are woven over and beneath the warp yarns. According to Figure 1.1, the warp in this method of weaving fiber is equivalent to silk fiber, and the weft is equivalent to sisal fiber. Three layers of a unidirectional mat made of silk and sisal fibers are used to create the composite.

Bidirectional mats made of silk-sisal fiber have been laid out in a cross-lay pattern, layering up in a 90° orientation. All of the reinforcement's layers were covered with epoxy resin. In order to ensure an improved interaction between the reinforcement and the matrix, facilitate a uniform resin distribution, and achieve the required thickness of 3 mm, rollers are then manually used to roll over the wet composite. The hybrid composites are then allowed to cure in a typical atmosphere.

1.3 Result and Discussion

By using a hand layup technique, a hybrid silk-sisal composite structure (300 mm × 300 mm) was developed. To determine the significance, all samples are prepared in accordance with the ASTM standard for mechanical testing, specifically the flexural strength and impact strength. In addition, the degradation characteristics were discussed to explore the selection of one-time usable biodegradable materials. Finally, a morphological study was performed to understand the fracture behavior of the developed composite structure.

1.3.1 Silk-Sisal on Mechanical Properties

1.3.1.1 Flexural Strength

One of the testing techniques used to calculate the rupture behavior of a composite structure is flexural strength [7, 15]. Granite nanofillers on silk-sisal hybrid samples were made in accordance with ASTM: D790, and a maximum bending load of 5 mm/min was applied to the samples until they fractured or broke. Table 1.2 lists the average flexural and impact values of pure (silk and sisal) and hybrid (silk, sisal, and granite filler) composites.

Table 1.2 Properties of developed silk-sisal composite.

S. no.	Name of the composite	Weight %	No. of samples	Flexural strength (MPa)	Impact strength (J)	Damping
1	Silk-sisal reinforced composite	No filler	1	84.60	0.90	0.37
			2	91.90	0.50	0.31
			3	97.60	0.80	0.37
			Average	**91.37**	**0.733**	**0.033**
2		2 wt.% of filler	1	117.90	0.70	0.04
			2	102.80	0.80	0.03
			3	81.64	0.90	0.03
			Average	**100.78**	**0.80**	**0.036**
3		4 wt.% of filler	1	114.90	0.60	0.06
			2	99.78	0.80	0.03
			3	123.90	0.70	0.03
			Average	**112.86**	**0.70**	**0.043**
4		6 wt.% of filler	1	101.5	0.60	0.03
			2	99.25	0.90	0.03
			3	97.28	0.70	0.40
			Average	**99.34**	**0.733**	**0.058**

The variations in flexural strength of silk-sisal as a function of the weight percentage of granite nanofiller content are investigated, and the average value of pure silk-sisal is 91.37 MPa, whereas granite nanofiller weight percentages of 2, 4, and 6 on silk-sisal were found to result in 100.78 MPa, 112.86 MPa, and 90.63 MPa, respectively. Figure 1.2(a) displays the stress-strain curve's reasonably responsible plot for each of the four compositions. It was noted that bell-shaped trends rather than increased flexural strength are represented by increasing the weight percentage of granite nanofillers.

However, adding granite nanofillers from 2 to 4 wt.%, which is about 10.7% higher, causes a drastic change in flexural strength, while changing from 4 to 6 wt.% causes a sharp decrease of about 19.6%. The experimental results evidenced that silk-sisal may be an alternative hybrid composite of banana-sisal to enhance the flexural strength, even without adding granite nanofillers. The current findings showed that pure silk-sisal had a 1.25 times lower flexural strength value exposed at a 4 wt.% content of granite nanofillers (without nanofillers). Similarly, for 2 and 6 wt.% of granite nanofiller content, respectively, flexural strength values are 1.1 times higher and remain unchanged. The desired flexural strength is found to be significantly influenced by the content of granite nanofillers, which is 4 wt.%.

1.3.1.2 Impact Strength

ASTM: D481 examines the impact competence of the unnotched hybrid composite under the unexpected applied load for four different sample compositions. Pure silk-sisal had an average impact strength of 0.75 J,

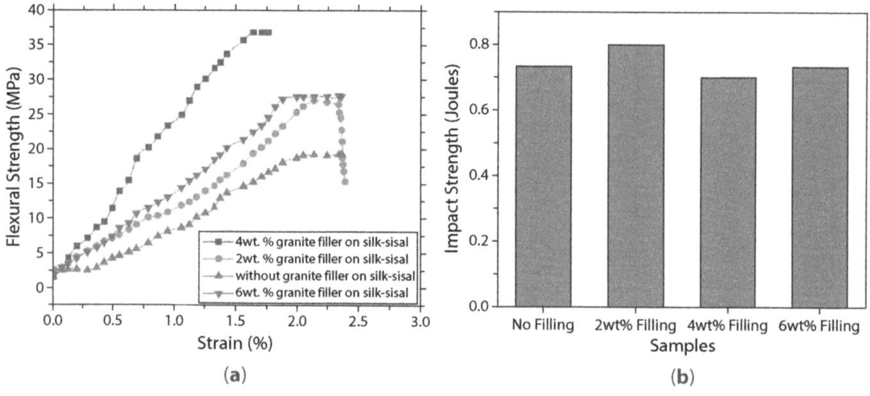

Figure 1.2 Stress vs. strain curve (a) flexural strength, (b) impact strength.

while granite nanofillers with weight percentages of 2, 4, and 6 on silk-sisal had values of 0.83 J, 0.6 J, and 0.6 J, respectively. Figure 1.2(b) displays the relatively responsible plots for all four compositions.

However, adding granite nanofillers results in a significant variation in impact strength, which is about 27.7% less by adding 2 to 4 wt.% compared to no change when adding 4 to 6 wt.%. On the other hand, there has been an improvement in the bond between the fiber mat and the matrix (resin + granite nanofiller). The impact results showed that pure silk-sisal had 1.1 times lower impact strength value than granite nanofillers at 2 wt.% content (without nanofillers). Similarly, for 2 and 6 wt.% of granite nanofiller content, 0.99 times lower and the same impact strength values are observed.

1.3.2 Damping Response

According to a literature review, cantilever samples are made to the ASTM: E756 standard, which is 250 × 25 × 3 mm [20]. An accelerometer attached to the free end of the hybrid composite structure was used to continuously monitor and record the free vibration responses, and the damping ratio (Zeta) was calculated using the experimental systems' logarithmic decrement [21]. The motion is based on Equation 1.1.

$$x(t) = X\, e^{-\zeta\omega_n t} sin\, sin(\omega_d t - \varphi) \tag{1.1}$$

According to the aforementioned equation, the hybrid composite structure oscillates at a frequency of $\omega_d = \left(\sqrt{1-\zeta}\right)\omega_n$. Zeta being less than 1, ω_d always being less than ω_n. Where ζ is the ratio of two successive oscillations, is the damping factor that needs to be calculated, n is the natural frequency of the developed hybrid composite structure, X and φ are the amplitude and phase angle response at initial conditions. Equation 1.1 can be used to express free vibration, and Equation 1.2 can be used to determine the actual damping ratio.

$$\delta = \frac{x_n}{x_{n+1}} = exp\left(\frac{2n\zeta\pi}{\sqrt{1-\zeta^2}}\right) \tag{1.2}$$

The experimental test was carried out by providing an initial displacement and recording the ensuing response; more information is shown in Figure 1.3(a). Based on the aforementioned equation, the chosen peak

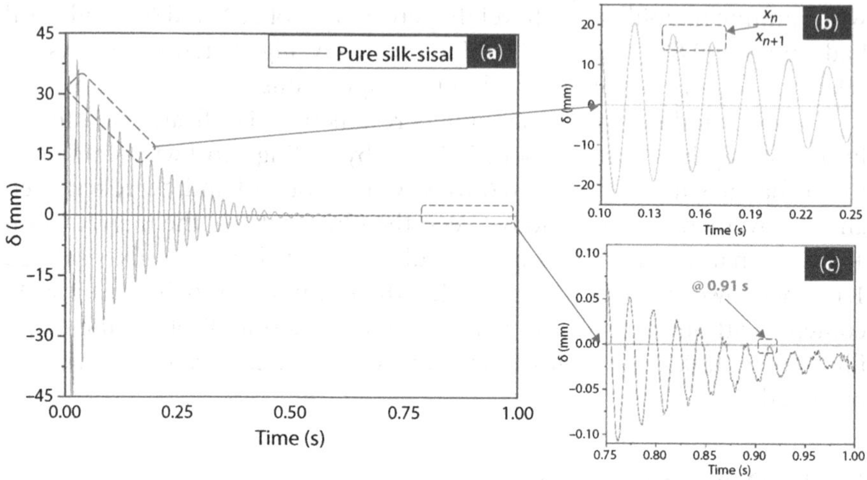

Figure 1.3 Typical vibration response plot for pure silk-sisal (a) experimental data standardized to peak with time series, (b) selected successive peaks for damping factor, (c) time at which frequency gets zero or negative.

with a constant time bound for all samples was highlighted in Figure 1.3(b) [8]. To reduce uncertainty, the average value was obtained from the three trails, and the damping ratio was calculated from the chosen two successive peaks. The time at which frequency enters the zero or negative region was shown in Figure 1.3(c).

The weight percentage of the granite nanofiller increases, so does the damping factor on the silk-sisal hybrid composite. However, a higher damping factor was obtained by incorporating 6 wt.% of granite nanofiller content, which is roughly 1.77 times more than pure silk-sisal (without nanofillers). Similar outcomes with 1.09 times and 1.29 times higher damping values are obtained for granite nanofiller contents of 2 and 4 wt.%, respectively. As a result, it was found that damping was more affected by the granite nanofiller's 6 wt.% content.

1.3.3 Fracture Morphology

The mechanical responses of flexural and impact strength of pure silk-sisal and different weight percentages of granite nanofiller compositions were studied and compared with previous research results on sisal-banana. Generally, sisal-banana compositions showed an improved tensile strength. Silk-sisal has clearly provided the ideal composition to replace sisal-banana. A clear trend was observed for silk-sisal composition to increase tensile

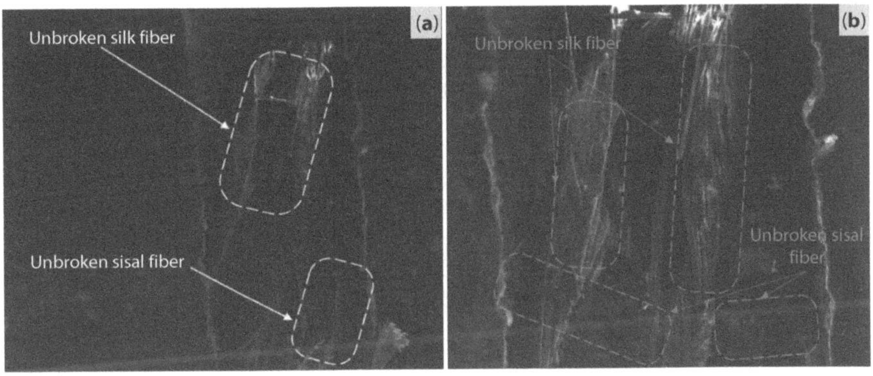

Figure 1.4 Flexural fracture microstructure (a) pure silk-sisal, (b) 6 wt.% granite nanofiller on silk-sisal.

strength than sisal-banana composition. The highest tensile breaking energy was also found in all granite nanofiller content compositions. Similarly, flexural strength was 1.59, 1.75, 1.98, and 1.73 times higher, while impact strength was 0.43, 0.79, 0.41, and 0.79 times lower. Based on the damping and mechanical responses of pure silk-sisal and 6% granite nanofiller compositions, the fracture morphology was discussed further.

The study's goal is to use existing research findings to replace sisal-banana in automobile applications. Flexural strength is an important factor in determining the functionality of any material. Flexural fractography images (pure silk-sisal and 6% granite nanofiller) are used for further analysis, as shown in Figure 1.4(a-b).

In general, silk and sisal have better adhesive properties with epoxy, and the silk-sisal composition has good mechanical properties. Crack propagation under the loading roller, on the other hand, could be associated with the flexural failure of pure silk-sisal and 6 wt.% granite nanofiller. Silk-sisal compositions with higher bending deformation and ductile failure nature are observed. Unbroken fibers and some broken silk fiber and voids were also observed.

1.3.4 Biodegradability

The samples were submerged in the soil for 144 hours while the weight and morphology of the films were continuously monitored. When the time spent in the soil was extended, it was discovered that the weight loss percentage of each silk-sisal fiber (with and without nanofillers) steadily increased, indicating the fiber's biodegradable qualities (Figure 1.5).

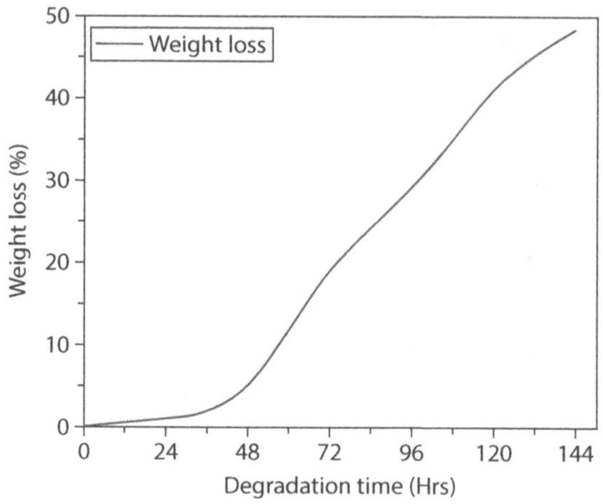

Figure 1.5 Weight reduction in the percentage.

Microorganisms found in soil and on the cutting-edge surfaces attack gelatin as it degrades. The amount of the antibacterial compound affected how quickly the strength of the silk-sisal composite was lost. Every sample had a smooth surface prior to the biodegradation test. According to the roughness and hole formation on the film surface, which were caused by the enzymatic activity of the microorganisms in soil, the samples started to degrade after 144 hours of soil burial testing. In addition, the silk-sisal hybrid composite will be consumed as food by invertebrates and insects. Termites', insects', and rodents' gnawing activities also cause mechanical damage.

1.4 Conclusions

On a silk-sisal hybrid composite, the impact of various granite nanofiller weight percentages is investigated. The conclusions are as follows:

- The developed silk-sisal hybrid composite's flexural strength increased as the weight percentage of granite nanofillers increased in comparison to pure silk-sisal. By adding 6 wt.% of granite nanofillers to silk-sisal, the flexural strength value was improved by about 9%.
- In addition, adding granite nanofillers to silk-sisal does not change the impact behavior in an acceptable way; it responds similarly to pure silk-sisal hybrid composite. This is because

the composition of silk and sisal has a higher stiffness value, which can reduce impact strength and control shearing behavior on fiber mats.

- By adding 6 wt.% of granite nanofiller content—which is roughly 75.8% higher than pure silk-sisal—a higher damping factor was achieved. Similar percentages of 9% and 30%, respectively, for 2 wt.% and 4 wt.% of granite nanofiller content as a result, it was discovered that the granite nanofiller content of 6 wt.% had a greater impact on the structure of the silk-sisal hybrid composite.
- Biodegradable tests were used to study the hybrid silk-sisal composite's physical characteristics. The biodegradability test yielded acceptable degradation according to the results. Therefore, one of the better options to improve the damping properties is a developed silk-sisal hybrid composite added with granite nanofillers.

References

1. Sujon, M.A.S., Habib, M.A., Abedin, M.Z., Experimental investigation of the mechanical and water absorption properties on fiber stacking sequence and orientation of jute/carbon epoxy hybrid composites. *J. Mater. Res. Technol.*, 9, 5, 10970–10981, 2020.
2. Doomra, A., Sandhu, S.S., Singh, B., Effect of post weld heat treatment on metallurgical and mechanical properties of electron beam welded AISI 409 ferritic steel. *Metall. Mater. Eng.*, 26, 3, 279–292, 2020.
3. Badrinath, R. and Senthilvelan, T., Comparative investigation on mechanical properties of banana and sisal reinforced polymer based composites. *Proc. Mater. Sci.*, 5, 2263–2272, 2014.
4. Thyla, P.R., Sripriyan, K., Prashanth, M., Ramu, M., Analysis of molten pool characteristics on hybrid GMAW process. *Asian J. Res. Soc. Sci. Humanit.*, 7, 10, 142–154, 2017.
5. Davoodi, M.M., Sapuan, S.M., Ahmad, D., A. Ali, Khalina, A., Jonoobi, M., Mechanical properties of hybrid kenaf/glass reinforced epoxy composite for passenger car bumper beam. *Mater. Des.*, 31, 10, 4927–4932, 2010.
6. Karuthapandi, S., Ramu, M., Thyla, P.R., Effects of the use of a flat wire electrode in gas metal arc welding and fuzzy logic model for the prediction of weldment shape profile. *J. Mech. Sci. Technol.*, 31, 5, 2477–2486, 2017.
7. Karuthapandi, S., Ramu, M., Palani, P.K., Study and analysis of the macrostructure characteristics in FCAW with the use of a flat wire electrode and by optimizing the process parameter using the Taguchi method and regression analysis. *High Temp. Mater. Process.*, 20, 3, 2016.

8. Kuppuraj, A. and Angamuthu, M., Investigation of mechanical properties and free vibration behavior of graphene/basalt nano filler banana/sisal hybrid composite. *Polym. Polym. Compos.*, 30, 1–12, 2022.
9. Muthalagu, R., Murugesan, J., Kumar, S.S., Babu, B.S., Tensile attributes and material analysis of kevlar and date palm fibers reinforced epoxy composites for automotive bumper applications. *Mater. Today Proc.*, 46, 433–438, 2021.
10. Yan, Z., Shi, B., Sun, L., Xiao, J., Surface defect detection of aluminum alloy welds with 3D depth image and 2D gray image. *Int. J. Adv. Manuf. Technol.*, 110, 3–4, 741–752, 2020.
11. Sanjeevi, S., Effects of water absorption on the mechanical properties of hybrid natural fibre/phenol formaldehyde composites. *Sci. Rep.*, 11, 1, 1–11, 2021.
12. Zhang, Y.Q., Zhang, H.Q., Li, J.F., Liu, W.M., Effect of heat input on microstructure and toughness of coarse grain heat affected zone in Nb microalloyed HSLA steels. *J. Iron Steel Res. Int.*, 16, 5, 73–80, 2009.
13. Negawo, T.A., Polat, Y., Akgul, Y., Kilic, A., Jawaid, M., Mechanical and dynamic mechanical thermal properties of ensete fiber/woven glass fiber fabric hybrid composites. *Compos. Struct.*, 259, 113221, 2021.
14. Akash, S., Avinash, S., Ramachandra, M., A study on mechanical properties of silk fiber reinforced epoxy resin bio-composite with siCas filler addition. *Mater. Today Proc.*, 5, 1, 3219–3228, 2018.
15. Arthanarieswaran, V.P., Kumaravel, A., Kathirselvam, M., Evaluation of mechanical properties of banana and sisal fiber reinforced epoxy composites: Influence of glass fiber hybridization. *Mater. Des.*, 64, 194–202, 2014.
16. Mirmohseni, A. and Zavareh, S., Preparation and characterization of an epoxy nanocomposite toughened by a combination of thermoplastic, layered and particulate nano-fillers. *Mater. Des.*, 31, 6, 2699–2706, 2010.
17. Sripriyan, K., Ramu, M., Thyla, P.R., Anantharuban, K., Weld bead characterization of flat wire electrode in GMAW process part II: A numerical study. *J. Mech. Sci. Technol.*, 35, 5, 1–8, 2021.
18. Sumesh, K.R. and Kanthavel, K., Effect of TiO_2 nano-filler in mechanical and free vibration damping behavior of hybrid natural fiber composites. *J. Braz. Soc. Mech. Sci. Eng.*, 42, 4, 2020.
19. Sripriyan, K., Ramu, M., Thyla, P.R., Anantharuban, K., Karthigha, M., Characteristic of weld bead using flat wire electrode in GMAW inline during the process: An experimental and numerical analysis. *Int. J. Press Vessel. Pip.*, 196, 2022.
20. Kumar, K.S., Siva, I., Jeyaraj, P., Jappes, J.T.W., Amico, S.C., Rajini, N., Synergy of fiber length and content on free vibration and damping behavior of natural fiber reinforced polyester composite beams. *Mater. Des.*, 56, 379–386, 2014.
21. Kumar, G.S., Chandrasekaran, M., Kumar, T.V., Vivek, P., Vibration and damping characteristics of sisal and glass fiber reinforced polyester composite. *Int. J. Eng. Technol.*, 7, 33, 503–509, 2018.

2

Effect of Plastic Particulate Addition on Polymer Composite Reinforced with *Prosopis juliflora* Fiber

Sakthi Balan G.[1], Aravind Raj S.[1]*, Jafrey Daniel James D.[2] and Ramesh M.[3]

[1]*Department of Manufacturing Engineering, School of Mechanical Engineering, Vellore Institute of Technology, Vellore, Tamil Nadu, India*
[2]*Department of Mechanical Engineering, K. Ramakrishnan College of Engineering, Tiruchirappalli, Tamil Nadu, India*
[3]*SRM Institute of Science and Technology, Ramapuram Campus, Ramapuram, Chennai, Tamil Nadu, India*

Abstract
The necessity of reusing waste products grows with each passing day. "Sustainability" is a term used in nearly all fields, particularly in the materials industry. Plastics are indispensable in the modern era, but the disposal of used plastics has the greatest impact on the environment. Recycling and reusing used plastic waste are one of the safest ways to address this problem. In this study, *Prosopis juliflora* fibers are used as reinforcement, while used plastic particles are used as filler in a polymer composite. The input factors that influence the composite's mechanical properties are selected and varied to determine the optimal fiber and filler composition for getting the desired composite's mechanical properties. The composites were evaluated for their water absorption, tensile, and hardness properties, and the responses were optimized using Plackett–Burman design and simple regression. The results indicate that the incorporation of recycled plastics affects the hardness and water absorption properties of composites while fibers enhance their tensile properties.

Keywords: Plackett–Burman, regression analysis, *Prosopis juliflora*, particulates of plastics, chemical treatment

**Corresponding author*: aravindsakthivel@hotmail.com

S. Thirumalai Kumaran and Tae Jo Ko (eds.) Sustainable Machining and Green Manufacturing, (13–32) © 2024 Scrivener Publishing LLC

2.1 Introduction

The population of the modern world is increasing daily. Due to the rise in population density, people must rely on nature to survive. The rate of depletion of nature's resources is increasing. Modernization and industrialization have created a great deal of pollution, which has wreaked havoc on nature. As people become increasingly dependent on plastics, their use continues to rise. Even though developed countries are taking numerous steps to combat pollution, the pollution rate continues to rise. Used plastics are dumped as waste in the oceans and seas. In some locations, it is burned, causing severe air pollution and health issues. The only way to reduce this is to stop using plastics and to reuse and recycle the plastics that have already been produced. This causes the degradation of the planet and a disruption of the ecological cycle. Due to the escalating growth rate of *Prosopis juliflora* (PF) plants, the water table is decreasing in numerous locations [1]. The majority of the area is covered with PF plants, and the only feasible use for it is to provide fossil fuel to the villagers in the surrounding communities [2]. Aspect ratio and fiber loading influenced the mechanical characteristics of composites manufactured from *Prosopis juliflora* fiber reinforced in a phenol formaldehyde matrix. The incorporation of *Prosopis juliflora* into the matrix was seen to result in an improvement of the material's mechanical properties [3]. Rice husk and PF fiber–reinforced biocomposites were developed to reveal their mechanical properties. Compared to an unreinforced composite, they discovered that the incorporation of rice husk improved the material's mechanical qualities by 45% [4]. The raw and alkali treatment of the PF fibers were characterized, and the thermal, mechanical, morphological, and electrical properties of the PF fibers were reported [5]. The mechanical properties of a PF fiber–reinforced hybrid composite can be modified by making changes to both the chemical treatment and the loading of fibers. They discovered that the modified *Prosopis Juliflora* fibre (PJF) contributed to enhanced matrix-reinforcement adhesion. This improvement was found to be significant. This composite material, which is relatively lightweight, was suggested for use in the automotive industry due to the advantages it affords in that regard [6]. It was found that chemical treatment improves the mechanical properties of PF hybrid composites [7]. The hybrid composite's mechanical properties were found to be suitable for structural applications. The hybrid composite is made up of PF and glass fiber that is reinforced in epoxy [8]. Fibers were extracted from mesquite plants, and it was revealed that these fibers might be utilized in the creation of composites, the manufacture of paper, and the textile sector [9]. The research on the moisture absorption characteristics of the jute fiber–reinforced composite filled with waste plastics found that the addition of waste plastic particles will enhance the material's moisture resistance abilities [10].

To create polymer-based composites with the assistance of fillers and reinforced particles, the hand layup process is a common and widely utilized approach [11]. Researchers developed porous carbon and reported that this can be used for storage batteries as a surface coating and that this could be useful to boost the charge capacity [12]. Kenaf fiber and banana fiber–reinforced polyester composite properties were examined, and it was discovered that the NaOH and SLS-treated samples showed improved properties and strong interfacial bonding than the untreated samples [13]. Jute, sisal, flax, hemp, pineapple, cotton, kenaf, E-glass, and carbon are just some of the many types of fibers that have been employed as reinforcements in the development of a new class of polymer-based materials that are intended to replace other conventional materials [14, 15].

In light of the aforementioned research, the current work is to develop a polymer-based composite by reinforcing waste plastics with PF fibers. Additionally, it will investigate the water absorption, hardness, and tensile strength of the composites. Plackett–Burman design and regression analysis have been utilized in order to maximize the impact that input parameters have on outcomes.

2.2 Materials and Methods

The typical method of obtaining PF fibers involves removing the moisture from the bark of *Prosopis juliflora* plants, which is then followed by drying the bark in the sunlight. Since it is made of natural fibers, it needs to be processed before it can be set to use as a reinforcing material. Therefore, it is processed using various chemicals like alkalis and silanes [16]. The extracted fibers are immersed in NaOH 5% (w/v) [17] for 1 hour and washed with HCl solution to attain a neutral pH value. After having a portion of the fiber immersed for an hour in triethoxy vinyl silane solution with a concentration of 5% (v/v), the solution is then rinsed away with deionized water and heated in an oven [18–20]. The fibers are removed from the bark of the *Prosopis juliflora* trees, then subjected to a chemical treatment that enhances its mechanical properties, and finally, it is used to strengthen a polymer composite material [15]. After being collected, crushed, and processed into plastic particulates, the waste plastics are then put to use as a filler in the composite. As a filler material, polypropylenes were selected from the wide variety of polymers available because they are resistant to degradation. After being recovered from the waste plastics, the items manufactured of polypropylene are separated out into a separate pile, where they are then cleaned to eliminate the sand or other foreign matter. After that, the waste plastics are pulverized and reduced to short particles. It undergoes repeated crushing in order to keep the same particle size throughout the final product [21, 22]. Resin and hardener are combined in the proportions specified by the producer of the product.

There are a plethora of tools and techniques available for use in the production of composites. Many factors, such as the types of fibers and fillers employed, the orientation of the fibers, the fiber volume, the number of layers, the matrix material, and the curing period, dictated the production process employed. Polymer matrix composites are often fabricated *via* hand layup, spray layup, compression molding, resin transfer molding, autoclave molding, or other comparable techniques. Because the resin can cure at room temperature and we use fiber in addition to fillers, the traditional hand layup procedure was used to develop the composites. In addition to the hand layup method, a vibrating base was used as the substrate onto which the mold was positioned. The mold is placed into the vibrating base once the hand layup process has been finished, and the mold was then subjected to vibration. The hand layup procedure must be finished and then immediately followed by the vibration action. Because of the vibrations, the resin and the fillers, in addition to the fibers, will disperse and mix together in a uniform manner. The air that could have been trapped inside the composite during the layup process was allowed to escape when the mold is vibrated. This allows the composite to be free of defects and ensures that the thickness of the composite is kept consistently across all of its dimensions. According to the instructions provided by the manufacturer, the epoxy resin and the hardener are combined in a proportion of exactly 50:50 [23]. After the resin that has been mixed has been evenly distributed throughout the surface of the mold with the assistance of a roller, the fibers and the plastic fillers are next dispersed across the matrix. The technique was repeated as many times as necessary until the desired thickness was achieved. After the mold has been covered and vibrated for an extended amount of time, it was then allowed to cure out completely. After the curing process was complete, the composite will be removed from the mold, and the appropriate dimensions will be employed to cut the specimens.

The tensile test for polymer composites was carried out in accordance with the guidelines of ASTM D3039-76 [24, 25]. The sample was cut to adhere to the appropriate dimensions for the standards, which are depicted in Figure 2.1. The Tinius Olsen universal testing machine was utilized to perform the evaluation. Both the speed of the cross head and the strain rate are predetermined as per the standards.

Microhardness of the composites was measured using the Vickers microhardness tester in line with ASTM E92 [26, 27]. The specimen is subjected to a force of 10 kg by means of a diamond pyramid with a square base. The Vickers pyramid number was the average of the measurements taken along the indentation's diagonals. In the region where the diamond pyramid rests on the specimen, there may be a filler, fiber, or matrix. The hardness of the specimens was measured in 10 different spots, and the average of those readings was recorded.

In order to conduct an effective water absorption test, specimens must be cut to exact specifications in line with ASTM D570 [23]. Prior to the examination, the specimens are weighed. Insulation tape was placed around the circumferences of the specimens to stop water from seeping inside. Once the samples are tagged, then dip them in water and leave them for 24 hours [22]. After 24 hours, the composites were taken out of the beakers and the excess water was cleaned out with a cotton cloth. In order to determine the proportion of water absorption, the specimens were reweighed, and the difference between the new and old weights was used to get the percentage of water absorption.

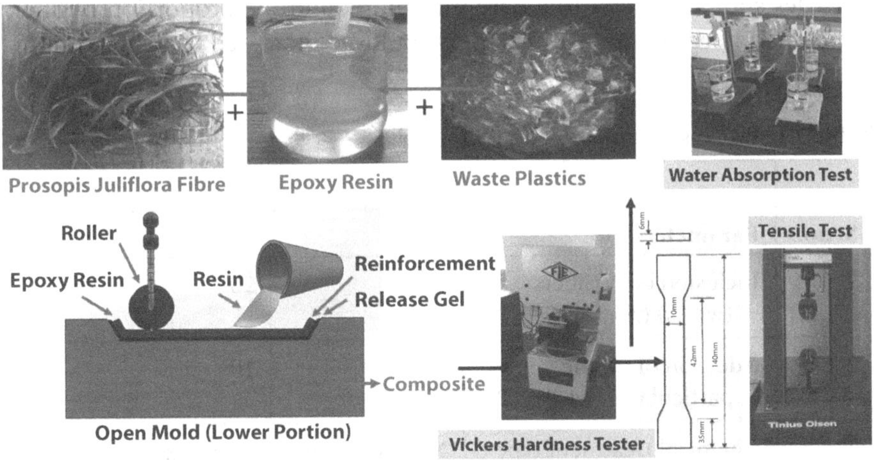

Figure 2.1 Experimental detail for the present work.

2.3 Results and Discussion

2.3.1 Influence of Process Parameters

The three most important factors within the scope of this study are the fiber composition, the waste plastic composition, and the chemical treatment given to the fibers. These three variables are considered at two different levels, and the specimens are developed in accordance with the structure provided by the Plackett–Burman experimental design [28–30]. Table 2.1 contains a list of the process parameters and their corresponding levels. The MINITAB software was used to perform the optimization. The results of the multiple tests are loaded into the software, and a regression equation was used to perform the analysis of the data. The outcomes of the experiment are presented in Table 2.2, as specified by the design.

Figures 2.2 and 2.3 show the results of the tensile test, while Figures 2.4 and 2.5 show the results of the hardness test and the water intake test. According to the findings of the tensile strength test, the 12th trial achieved the highest possible strength when compared to the findings of the other trials. In the 12th experiment, the percentage of fiber was increased to its maximum level, the quantity of plastic filler was decreased to its lowest level, and the fiber was treated with silane. It was clear from examining the other responses that the incorporation of the filler results in a reduction of the tensile strength of the composite.

When designing the composites, it was preferable to achieve a reduced hardness while simultaneously increasing the tensile strength. It was clear that in order to achieve the desired level of softness, the addition of plastic particles should be kept to a bare minimum, and the proportion of fiber, which had been treated with silane, should be increased to its maximum.

Table 2.1 Process parameters and their levels.

S. no.	Parameters	Levels	
		1	**2**
1	Inclusion of PF fibers (wt. %) (A)	10	20
2	Addition of plastic particulates (wt. %) (B)	5	10
3	Chemical treatment type (C)	NaOH (Alkali)	Triethoxy vinyl silane

Table 2.2 Experimental results for the Plackett–Burman design trials.

Trials	Input parameters			Output parameters		
	Fiber percentage (A)	Plastic percentage (B)	Chemical treatment (C)	Tensile strength (MPa)	Hardness (HV)	Water absorption (%)
1	20	5	Silane	76.65	30.45	1.67
2	10	5	NaOH	65.95	31.37	1.81
3	10	5	Silane	69.31	31.85	1.62
4	10	10	Silane	66.82	45.67	1.33
5	10	10	NaOH	60.38	46.59	1.53
6	20	10	Silane	71.41	40.13	1.59
7	20	10	NaOH	71.56	40.63	1.67
8	10	5	NaOH	66.34	32.67	1.84
9	20	5	NaOH	70.59	31.48	1.87
10	10	10	Silane	67.34	45.92	1.35
11	20	10	NaOH	70.82	43.28	1.61
12	20	5	Silane	76.77	30.97	1.74

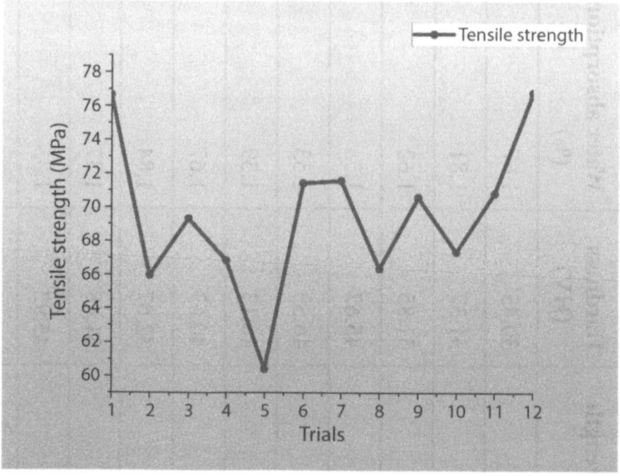

Figure 2.2 Tensile strength results for 12 trials.

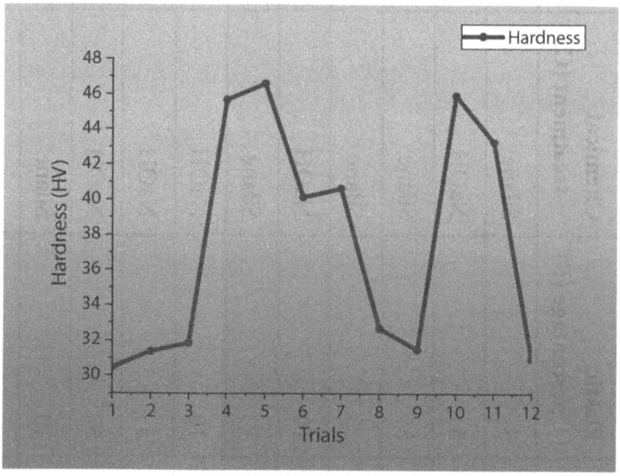

Figure 2.3 Microhardness results for 12 trials.

The fifth experiment yielded the highest value of hardness, which was created with 10 weight percent of plastic fillers and 10 weight percent fiber additives. This result was observed for the fifth trial.

The fourth study reported the lowest proportion of water intake, which also included the lowest percentage of fiber and the highest amount of plastic particles. It was discovered that the incorporation of waste plastic

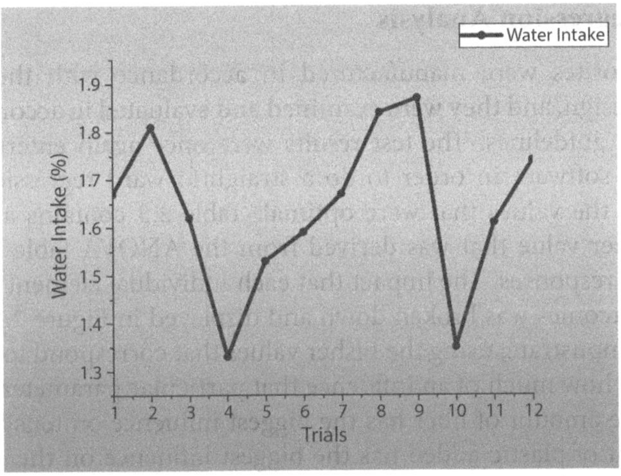

Figure 2.4 Water intake responses for 12 trials.

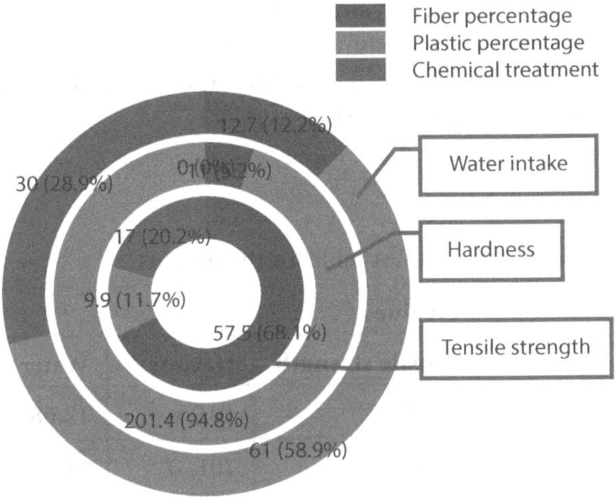

Figure 2.5 Contribution levels of each factor on the output responses.

particles reduced the amount of water that was absorbed. The water absorption percentage that was measured was highest for the ninth trial, which had a greater fiber percentage and a lower plastic filler percentage. When filler is added to polymer composites, the properties of the composites are altered to a greater extent [4, 31, 32].

2.3.2 Regression Analysis

The composites were manufactured in accordance with the Plackett-Burman design, and they were examined and evaluated in accordance with the ASTM guidelines. The test results were once again entered into the MINITAB software in order to do a straightforward regression analysis and locate the values that were optimal. Table 2.3 contains a tabulation of the Fisher value that was derived from the ANOVA table for each of the output responses. The impact that each individual element had on the various outcomes was broken down and displayed in Figure 2.5. It is possible to demonstrate, using the Fisher values that correspond to each input parameter, how much of an influence that particular parameter has on the results. The amount of fiber has the biggest influence on tensile strength, the amount of plastic added has the biggest influence on the composite's hardness, and the amount of plastic filler has the biggest influence on how much water the material absorbs.

Table 2.4 contains a summary of the models for each output factor. If a model's regression coefficient value is larger than 90%, it can be inferred that the model is statistically significant. In order to simplify its presentation and make the regression coefficient more readily understandable, Figure 2.6 presents it as a bar chart.

Table 2.3 F-values from ANOVA table for the outputs.

	F-value		
Source	**Tensile strength**	**Hardness**	**Water absorption**
Fiber percentage	57.45	10.96	12.68
Plastic percentage	9.88	201.37	61.04
Chemical treatment	17	0.04	29.97

Table 2.4 Model summary for the various outputs.

Output responses	**R-square**	**R-square (adjusted)**	**R-square (predicted)**
Tensile strength	91.33%	88.09%	80.50%
Hardness	96.37%	95.01%	91.83%
Water absorption	92.84%	90.15%	83.88%

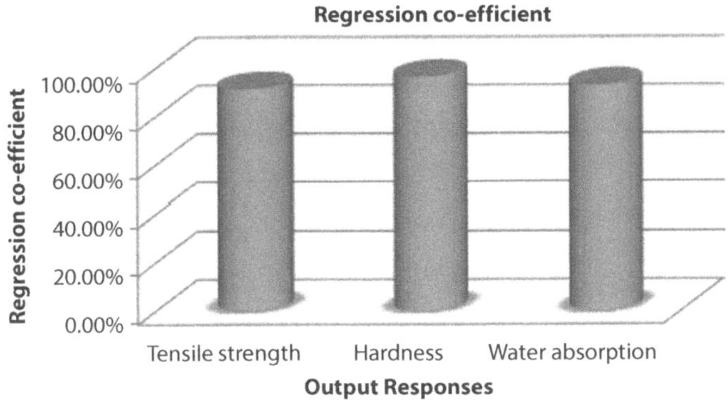

Figure 2.6 Regression coefficient for the outputs.

The plots for the tensile test outcomes are depicted in Figures 2.7–2.9. These figures show the main effect, the interaction, and the cube plot. When it related to tensile strength, the main effect plot was constructed using the "larger is better" core principle. The interaction plot illustrates how the influence of each input variable contributes to the final results. The cube plot makes it abundantly clear that in order to get the highest possible tensile strength, the fiber addition must be as high as possible, the amount of plastic filler must be kept to a minimum, and the fiber must be silane-treated [33].

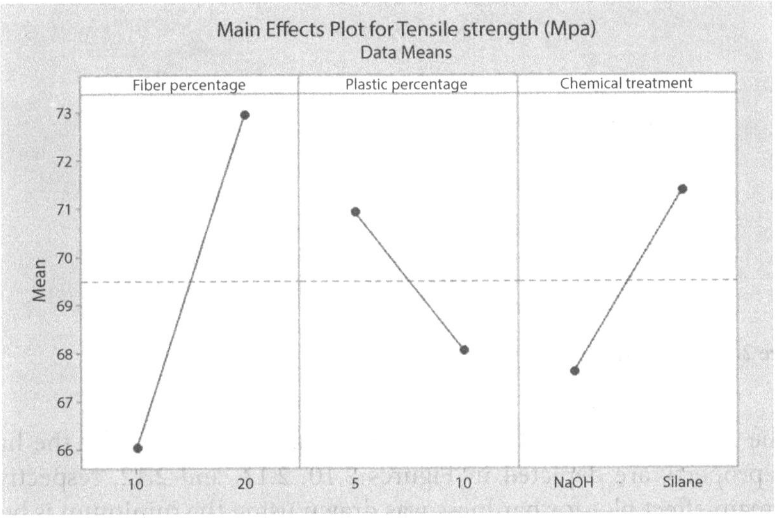

Figure 2.7 Tensile strength – Main effect plot.

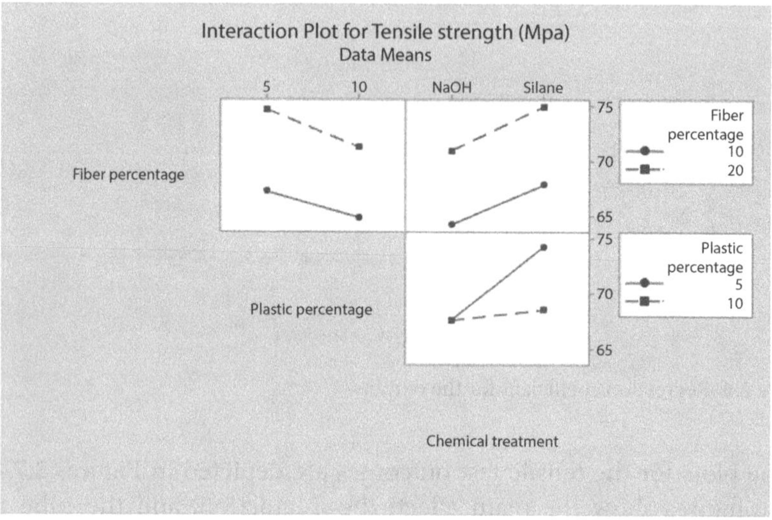

Figure 2.8 Tensile strength – Interaction plot.

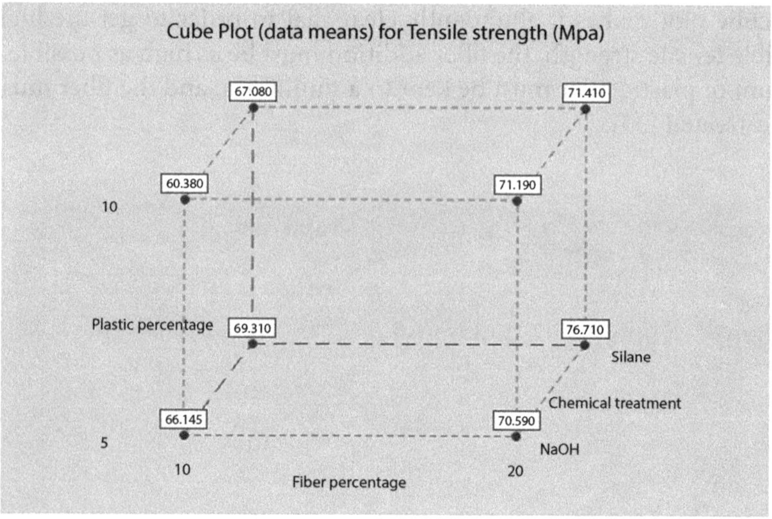

Figure 2.9 Tensile strength – Cube plot.

The main effect plots, interaction plots, and cube plots for the hardness property are depicted in Figures 2.10, 2.11, and 2.12, respectively. The main effect plot for hardness was drawn using the minimum is better principle, while the interaction plot clearly showed the correlation between

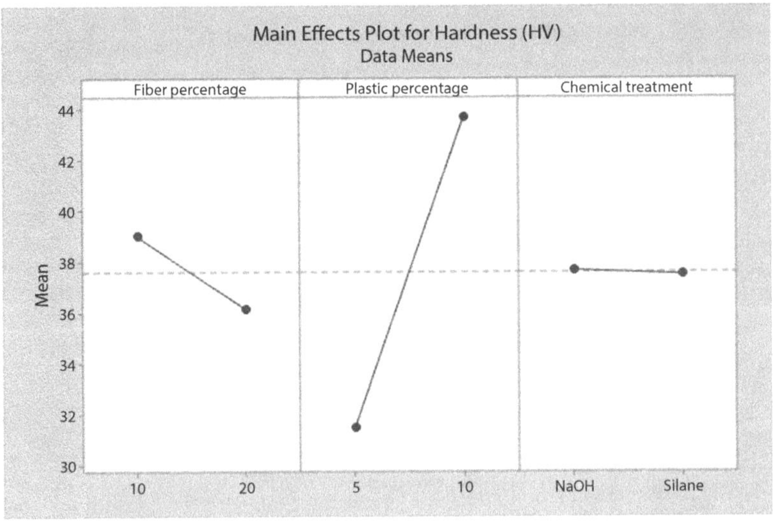

Figure 2.10 Hardness – Main effect plot.

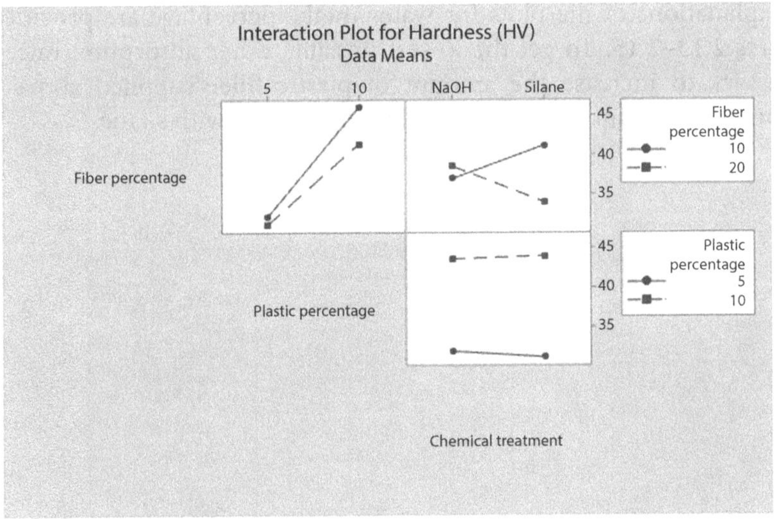

Figure 2.11 Hardness – Interaction plot.

fiber inclusion and chemical treatment. The main effect plot for hardness was also plotted for this principle. According to the cube plot, in order to achieve the lowest possible hardness, the input parameters that were used to get the highest possible tensile strength need to be retained.

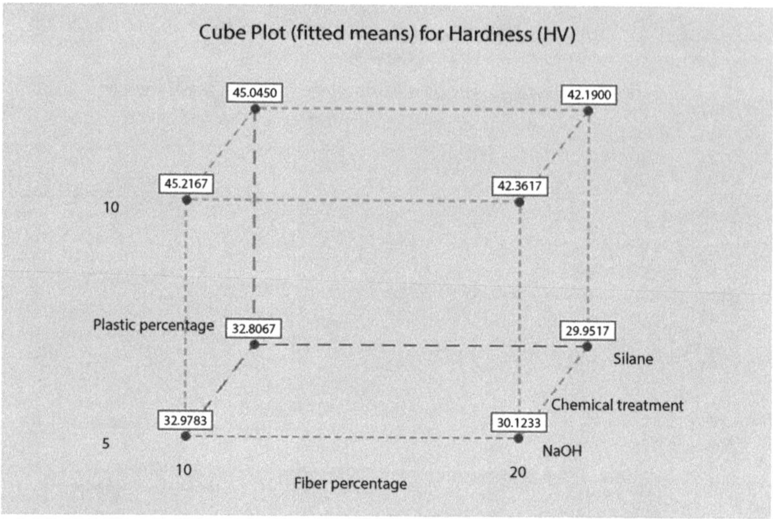

Figure 2.12 Hardness – Cube plot.

Explanations of the plots for water intake percentage are provided in Figures 2.13–2.15. To get the lowest possible water absorption rate, it is necessary to increase the amount of plastic filler supplied above and beyond the amount of fibers and to treat the fibers with silane.

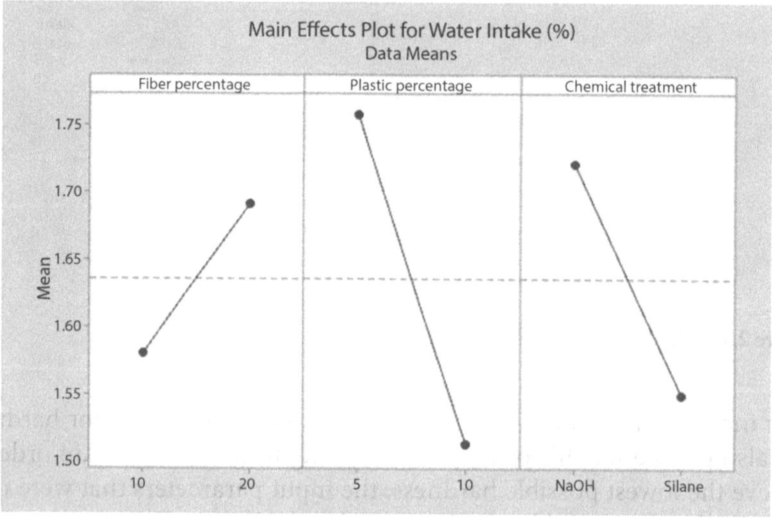

Figure 2.13 Water intake – Main effect plot.

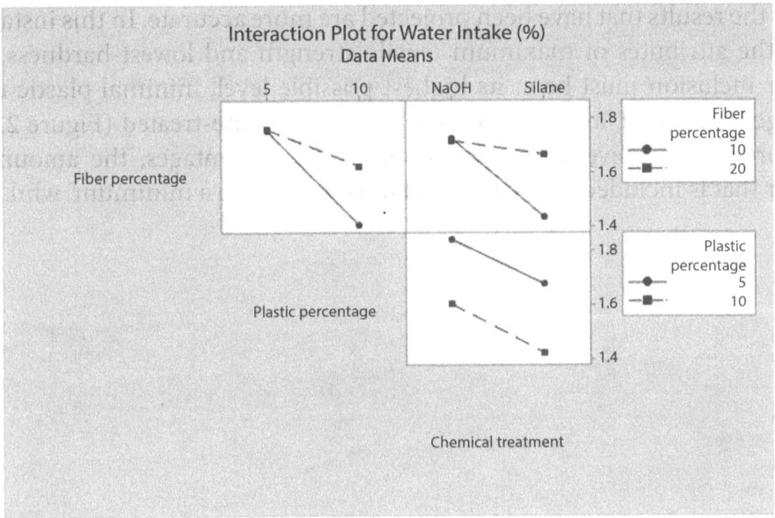

Figure 2.14 Water intake – Interaction plot.

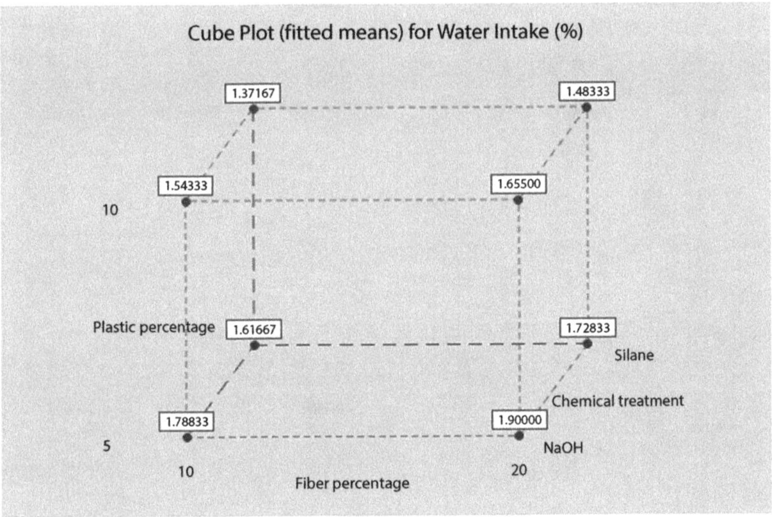

Figure 2.15 Water intake – Cube plot.

2.3.3 Optimized Responses

The software predicted based on the outcomes of the experiments as to which input parameters need to be optimized in order to obtain the maximum outputs. When the desirability is closer to the unit value, it indicates that the results that have been projected are more accurate. In this instance, for the attributes of maximum tensile strength and lowest hardness, the fiber inclusion must be at its highest possible level, minimal plastic filler usage is required, and the fiber itself must be silane-treated (Figure 2.16). In order to achieve minimum water intake percentages, the amount of fiber that is included in the material must be kept to a minimum, while the

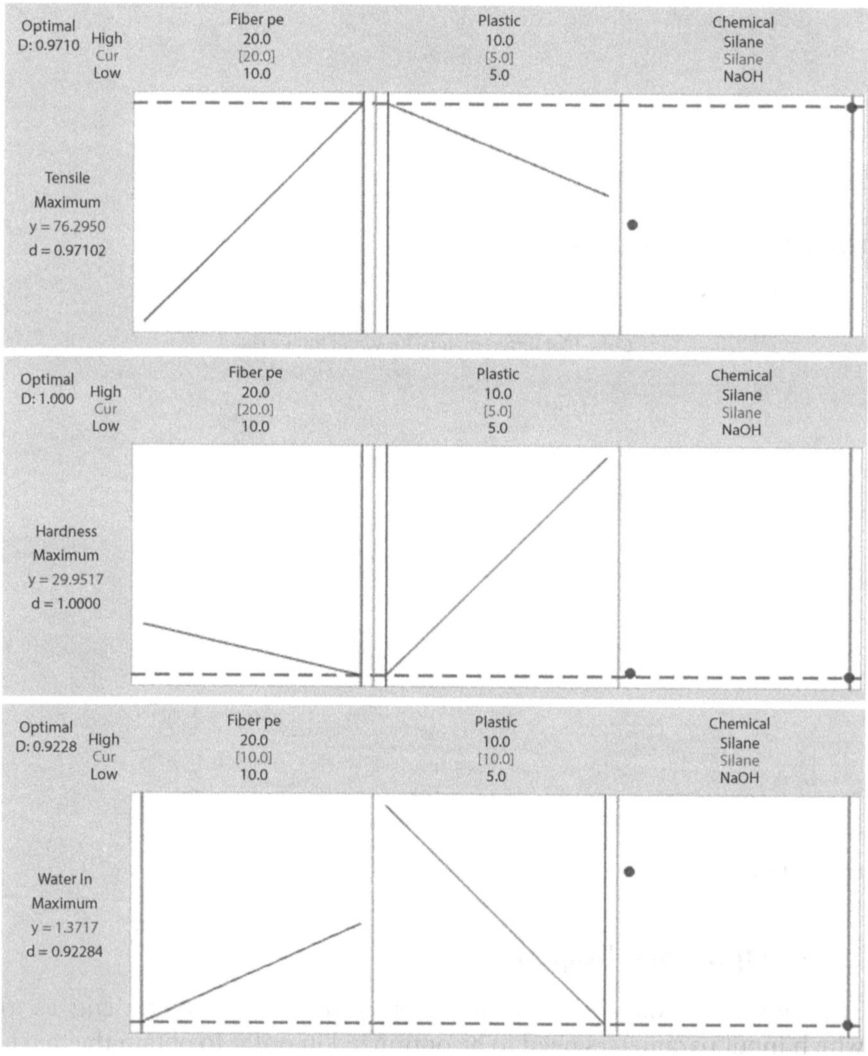

Figure 2.16 Optimized and predicted values for each parameter.

Table 2.5 Validation test results.

Parameters	Units	Optimal control parameters		Error (%)
		Predicted levels	Experimental results	
Maximum tensile strength	MPa	A2B1C2	75.52	1
Minimum hardness	HV	A2B1C2	30.45	1.64
Minimum water intake	%	A1B2C2	1.40	2.14

amount of plastic filler that was added must be increased using fibers that have been treated with silane.

In the end, the additional specimens were produced based on the predicted input factors, and they were put through validation testing. Table 2.5 displays the results of a comparison between predicted and experimental input levels, along with the error percentages. The amount of inaccuracy that occurs when comparing the predicted values to the experimental data was within the acceptable range.

2.4 Conclusion

The hand layup method was used to manufacture the composites that incorporate waste polymers and are reinforced with *Prosopis juliflora*. The specimens that were fabricated are evaluated in accordance with the standards. In accordance with the predicted levels, the confirmation tests were also carried out, and the findings are within the range of values that are considered acceptable. According to the findings, in order to achieve maximum tensile strength and minimum hardness, the input levels need to be A2B1C2, while in order to achieve minimum water absorption, the parameter levels need to be A1B2C2, which means that 10 weight percent of PF fiber addition, 10 weight percent of waste plastics, and the fibers need to be silane-treated. According to the findings, it was fairly evident that the incorporation of PF fibers and the chemical treatment that was applied to them have a significant impact on the tensile properties. The addition of a higher percentage of waste plastic fillers has an impact on the water intake characteristics as well as the hardness of the material.

References

1. Chapman, D., Prosopis juliflora (Sw.) DC, *Bulletin OEPP/EPPO Bulletin*, 49, 290–297, 2019. doi:10.1111/epp.12531.
2. Mbaabu, P.R., Olago, D., Gichaba, M., Eckert, S., Eschen, R., Oriaso, S., Choge, S.K., Linders, T.E.W., Schaffner, U., Restoration of degraded grasslands, but not invasion by Prosopis juliflora, avoids trade-offs between climate change mitigation and other ecosystem services. *Sci. Rep.*, 10, 1–13, 2020.
3. Sujin Jose, A., Athijayamani, A., Ramanathan, K., Sidhardhan, S., Effects of aspect ratio and loading on the mechanical properties of prosopis juliflora fibre-reinforced phenol formaldehyde composites. *Fibres Text. East. Eur.*, 25, 59–64, 2017.
4. Santhosh, M.S., Karthikeyan, G., Sasikumar, R., Hariharan, R., Mohanraj, R., Mechanical and morphological behaviour of rice husk/prosopis juliflora reinforced bio composites. *Mater. Today Proc.*, 27, 556–560, 2020.
5. Madhu, P., Pradeep, S., Sanjay, M.R., Siengchin, S., Characterization of raw and alkali treated prosopis juliflora fibers for potential polymer composite reinforcement. *IOP Conf. Ser. Mater. Sci. Eng.*, 653, 2019.
6. Rajan, B.S., Balaji, M.A.S., Saravanakumar, S.S., Effect of chemical treatment and fiber loading on physico-mechanical properties of Prosopis juliflora fiber reinforced hybrid friction composite. *Mater. Res. Express*, 6, 2019.
7. Krishnudu, D.M., Sreeramulu, D., Reddy, P.V., Rao, H.R., Effect of alkali treatment on mechanical properties of prosopis juliflora hybrid composites. *Int.. J. Appl. Eng. Res.*, 13, 2933–2935, 2018.
8. Manoj Kumar, G., Uthranarayan, C., Joseph Jebaraj, D.J., Keerthana, S., Ganesh, N., Exploration of tensile, flexural and hardness test properties of prosopis juliflora/glass/epoxy hybrid composite laminates. *J. Phys. Conf. Ser.*, 1362, 2019.
9. Howayda, M.M., Taiseer Hassan, M., Ahmed, A., Youssif, A., Production of fiber from mesquite plant (prosopis juliflora. L). *Int. J. Eng. Appl. Sci.*, 6, 52–56, 2019.
10. Balan, G.S. and Ravichandran, M., Study of moisture absorption characteristics of jute fiber reinforced waste plastic filled polymer composite. *Mater. Today Proc.*, 27, 712–717, 2020.
11. Balan, G.S., Balasundaram, R., Chellamuthu, K., Gopan, N.S., Dinesh, S., Vijayan, V., Sathish, T., Rajkumar, S., Flame resistance characteristics of woven jute fiber reinforced fly ash filled polymer composite. *J. Nanomater.*, 2022, 2022. https://doi.org/10.1155/2022/9704980.
12. Shanmugapriya, S., Surendran, S., Lee, Y.S., Selvan, R.K., Improved surface charge storage properties of Prosopis juliflora (pods) derived onion-like porous carbon through redox-mediated reactions for electric double layer capacitors. *Appl. Surf. Sci.*, 492, 896–908, 2019.

13. Alavudeen, A., Rajini, N., Karthikeyan, S., Thiruchitrambalam, M., Venkateshwaren, N., Mechanical properties of banana/kenaf fiber-reinforced hybrid polyester composites: Effect of woven fabric and random orientation. *Mater. Des.*, 66, 246–257, 2015.
14. Saba, N., Paridah, M.T., Jawaid, M., Mechanical properties of kenaf fibre reinforced polymer composite: A review. *Constr. Build. Mater.*, 76, 87–96, 2015.
15. Balasundaram, R., Devi, S.S., Balan, S.G., Machine learning approaches for prediction of properties of natural fiber composites: Apriori algorithm. *Aust. J. Mech. Eng.*, 1–16, 2022.
16. Balan, S.G., Ravichandran, M., Kumar, S.V., Study of ageing effect on mechanical properties of prosopis juliflora fibre reinforced palm seed powder filled polymer composite. *Aust. J. Mech. Eng.*, 18, 1–13, 2020.
17. Balan, S.G., Sridharan, M., Balasundaram, R., Sasikaran, A., Sagar, M., Dinesh, S., Vijayan, V., Rajkumar, S., Degradation analysis of jute fiber reinforced waste tile powder-filled polymer composite on wear characteristics. *Int. J. Polym. Sci.*, 2021, 1–13, 2021. https://doi.org/10.1155/2021/8587383.
18. Madhu, P., Sanjay, M.R., Senthamaraikannan, P., Pradeep, S., Siengchin, S., Jawaid, M., Kathiresan, M., Effect of various chemical treatments of prosopis juliflora fibers as composite reinforcement: Physicochemical, thermal, mechanical, and morphological properties. *J. Nat. Fibers*, 17, 833–844, 2020.
19. Reddy, P.V., Krishnudu, M.D., Prasad, R.P., Reddy, V.S.R., A study on alkali treatment influence on prosopis juliflora fiber-reinforced epoxy composites. *J. Nat. Fibers*, 18, 1094–1106, 2021.
20. Rajan, S.B., Saibalaji, M.A., Mohideen, R.S., Tribological performance evaluation of epoxy modified phenolic FC reinforced with chemically modified Prosopis juliflora bark fiber. *Mater. Res. Express*, 6, 2019.
21. Balan, G.S., Kumar, V.S., Raj, S.A., Study of water absorption characteristics of waste plastics and GFRP composites. *Mater. Today Proc.*, 22, 3116–3125, 2019.
22. Balan, G.S., Kumar, V.S., Rajaram, S., Ravichandran, M., Investigation on water absorption and wear characteristics of waste plastics and seashell powder reinforced polymer composite. *J. Tribol.*, 27, 57–70, 2020.
23. Ganapathy, S.B., Sakthivel, A.R., Sultan, M.T.H., Shahar, T., Sebaey, T.A., Effect of prosopis juliflora thorns on mechanical properties of plastic waste reinforced epoxy composites. *Polym. (Basel)*, 14, 2022.
24. Palanivendhan, M., Chandradass, J., Kaviyarasu, T., Philip, J., Fabrication and characteristics of hybrid glass fiber/Prosopis Juliflora reinforced epoxy composite. *Mater. Today Proc.*, 45, 6833–6837, 2021.
25. Madhu, P., Rangappa, S.M., Khan, A., Otaibi, A.A., Al-Zahrani, S.A., Pradeep, S., Gupta, M.K., Boonyasopon, P., Siengchin, S., Experimental investigation on the mechanical and morphological behavior of Prosopis juliflora bark fibers/E-glass/carbon fabrics reinforced hybrid polymeric composites for structural applications. *Polym. Compos.*, 41, 4983–4993, 2020.

26. Balan, G.S., Krishnan, A.M., Saravanavel, S., Ravichandran, M., Investigation of hardness characteristics of waste plastics and egg shell powder reinforced polymer composite by stirring route. *Mater. Today Proc.*, 33, 4090–4093, 2020.
27. Balan, G.S., Krishnan, A.M., Saravanavel, S., Ravichandran, M., Investigation on properties of bahunia racemosa fiber and egg shell powder reinforced polymer composite. *Mater. Today Proc.*, 33, 4084–4089, 2020.
28. Boudraa, H., Kadri, N., Mouni, L., Madani, K., Microwave-assisted hydrodistillation of essential oil from fennel seeds: Optimization using Plackett–Burman design and response surface methodology. *J. Appl. Res. Med. Aromat. Plants*, 23, 100307, 2021.
29. Naysmith, A., Mian, N.S., Rana, S., Development of conductive textile fabric using plackett–burman optimized green synthesized silver nanoparticles and *in situ* polymerized polypyrrole. *Green Chem. Lett. Rev.*, 16, 29, 2023, doi: 10.1080/17518253.2022.2158690.
30. Laime-Oviedo, L.A., Soncco-Ccahui, A.A., Peralta-Alarcon, G., Arenas-Chavez, C.A., Pineda-Tapia, J.L., Diaz-Rosado, J.C., Alvarez-Risco, A., del-Aguila-Arcentales, S., Davies, N.M., Yanez, J.A., Vera-Gonzales, C., Optimization of synthesis of silver nanoparticles conjugated and response surface methodology-preliminary antibacterial activity. *Processes*, 10, 1727, 2022.
31. Reddy, P.V., Prasad, P.R., Krishnudu, D.M., Hussain, P., Influence of fillers on mechanical properties of prosopis juliflora fiber reinforced hybrid composites. *Mater. Today Proc.*, 19, 384–387, 2019.
32. Balan, G.S., Prakash, B., Raj, S.A., Karthik, S., Kumar, V.M., Analysis on the impact of recycled ceramic waste, fiber format, and manufacturing techniques on polymer composite properties. *Eng. Res. Express*, 4, 2022.
33. Benin, S.R., Kannan, S., Moses, A.J., Bright, R.J., Mechanical characterization of prosopis juliflora reinforced polymer matrix composites with filler material. *Mater. Today Proc.*, 33, 1110–1115, 2020.

3

Effect of Various Manufacturing Techniques on Mechanical Properties of Biofiber-Reinforced Composites

M. Sasi Kumar[1]*, S. Sathish[2], M. Makeshkumar[2], S. Gokulkumar[2], L. Prabhu[2], S. Hemalatha[1], S. Ponnavan[1] and Nancy Chopra[1]

[1]*Department of Aeronautical Engineering, KIT-KalaignarKarunanidhi Institute of Technology, Tamil Nadu, India*
[2]*Department of Mechanical Engineering, KPR Institute of Engineering and Technology, Tamil Nadu, India*

Abstract
Bio-based fiber–reinforced composites (BFRCs) have been used in many industrial applications such as automobile, aircraft, sport, musical, furniture, marine, and medical fields due to cost-effectiveness, easy availability, resistance to fatigue and impact, ease of extraction, ease of fabrication, strength-to-weight ratio, eco-friendliness, biodegradability, and high stiffness. BFRCs are fabricated through various manufacturing techniques such as hand layup, injection molding, resin transfer molding, compression molding, vacuum bag molding, extrusion, melt compounding, pultrusion, spray layup, and blow molding. One of their drawbacks is poor adhesion between the hydrophilic reinforcement and the hydrophobic resin. Due to incompatibility between biofiber and biopolymer and moderate mechanical properties, BFRCs have been found to be less used in high-strength applications. This review provides an insight on various effects of manufacturing techniques on the mechanical properties of BFRCs to create sustainable and environmentally friendly systems in the materials world.

Keywords: Biofiber, biopolymer, manufacturing techniques, mechanical properties, eco-friendly

**Corresponding author*: sasikumar.mak@gmail.com

S. Thirumalai Kumaran and Tae Jo Ko (eds.) Sustainable Machining and Green Manufacturing, (33–64) © 2024 Scrivener Publishing LLC

3.1 Introduction

Composite materials are materials created by combining two or more distinct materials. The aerospace and automobile industries are the largest consumers of composite materials. This is because composite materials have less weight and have mechanical qualities comparable to those of commonly used materials. Properties of natural fibers are determined by their cellulose, hemicellulose, and lignin contents, which vary depending on where the fibers are grown [1]. In a fast-developing civilization, materials

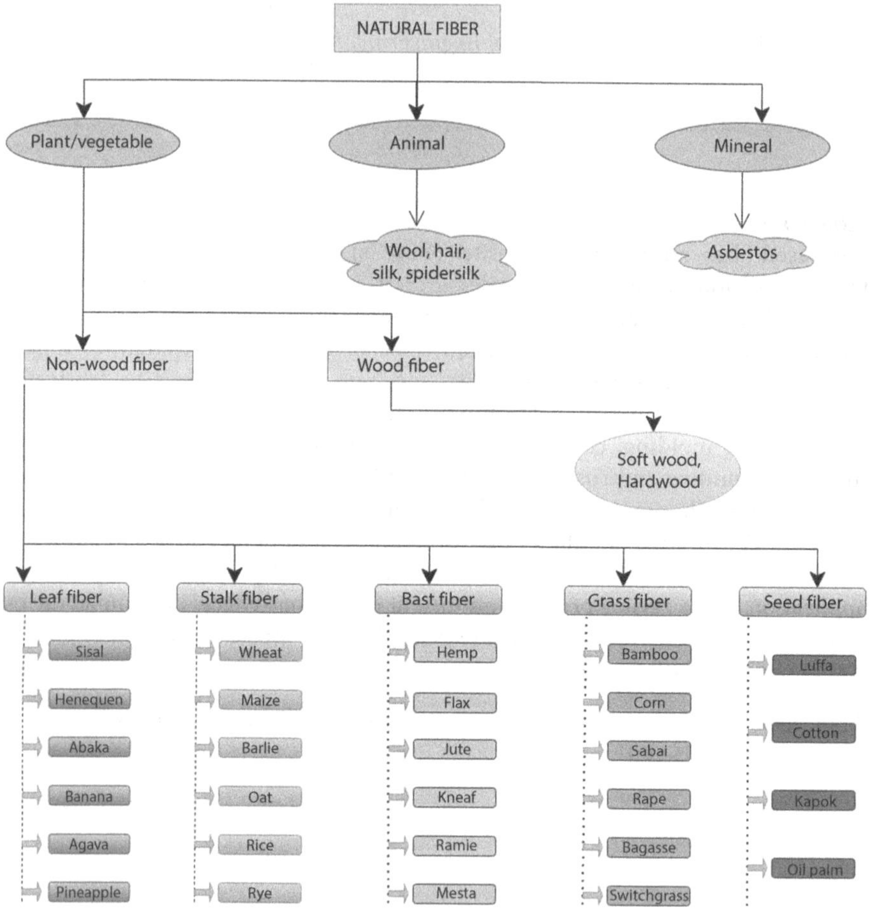

Figure 3.1 Classification of natural fibers.

with exceptional characteristics are needed. Metal alloys, ceramics, and polymers cannot meet this need. Modern technologies require strength and high-performance service materials. Material science research has fulfilled the above requirements. The volume fraction of fiber reinforcements contained and the orientation of fibers determine the mechanical strength of the composite material. In this paper, it is investigated how the mechanical properties of biofiber-reinforced composites are affected by different manufacturing methods such as hand layup, compression molding, injection molding, resin transfer molding, vacuum bag molding, and spray lamination. Figure 3.1 represents the classification of natural fiber from various sources.

3.2 Manufacturing Methods

Researchers study how biofiber-reinforced composite manufacturing processes affect mechanical performance. Composite materials are manufactured by various techniques that include injection molding, hand layup, filament molding, resin transfer, compression molding, and spray molding, as shown in Figure 3.2.

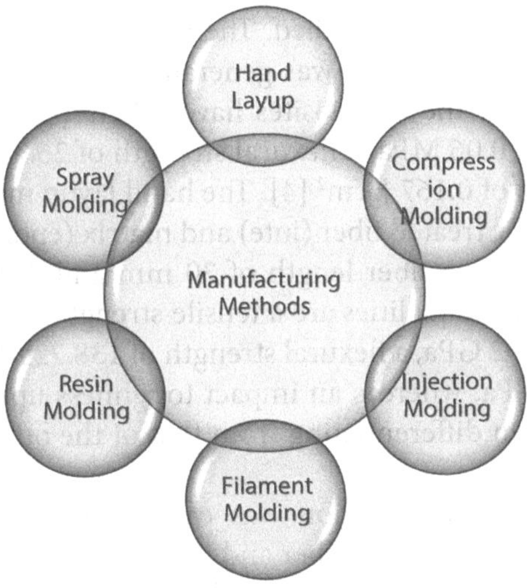

Figure 3.2 Manufacturing techniques.

3.3 Hand Layup Technique

Composite materials have become very important in modern life, so the hand layup process is the basic and traditional open-molding method for making composite materials. After the gel is put on the open mold, the matrix is poured into it, and then the fiber is put on the mold. The moving rollers are used to get rid of air bubbles in the cavity. It is cured under room temperature. We discuss the mechanical properties of different fibers, namely, coir, palm, sisal, carbon, pulp, bamboo, jute, coconut coir, *Carica* papaya stem, and *Dichrostachys cinerea* bark reinforced with the matrix (polyester, epoxy, polyvinyl chloride, vinyl ester) while using the hand layup technique procedure.

The hand layup method was carried out for the experimental investigation of *Cissus quadrangularis* stem and polyester matrix with a fiber length of 40 mm and fiber weight up to 30%. The tensile modulus and tensile strength are 1.400 GPa and 90.20 MPa, respectively, the flexural modulus and flexural strength are 2.250 GPa and 103.040 MPa, respectively, whereas impact toughness is 10.7 J/cm^2 and actual hardness is 92 HRRW when compared to the other manufacturing methods [2]. The mechanical characteristics of composites fabricated through the hand layup method are listed in Table 3.1.

The mechanical features of the composite of *Calotropis* reinforcement and epoxy matrix were then studied. The hand layup took 30-mm length of fiber and 30 wt% of fiber. It was generally observed that the mechanical characteristics of the composites have a better result, with a tensile strength of 11.43 ± 0.06 MPa, a flexural strength of 26.38 ± 0.47 MPa, and an impact strength of 0.167 J/cm^2 [4]. The hand layup method was carried out to investigate the treated fiber (jute) and matrix (epoxy) by experimental investigation along a fiber length of 30 mm and 20% weight of fiber. Among its mechanical qualities are a tensile strength of 81.65 MPa, a tensile modulus of 0.792 GPa, a flexural strength of 138.72 MPa, and a flexural modulus of 2.75 GPa, whereas an impact toughness up to 3.41 J/cm^2 has been observed when differentiating from that of the other manufacturing methods [8].

The experimental investigation was carried out for jute and epoxy where the 30-mm length of the fiber and weight up to 20 wt% by the hand layup process were used. The mechanical results for tensile modulus, tensile strength, flexural modulus, and flexural strength are 0.638 GPa, 75.15 MPa, 2.61 GPa, and 124.01 MPa, respectively, and the impact strength is 3.25 J/cm^2 [8]. An experimental investigation was carried out for treated *Dichrostachys cinerea* bark reinforcement and epoxy resin with 30 wt% of

Table 3.1 Mechanical properties of composites made using the hand layup method.

Reinforcement	Matrix	Manufacturing methods	Tensile strength	Flexural strength	Impact	References
Banana fiber	Polyester resin	Hand layup method	19.91 MPa	92.761 MPa	-	[3]
Sugar tree (*Grewia hirsuta*)	Epoxy resin	Hand layup method	56.107 N/ mm^2	68.039 N/mm^2	0.0512 J/mm^2	[4]
Flax fibers	Epoxy resin	Hand layup method	65.07 MPa	-	-	[5]
Banana and sisal fibers	Epoxy resin	Hand layup method	14 MPa	-	3.616 J/ mm^2	[6]
Sisal, jute, and *Sorghum bicolor* fibers	Polyester resin	Hand layup method	241.2 MPa	86.8 MPa	26.3 J/m^2	[7]

fiber through the hand layup process. Mechanical properties such as tensile strength (measured in MPa) and modulus (in GPa) (50.75 MPa and 0.523 GPa, respectively), flexural strength and modulus (74.06 MPa and 4.764 GPa, respectively), and impact strength (8.4665 J/cm^2) have been measured and analyzed [9]. *Dichrostachys cinerea* bark reinforcement and epoxy resin were used for the hand layup method for fabrication with 0.07 mm of length and 30 wt% of fiber. They found a tensile strength of 46.69 MPa, an elastic modulus of 0.484 GPa, a bending modulus of 4.456 GPa, and a flexural toughness up to 68.89 MPa, whereas the impact toughness was up to 7.577 J/cm^2 [9]. The hand layup method was carried out for the experimental investigation of *Calotropis gigantea* and epoxy matrix wherein the fiber weight was 10%. It has been observed that mechanical characteristics showed a tensile stiffness up to 48.73 MPa, a flexural stiffness up to 195.19 MPa, an impact stiffness up to 6.34 J/cm^2, and a hardness up to 67 HRRW when compared to the other manufacturing methods [10]. The treated *Carica papaya* stem fiber and epoxy matrix were used for fabrication with 20% fiber weight by hand layup method. Tensile stiffness, elastic modulus, and impact stiffness are 122 MPa, 118.9 MPa, and 7.8 J/cm^2, respectively [11]. The hand layup method was used for the experimental investigation of the treated tapsi fiber and epoxy matrix with 15 wt% of fiber. Mechanical properties like the ultimate tensile stiffness ranged up to 43.02 MPa whereas flexural stiffness ranged up to 167.19 MPa when compared to the other manufacturing methods [12]. The effects of natural fiber composites fabricated by hand are shown in Table 3.2.

The mechanical characteristics of the filler named *Shorea robusta* with resin (polyester) were evaluated. They used the hand layup method for this experimental investigation. After the experiment, they found better mechanical characteristics like a tensile stiffness of 15 MPa and a flexural stiffness of 26 MPa [13]. Then, sugar palm fiber and rosella fiber reinforced with a matrix named vinyl ester were prepared [14] in which the vinyl ester matrix is used for the hand layup process. It is found to have improved mechanical properties such as a tensile stiffness of 24 MPa and a flexural stiffness of 110.5 MPa. The mechanical attributes of a hybrid composite of wood flour and pulp fiber reinforced with polyvinyl chloride as a resin were assessed using the hand layup technique. Thus, the mechanical features of reinforcement have been obtained in that the ultimate tensile stiffness was up to 17 MPa and the flexural stiffness was up to 42 MPa [15]. Table 3.3 presents several composites made by the hand layup technique.

Table 3.2 Effect of hand-laid natural fiber-reinforced polymer matrix composites.

Reinforcement	Matrix	Fabrication methods	Effect/Influences	References
Banana fiber	Polyester resin	Hand layup	Composites with a coconut shell content of 3 volume percent showed the highest tensile and flexural strengths.	[3]
Flax fiber	Epoxy	Hand layup	Four-layer fiber composites have a higher tensile strength at a volume percentage of 42%.	[5]
Banana and sisal fiber	Epoxy	Hand layup	Fiber-matrix interfacial bonding improved mechanical characteristics.	[6]
Sisal/jute/ *Sorghum* fibers	Polyester resin	Hand layup	Maximum tensile strength is thermally stable and decomposes.	[7]

3.4 Compression Techniques

Composites are rapidly becoming the material of choice in many technical applications, and some molding processes are playing an extremely significant part in this trend. One of them is known as compression molding, and it has been demonstrated to be a helpful approach for the production of complicated fiberglass-reinforced polymers. This is mostly owing to the fact that it functions on a quick cycle time due to the large-volume and high-pressure methodology. There are several types of compression

Table 3.3 Composites made by the hand layup technique.

Reinforcement/Matrix	
Novel *Acacia tortilis* fiber/ polyester composite [16]	Flax, linen, and bamboo fabric fiber/ epoxy resin [22]
Coir fiber/glass fiber as fiber/ epoxy composite [17]	Banana and flax fibers/epoxy resin [23]
Luffa fiber/epoxy matrix [18]	Bamboo fiber/polymer composite [24]
Jute as fiber/epoxy as a matrix [19]	Hybrid fiber with Nano SiC particles composite, palm-wood-SiC/epoxy resin [25]
Lantana camara fiber/epoxy composite [20]	Natural sisal, jute fiber/epoxy resin composites [26]
Jute fiber and *Cyperus pangorei* fiber/epoxy composites [21]	Glass-jute fiber/polymer composites [27]

molding based on the type of material to be molded [28]. Compression molding is a typical method of producing thermosets, elastomers, thermoplastics, and natural rubbers. There are several advantages of compression molding, including producing high-volume, dimensional precision, reduced-weight, corrosion-resistant, high-strength, and temperature-resistant products with superior surface quality [28, 29].

3.4.1 Mechanical Properties of Products Made by Compression Molding Techniques

The red banana peduncle fiber and polyester matrix were used for compression molding for fabrication of 400 mm of fiber. Results showed a tensile property of 11.61 MPa, a flexural strength of 2 MPa, and an impact strength of 4.6 J/cm^2 [30]. The composite plate's fabrication epoxy resin is combined with sisal fiber to create this. In order to test its mechanical qualities, it is manufactured using a compression molding process. First, the fiber is treated with a NaOH solution, which creates a strong link between the fiber and the matrix; then, it is discovered that the mechanical qualities improved [31]. The compression molding method was carried out for the experimental investigation of treated *Luffa* fiber and epoxy matrix with a fiber length of 30 mm and fiber weight of 6%. From further observation, the elastic strength was 18.3 MPa, the elastic modulus was 0.0101 GPa, the flexural stiffness was up to 220 MPa, and the impact stiffness was up to 3.4 J/m when compared to some other manufacturing methods [32]. Compression molding was used to successfully create both single-fiber and hybrid composites, and their mechanical, thermal, and tribological characteristics were tested. Compared to other composites studied, the mechanical characteristics of MC2 composites are superior. Some studies have suggested that the increased synergy of banana and jute fiber at 20 wt% is to blame. Overall, the proportion of fiber weight has a bigger effect on composites' tribological features than any of the other parameters studied here. The NaOH treatment improved the mechanical and tribological qualities [33]. To evaluate the mechanical performance of the banana/flax/waste tea leaf fiber with epoxy resin, fabrication is done by using the compression molding technique. By using 5% weight of NaOH solution, the banana and waste tea leaf fiber are treated for strong bonding. As a result, the mechanical performance can be increased by the hybridization of 25% of banana fiber with weight percentage of 5 of waste tea leaf fiber [34]. The mechanical characteristics of the seed powder *Polyalthia longifolia* mixed with the matrix vinyl ester were evaluated. For the fine adhesion linkage of fiber, they used the vinyl ester matrix. They observed fine mechanical properties of combining the fibers after completing the experiment like tensile, flexural, and impact strengths of 31 MPa, 125 MPa, and 31 KJ/m^2, respectively, whereas hardness is 36 [35]. Then, the mechanical attributes of coir fiber and chitosan powder reinforced with vinyl ester used as a matrix were studied [36]. The tea-glass-sisal fiber is used here with the epoxy to examine the mechanical parameters. Compression molding is used to fabricate the composite plates. Both sisal and tea fibers were first treated with 5% alkaline solution. Note that the hybridization of the sisal and glass showed better performance for flexural and impact strengths as compared to the other fibers [37]. Table 3.4 shows how compression molding affects the mechanical properties of composites.

The compression molding method of fabrication was carried out for coir fiber. The results concluded some mechanical properties of fibers such that properties of tensile and flexural are 22 MPa and 29 MPa, respectively. For fibers like bagasse and coir reinforced with vinyl ester, the mechanical properties are investigated [42]. The authors examined the mechanical performance using *Ananas comosus-Camellia sinensis*-glass fiber hybridization with epoxy resin. Compression molding is the technology used in the production of the composite materials. Results show that impact, tensile, and flexural strengths may be improved [43]. From the compression molding method, they have found better mechanical properties of bagasse fiber. After the experiment, they acquired the mechanical characteristics for the

Table 3.4 Mechanical characteristics of composites produced via compression molding.

Reinforcement	Matrix	Manufacturing methods	Tensile strength	Flexural strength	Impact	References
Madar fiber/sugarcane bagasse	Epoxy	Compression molding	61 MPa	147 MPa	54 KJ/m^2	[38]
Hedysarum coronarium-based green fiber	Polymer	Compression molding	33.4 MPa	-	-	[39]
Hemp/flax/noil hemp/noil flax fiber	Epoxy	Compression molding	-	30.04 MPa	-	[40]
Pineapple/flax	Epoxy	Compression molding	37.89 MPa	70.28 MPa	96.99 J/m	[41]

hybrid composite like 43 MPa and 53 MPa of tensile and flexural values. Thus, an experimental investigation of a variety of bagasse fiber reinforced with vinyl ester was carried out [44]. By using the flax/jute fiber with the epoxy matrix, the fabrication of the laminates was done and mechanical performance was examined. Firstly, the fibers are chemically treated with alkaline with 5% NaOH to make the bond strong between the fiber and the matrix. Then, compression molding technique is used for the fabrication. Finally, it is noted that the mechanical strengths can be improved through hybridization [45]. The compression molding process was used on the bagasse fiber to carry out the experimental investigation. After completing the experiment, they found that the mechanical characteristics for the hybrid composite are as follows: tensile strength is 51 MPa and flexural stiffness is 63 MPa. With the compression molding technique, the composite materials are fabricated by using kenaf-glass-waste tea leaf fiber with the epoxy. The mechanical properties can be evaluated by using 40% weight of fiber with 60% weight of matrix. The improvement is observed in the mechanical properties like tensile strength, flexural strength, and impact strength [37]. The results of compression composites made from natural fibers are shown in Table 3.5.

For the manufacture of laminates, the compression molding method is the technique that is employed. In this study, the hybridization of jute-glass-tea leaf fiber with epoxy resin was utilized in order to investigate various mechanical characteristics. These included flexural, tensile, and impact strengths. Therefore, the mechanical qualities can be improved by including 10% by weight of glass fiber, 20% by weight of jute fiber, and 10% by weight of tea leaf fiber into the hybridization process [51]. The natural fiber coir and the filler termite mound powder were reinforced with vinyl ester. Here, some mechanical characteristics for the natural fiber coir and rice hush (filler) were studied. In this experiment, they used vinyl ester as a matrix to obtain better results. After this investigation, the physical characteristics such as tensile, flexural, and impact strengths of 32 MPa, 39 MPa, and 53 KJ/m^2 [49], respectively, were obtained. Great attention is given to look into the mechanical properties using flax fiber, SiC and Al_2O_3 fillers, and epoxy as a resin. The fabrication is done using the compression molding technique. Finally, it is revealed that the addition of flax fiber up to 20% weight and the addition of SiC filler up to 8% weight can improve the various mechanical properties of the composite material [52]. Table 3.6 presents several composites made by the compression molding technique.

Table 3.5 Compression molding processes and their effect on natural fiber-reinforced polymer matrix composites.

Reinforcement	Matrix	Fabrication methods	Effect/influences	References
Flax fiber/ waste tire rubber (RR)	Recycled HDPE	Compression molding	Higher tensile and impact strengths	[46]
Abutilon indicum plant leaves/ glass fiber	Epoxy	Compression molding	The addition of bio-reduced graphene oxide to a polymer matrix increases the matrix material's tensile, flexural, and impact strengths by 28%, 55%, and 80%, respectively.	[47]
Banana and coir fiber	Epoxy	Compression molding	The combination of banana/coir fiber with epoxy matrix achieved the maximum and improved the mechanical properties.	[48]
Jute and glass fiber	Epoxy	Compression molding	42% more flexural strength over untreated and without microfiber-injected composites.	[49]
Sisal and hemp fibers	Polypropylene	Compression molding	Recycled PP surgical masks are used as a matrix for sisal and hemp reinforcement. The hot compression molding method sandwiches PP masks with natural fibers. The tensile strength of sisal and hemp fiber composites increases by 197% and 305%, respectively. Tensile elongation increased to 574% for sisal fiber composites.	[50]

Table 3.6 Composites made by the compression molding technique.

Reinforcement/Matrix	
Jute, hemp fibers, cross-ply composite panels [53]	Jute fiber/polypropylene [60]
Air-laid fibers/polymer-based composites [54]	Kenaf-cornstarch fiber/ polymer composites [57]
Flax fiber/high-density polyethylene [55]	Kenaf fiber/polypropylene [61]
Coconut fibers/polymer-based composites [56]	Kenaf-soy flour fiber/polymer composites [62]
Bagasse-cornstarch/polymer composites [57]	Wood/polyethylene [63]
Banana fiber/low-density polyethylene [58]	Ramie fiber/polylactic acid [64]
Hemp fiber/polylactic acid [59]	

3.5 Injection Technique

Molten materials are injected into molds. It produces thousands of identical goods in mass manufacturing. Injection molding uses thermoplastic and thermosetting polymers, although metals, glassware, elastomers, and confections are also utilized. Injection molding is a common method of manufacturing thermoplastic. It means once it is melted, it can be reused by remelting, reshaping, and refreezing where all are serving different purposes. Listed are some materials used such as acrylonitrile butadiene styrene (ABS), poly(methyl methacrylate) (PMMA), thermoplastic rubber (TPR), thermoplastic polyurethane (TPU), and polypropylene (PP).

The parts included in this process are hopper, heater, screw, motor, hydraulic mechanism, etc. In a traditional injection molding process, pelletized fiber composites are sustained into a hop-picker, then it is transported through heated screw mass. Through a nozzle, the screw pours the material in the mold once the material has been melted to the desired amount in a barrel. After the mold cools, it acquires the desired shape. Injection molding has numerous applications in a wide variety of industries from the automotive to the medical industry. Some major advantages of using injection molding are high efficiency, enhanced strength, detailed features, reusage of material, and allowing for complex geometries with tight tolerances. It totally depends on the material properties to find the right balance, and there is a variety of options when it comes to color because it is well-suited with a large set of materials as well as colors.

The injection molding method was carried out for the experimental investigation of the fiber treated *Areca* sheath and matrix polyethylene terephthalate where the fiber length is 3 mm and the weight of the fiber is 20%. Mechanical properties are as follows: hardness of 27.74 VHN and flexural, tensile, and impact of 52.88 MPa, 57.28 MPa, and 2.254 J/cm^2, respectively. It is then compared to the other manufacturing methods [65]. Thermoplastic composites were injected into molds. Exhibiting a bending strength of 477 MPa, flexural modulus of 25.7 GPa, and fiber volume content of 67% at a manufacturing temperature of 270°C, the material has exceptional mechanical characteristics [66]. The mechanical characteristics of composites manufactured using the injection molding method are listed in Table 3.7. Table 3.8 shows how injection molding affects natural fiber composites.

Injection molding was used to conduct experiments with a reinforcement length of 3 mm and a fiber weight percentage of 20% made from processed *Moringa oleifera* fruit fiber and polyethylene terephthalate matrix. The mechanical properties are as follows: a tensile strength of up to 65.92 MPa, a hardness of 30.62 VHN, a flexural strength of up to 98.49 MPa, and an impact strength of up to 2.866 J/cm^2 [65]. Table 3.9 presents several composites made by the injection molding technique.

3.6 Filament Techniques

To make a composite structure, filament winding is the process of wrapping resin-soaked glass or advanced fiber around a rotating mandrel. This provides the strength-to-weight ratio, optimal mechanical properties, and controlled fiber deposition. Furthermore, the disadvantages of the process include being restricted to convex-shaped components and the mandrel cost will be high for the large components.

This experimental investigation was carried out for flax yarn and polypropylene with 72% of fiber loading using filament winding. From the observations, the optimum results in mechanical characteristics are as follows: tensile parameter and modulus parameter are 321 MPa and 29 GPa, respectively [84]. The effects of natural fiber composites manufactured using the filament molding technique are presented in Table 3.10. Table 3.11 presents several composites made by the filament winding technique.

3.7 Vacuum-Assisted Resin Transfer Molding Technique

The vacuum technique method was found to be the most common composite manufacturing technology used to create laminated structures. It is

Table 3.7 Mechanical properties of composite materials manufactured using the injection molding technique.

Reinforcement	Matrix	Manufacturing methods	Tensile strength	Flexural strength	Impact	References
Hemp fiber	Polylactic acid	Injection molding	54.31 ± 7.25	99.25 ± 9.42	8.99 KJ/m^2	[67]
Jute	Polypropylene	Injection molding	74	112	-	[68]
Ethiopian fibers and sisal fiber	Polyethylene	Injection molding	66	11.94	-	[69]
Sisal and flax fiber	Polypropylene	Injection molding	34.64	71.38	-	[70]
Hemp fiber	(BioHDPE)	Injection molding	15.5 ± 1.4	-	3.9 ± 0.2 KJ/m^2	[71]

Table 3.8 The impact of natural fiber-reinforced composite polymer matrix molded via the injection molding process.

Reinforcement	Matrix	Fabrication methods	Effect/Influences	References
Hemp fiber	Polylactic acid	Injection molding	Injection molding process results in better mechanical properties.	[67]
Jute	Polypropylene	Injection molding	Higher mechanical strength compared to others manufacturing methods.	[68]
Ethiopian fibers and sisal fiber	Polyethylene	Injection molding	This study found that sisal and Ethiopian fibers are effective load-bearing and structural materials for various applications.	[69]
Sisal and flax fibers	Polypropylene	Injection molding	Polypropylene matrix-sisal fiber-flax fiber biocomposites have improved tensile and flexural properties.	[70]
Hemp fiber	BioHDPE	Injection molding	Stiffness, hardness, compatibility, and thermal stability have improved.	[71]

widely used in the aircraft sector. The vacuum bagging procedure facilitates hand lamination while creating a component with improved characteristics due to compaction. Following the hand layup process, the laminate is inserted between the vacuum bag and mold to guarantee even fiber infusion into the matrix material. Vacuum bagging eliminates air spaces while increasing the fiber-to-epoxy ratio. Overall, it is an excellent procedure for improving composite laminates. Sealant tape, release fiber, release film, breather fabric, vacuum pot, vacuum bag, vacuum line, vacuum gauge, and other components are necessary for vacuum bagging. Vacuum bags are also employed in industrial settings for laminating parts with complex geometries, such as rounded or asymmetrical surfaces. To create a vacuum at little expense, you can use a Venturi vacuum and an air compressor.

Table 3.9 Composites made by the injection molding technique.

Reinforcement/Matrix	
Flax, pulp, wood powder/polymer composites [72]	Hemp fiber/polypropylene composites [78]
Kenaf fiber, reinforced composites/polypropylene and polylactic acid [73]	Rice husk flour/polypropylene composites [79]
Flax fiber/polypropylene composites [74]	Kenaf, jute, bamboo, sugarcane, banana fibers/polypropylene composites [80]
Kenaf fibers/polypropylene composites [75]	Jute fiber/polypropylene composites [81]
Kenaf fiber/polypropylene composites [76]	Glass fiber/polypropylene composites [82]
Kraft pulp and hemp fiber/ polypropylene composites [77]	Cellulose fibers/polypropylene composites [83]

The fundamental benefit of vacuum forming over other big molding methods is the cheaper cost of tooling. Moving on, vacuum forming is frequently a cost-effective method of production because of the short time span between the concept and the finished product. It may also enhance open-mold laminates' mechanical properties. This enables a broader range and mix of materials and a stronger link between them. Some examples used in creating the vacuum bags are polyamide and polyethylene. Other material examples are as follows: epoxy, porous wood, steel, lead, aluminum, hardwood, polyester, and plastic. Figure 3.3 represents some vacuum assisted techniques.

Fiber loading of 39% was used in the experimental study of flax and unsaturated polyester using a technique termed resin transfer. Mechanical properties such as a tensile stiffness of 61 MPa, a Young's elastic modulus of 6 GPa, a flexural stiffness up to 91 MPa, and a flexural modulus of 5 GPa were measured for the 39% fiber loading and found to be optimal [89]. Using the resin transfer technique, an experimental examination was conducted for 35% fiber loading of jute and unsaturated polyester. Based on this information, the mechanical properties of the material at its fine optimum are as follows: a tensile strength of 50 MPa, a flexural strength of 103 MPa, a tensile modulus of 8 GPa, and a flexural modulus of 7 GPa [89]. The resin transfer molding method was carried out for the experimental investigation of cellulose fiber and bio-epoxy. Optimum results in mechanical

Table 3.10 Influence of filament molding on natural reinforced composites.

Reinforcement	Matrix	Fabrication methods	Effect/Influences	References
Bamboo fiber	Epoxy	Filament winding	There was an increase in the material's tensile, shear, and bending strengths.	[85]
Abaca/jute/sisal	Epoxy	Filament winding	Excellent tensile strength was obtained.	[86]
Abaca, jute, and glass fiber	Epoxy	Filament winding	Improved abaca fiber mechanical qualities.	[87]
Carbon fiber	Epoxy	Filament winding	High creep test mechanical, storage modulus, and strain.	[88]
Glass and carbon fiber	Epoxy	Filament winding	The manufacturing, mechanical, and interfacial characteristics of carbon fiber filament wound composites were examined by introducing liquid-like MWCNT reinforcements to the epoxy matrix. The resin matrix with liquid-like MWCNT reinforcements demonstrated good filament wound composite processing capabilities based on viscosity and pot life tests.	[89]

Table 3.11 Composites made by the filament winding technique.

Reinforcement/Matrix	
Carbon fiber/epoxy composites [90]	Kenaf fiber/polymer composites [95]
Jute fiber and glass fiber, abaca fiber/ epoxy composite [91]	Glass fibers/epoxy resin [96]
Sisal fiber/high-density polyethylene [92]	Kenaf fiber/epoxy composite [97]
Kenaf fiber/glass fiber/epoxy [93]	Glass fiber/epoxy composite [98]
Carbon fiber/epoxy composites [94]	

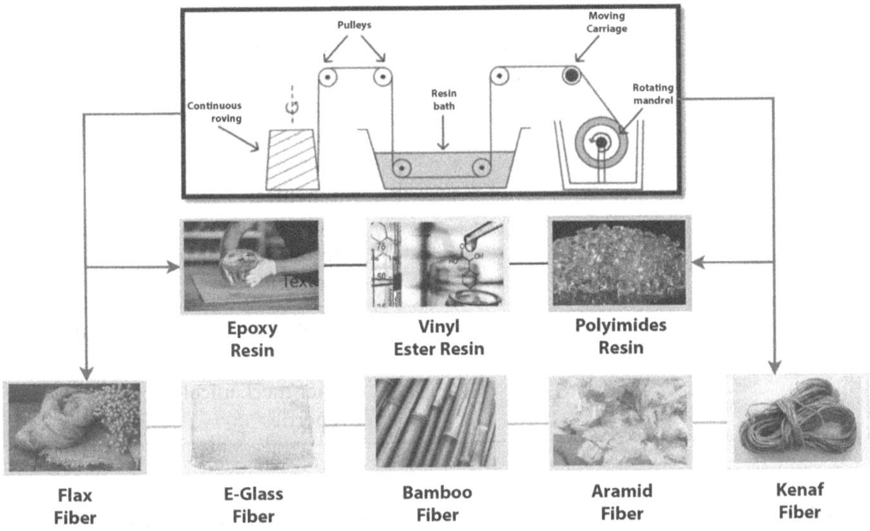

Figure 3.3 Vacuum-assisted techniques.

characteristics are observed, such as a tensile strength of 92 MPa, a flexural strength of 72 MPa, a tensile modulus of 9 GPa, and a flexural modulus up to 27 GPa. The sisal and epoxy matrix used for preparing the composite from fiber were 48% and 37%, respectively, using the resin transfer molding method for fabrication. They observed some outcomes such as a tensile strength and a modulus of 211 MPa and 20 GPa, respectively, for 48% of fiber loading and found outcomes for tensile stiffness and modulus of 183 MPa and 15 GPa for 37% of fiber load; 48% of fiber loading showed better results when compared to 37% of fiber load [99]. Table 3.12 presents several composites made by resin transfer molding.

Table 3.12 Effect of natural fiber-reinforced polymer matrix composites by the vacuum-assisted resin transfer molding.

Reinforcement	Matrix	Fabrication methods	Effect/Influences	References
Glass	Epoxy	Resin transfer molding	VARTM specimens sustained higher load than hand layup specimens.	[100]
Flax	Vinyl ester	Resin transfer molding	Mechanical features were used to design and build agricultural vehicle chemical storage tanks.	[101]
Ramie	Epoxy	Vacuum-assisted resin transfer molding	Hot compression with high pressure and temperature increases composite fiber content. Ramie fabric composites' mechanical characteristics improve with hot compaction.	[102]
Bamboo	Vinyl ester	Resin transfer molding	Treated fiber composites show higher mechanical properties	[103]

We used resin transfer molding to conduct experiments with flax and epoxy with fiber loadings of 46% and 54%, respectively. Findings include mechanical attributes such as a tensile stiffness of 280 MPa and a Young's modulus up to 35 GPa for 46% of fiber loading and mechanical attributes such as a tensile stiffness of 279 MPa and a Young's elastic modulus up to 39 GPa for 54% of fiber loading [99]. The mechanical characteristics of the composite material adapted from fiber named flax yarn and vinyl ester used as a matrix with 35% of fiber loading were studied using resin transfer molding. It is found to have better mechanical characteristics such as a tensile strength of 111 MPa, a flexural strength of 128 MPa, a tensile modulus of 10 GPa, and a flexural modulus of 10 GPa [101]. The flax yarn fiber and vinyl ester matrix were used to prepare composites from fiber containing 24% by using the resin transfer molding method for fabrication. They observed fine results of tensile stiffness and modulus of up to 248 MPa and 24 GPa, respectively [104]. The mechanical characteristics to produce composite material from the fiber like flax yarn and unsaturated polyester matrix with 34% of fiber loading were studied using resin molding. It has improved mechanical properties such as a tensile strength of 143 MPa, a flexural strength of 198 MPa, and a tensile modulus of 14 GPa [104]. The experimental work was carried out for unidirectional blast fiber with vacuum-assisted techniques. Improvements have been made to the material's mechanical properties, such as its interlaminar shear strength and interlaminar fracture toughness [105].

3.8 Spray Molding Technique

Spray-up, or chopping, is an open-mold process like hand layup that is used for constructing boats, containers, vehicle equipment, and bath sets in all forms and sizes. Chopped laminates have strong conformability and may be created faster than hand-laid parts for complicated forms. The operator controls the thickness and uniformity during the spray-up process, making it more operator-dependent than manual layup. Despite the modest manufacturing volume per mold, several molds can be used to achieve substantial production numbers. Simple inexpensive tools and straightforward processing are used in this technique. On-site manufacturing is possible with practically no component size restrictions—thanks to portable equipment. The procedure could be computerized.

Like hand layup, the mold is prepared by applying gel coat and curing it. The resin-soaked "chop" is deposited on the mold after being fed via a chopper gun together with continuous strand glass roving and started resin. After the laminate has been rolled, the glass strands are saturated and the chop is compacted. Chop laminate is stacked to the desired thickness by adding further layers. The chopped laminates can be reinforced with woven roving or knitted textiles from roll stock. It is simple to add core materials of the same sort as those used in hand layup. The molds used in hand layup are identical to these: they are single-cavity, basic fiberglass composite molds. Composites' inexpensive molds span the gamut from little to massive.

Using a spray coating approach, carbon nanotubes (CNTs) have been successfully deposited onto carbon fiber prepregs, allowing for precise regulation of network creation and promising localization. This method has advantages in terms of simplicity, adaptability, and the possibility for industrial scale-up when related to further conventional techniques of inserting CNTs into epoxy matrix for increasing characteristics of composite [106]. Spray-up molding techniques are used for jute fiber composite materials. Based on the findings, a jute/carbon spray-up robot may be fed the motion data of an expert. It can help reduce mistakes made while making jute composite panels and improve the mechanical properties [107].

Hybrid biomaterials of pineapple leaf fiber and glass-vinyl ester are investigated in this study with the use of an automated spray-up approach. This technique is adaptable for use with car and aviation interior parts. The goal of this study was to figure out how to use a manual chop spray cannon attached to an industrial robotic arm to produce large quantities of random discontinuous fiber composite (DFC). The result shows the improvement in tensile and thermal properties [108]. Carbon fiber composite materials including a thermoplastic matrix were employed, and aluminum particles were sprayed for metallization. As part of the research into the process viability, many parameter combinations for deposition have been tried. The morphology of the coating's surface and the particles' and coating's interface were analyzed using SEM. The outcome demonstrated the efficacy of the cold spray approach for the treatment of aluminum coatings on polymeric matrix composites.

3.9 Conclusion

Composites have risen in favor in both the academic and industrial areas because they combine the best features of their component parts. Most of these techniques were successful in enhancing material qualities for potential lightweight applications in crucial areas such as the medical, road transport, marine, shipbuilding, civil, and sports industries. There are many methods for producing composites, and each type of composite has its own special uses. Each material has unique physical properties including flexural and tensile strength, hardness, impact, stiffness, and tensile and flexural modulus. Many studies have been conducted to investigate the characteristics and phase transition of composite materials, and researchers have attempted to mix numerous important ingredients for their manufacturing. Researchers enhanced the proportion of fiber, matrix content, and a few other elements that work together to improve the composite to improve material efficiency. The weight percentage and kind of matrix or fiber material that is employed in a manufacturing technique will impact the level of success that the approach achieves. Because of this, production techniques are different for each type of material that is utilized.

References

1. Sumesh, K.R. and Kanthavel, K., Optimizing various parameters influencing mechanical properties of banana/coir natural fiber composites using grey relational analysis and artificial neural network models. *J. Ind. Text.*, 51, 4, 6705S–6727S, 2022.
2. Indran, S., Edwin Raj, R.D., Daniel, B.S.S., Binoj, J.S., Comprehensive characterization of natural cissus quadrangularis stem fiber composites as an alternate for conventional FRP composites. *J. Bionic Eng.*, 15, 5, 914–923, 2018.
3. Kannan, G. and Thangaraju, R., Evaluation of tensile, flexural and thermal characteristics on agro-waste based polymer composites reinforced with banana fiber/coconut shell filler. *J. Nat. Fibers*, 20, 1, 2154630, 2023.
4. Prasad, V., Muhammed Hunize, C.V., Abhiraj, R., II, Jospeh, M.A., Sekar, K., Ali, M., Mechanical properties of flax fiber reinforced composites manufactured using hand layup and compression molding—A comparison, in: *Lecture Notes in Mechanical Engineering*, pp. 781–789, Pleiades Journals, Springer, Singapore, 2019.
5. Saravana Kumar, A., MaivizhiSelvi, P., Rajeshkumar, L., Delamination in drilling of sisal/banana reinforced composites produced by hand lay-up process. *Appl. Mech. Mater.*, 867, 29–33, 2017.
6. Kumar, S.S., Effect of natural fiber loading on mechanical properties and thermal characteristics of hybrid polyester composites for industrial and construction fields. *Fibers Polym.*, 21, 7, 1508–1514, 2020.
7. Yoganandam, K., Ganeshan, P., Nagarajaganesh, B., Raja, K., Characterization studies on Calotropis procera fibers and their performance as reinforcements in epoxy matrix. *J. Nat. Fibers*, 17, 12, 1706–1718, 2020.
8. Ramakrishnan, S., Krishnamurthy, K., Rajasekar, R., Rajeshkumar, G., An experimental study on the effect of nano-clay addition on mechanical and water absorption behaviour of jute fibre reinforced epoxy composites. *J. Ind. Text.*, 49, 5, 597–620, 2019.
9. Baskaran, P.G., Kathiresan, M., Pandiarajan, P., Effect of alkali-treatment on Structural, thermal, tensile properties of dichrostachys cinerea bark fiber and its composites. *J. Nat. Fibers*, 19, 2, 433–449, 2022.
10. Vinod, A., Vijay, R., Singaravelu, D.L., Thermo mechanical characterization of calotropis gigantea stem powder-filled jute fiber-reinforced epoxy composites. *J. Nat. Fibers*, 15, 5, 648–657, 2018.
11. Kumaar, A.S., Senthilkumar, A., Saravanakumar, S.S., Senthamaraikannan, P., Loganathan, L., Muthu Chozha Rajan, B., Mechanical properties of alkali-treated carica papaya fiber-reinforced epoxy composites. *J. Nat. Fibers*, 19, 1, 269–279, 2022.
12. Reddy, K.H., Reddy, R.M., Ramesh, M., Mohana Krishnudu, D., Reddy, B.M., Rao, H.R., Impact of alkali treatment on characterization of tapsi (sterculia urens) natural bark fiber reinforced polymer composites. *J. Nat. Fibers*, 18, 3, 378–389, 2021.

13. Vimalanathan, P., Venkateshwaran, N., Santhanam, V., Mechanical, dynamic mechanical, and thermal analysis of shorearobusta-dispersed polyester composite. *Int. J. Polym. Anal. Charact.*, 21, 4, 314–326, 2016.
14. Razali, N., Sapuan, S.M., Razali, N., Mechanical properties and morphological analysis of roselle/sugar palm fiber reinforced vinyl ester hybrid composites, in: *Natural Fiber Reinforced Vinyl Ester and Vinyl Polymer Composites: Development, Characterization and Applications*, pp. 169–180, Elsevier, Woodhead Publishing, 2018.
15. Ashori, A., Kiani, H., Mozaffari, S.A., Mechanical properties of reinforced polyvinyl chloride composites: Effect of filler form and content. *J. Appl. Polym. Sci.*, 120, 3, 1788–1793, 2011.
16. Dawit, J.B., Lemu, H.G., Regassa, Y., Akessa, A.D., Investigation of the mechanical properties of Acacia tortilis fiber reinforced natural composite. *Mater. Today Proc.*, 38, 2953–2958, 2020.
17. Ashik, K.P., Sharma, R.S., Guptha, V.L.J., Investigation of moisture absorption and mechanical properties of natural/glass fiber reinforced polymer hybrid composites. *Mater. Today Proc.*, 5, 1, 3000–3007, 2018.
18. Krishnakumari, A., Devaraju, A., Saravanan, M., Evaluation of mechanical properties of hybrid rootfiber reinforced polymer composites. *Mater. Today Proc.*, 5, 6, 14560–14566, 2018.
19. Mishra, V., Biswas, S., Physical and mechanical properties of bi-directional jute fiber epoxy composites. *Procedia Eng.*, 51, 561–566, 2013.
20. Deo, C. and Acharya, S.K., Effect of moisture absorption on mechanical properties of chopped natural fiber reinforced epoxy composite. *J. Reinf. Plast. Compos.*, 29, 16, 2513–2521, 2010.
21. Vijay, R. and Singaravelu, D.L., Experimental investigation on the mechanical properties of Cyperus pangorei fibers and jute fiber-based natural fiber composites. *Int. J. Polym. Anal. Charact.*, 21, 7, 617–627, 2016.
22. Yan, L., Chouw, N., Yuan, X., Improving the mechanical properties of natural fibre fabric reinforced epoxy composites by alkali treatment. *J. Reinf. Plast. Compos.*, 31, 6, 425–437, 2012.
23. Srinivasan, V.S., Rajendra Boopathy, S., Sangeetha, D., Vijaya Ramnath, B., Evaluation of mechanical and thermal properties of banana-flax based natural fibre composite. *Mater. Des.*, 60, 620–627, 2014.
24. Lokesh, P., Surya Kumari, T.S.A., Gopi, R., Loganathan, G.B., A study on mechanical properties of bamboo fiber reinforced polymer composite. *Mater. Today Proc.*, 22, 897–903, 2020.
25. Barua, A., Jeet, S., Bagal, D.K., Satapathy, P., Agrawal, P.K., Evaluation of mechanical behavior of hybrid natural fiber reinforced nanosic particles composite using hybrid Taguchi-Cocoso method. *IJITEE*, 8, 10, 3341–3345, 2019.
26. Premnath, A.A., Impact of surface treatment on the mechanical properties of sisal and jute reinforced with epoxy resin natural fiber hybrid composites. *J. Nat. Fibers*, 16, 5, 718–728, 2019.

27. Magarajan, U., Kumar, S.D., Arvind, D., Kannan, N., Hemanandan, P., A comparative study on the static mechanical properties of glass fibre vs glass-jute fibre polymer composite. *Mater. Today Proc.*, 5, 2, 6711–6716, 2018.
28. Rajak, D.K., Pagar, D.D., Menezes, P.L., Linul, E., Fiber-reinforced polymer composites: Manufacturing, properties, and applications. *Polym.*, 11, 10, 1667, 2019.
29. Rajak, D.K., Pagar, D.D., Kumar, R., Pruncu, C., II, Recent progress of reinforcement materials: A comprehensive overview of composite materials. *J. Mater. Res. Technol.*, 8, 6, 6354–6374, 2019.
30. Manimaran, P., Jeyasekaran, A.S., Purohit, R., Pillai, G.P., An Experimental and numerical investigation on the mechanical properties of addition of wood flour fillers in red banana peduncle fiber reinforced polyester composites. *J. Nat. Fibers*, 17, 8, 1140–1158, 2020.
31. Kumaresan, M., Sathish, S., Karthi, N., Effect of fiber orientation on mechanical properties of sisal fiber reinforced epoxy composites. *J. Appl. Sci. Eng.*, 18, 3, 289–294, 2015.
32. Dharmalingam, S., Meenakshisundaram, O., Elumalai, V., Boopathy, R.S., An investigation on the interfacial adhesion between amine functionalized luffa fiber and epoxy resin and its effect on thermal and mechanical properties of their composites. *J. Nat. Fibers*, 18, 12, 2254–2269, 2021.
33. Karthi, N., Kumaresan, K., Rajeshkumar, G., Gokulkumar, S., Sathish, S., Tribological and thermo-mechanical performance of chemically modified musa acuminata/corchorus capsularis reinforced hybrid composites. *J. Nat. Fibers*, 19, 12, 4640–4653, 2022.
34. Prabhu, L., Krishnaraj, V., Sathish, S., Gokulkumar, S., Karthi, N., Rajeshkumar, L., Balaji, D., Vigneshkumar, N., Elango, K.S., Karpagam, J., Vijayalakshmi, V.J., Gowarthan, E.R., Jayakumar, H., Experimental investigation on mechanical properties of flax/banana/industrial waste tea leaf fiber reinforced hybrid polymer composites. *Mater. Today Proc.*, 45, 8136–8143, 2021.
35. Stalin, B., Nagaprasad, N., Vignesh, V., Ravichandran, M., Rajini, N., Ismail, S.O., Mohammad, F., Evaluation of mechanical, thermal and water absorption behaviors of Polyalthia longifolia seed reinforced vinyl ester composites. *Carbohydr. Polym.*, 248, 116748, 2020.
36. Rajamuneeswaran, S., Jayabal, S., Kalyana Sundaram, S., Balaji, N.S., Ramkumar, P., Effect of chitosan particle addition on the tensile and flexural strength of coir fiber reinforced polyester composites. *Appl. Mech. Mater.*, 813–814, 30–33, 2015.
37. Prabhu, L., Krishnaraj, V., Gokulkumar, S., Sathish, S., Sanjay, M.R., Siengchin, S., Mechanical, chemical and sound absorption properties of glass/kenaf/waste tea leaf fiber-reinforced hybrid epoxy composites. *J. Ind. Text.*, 51, 10, 1674–1700, 2022.
38. Chandrika, V.S., Anamika, A., Jeeva, C., Perumal, B., Kumar, S.S., Roseline, J.F., Raghavan, I.K., Natural fiber incorporated polymer matrix composites

for electronic circuit board applications. *Adv. Mater. Sci. Eng.*, 2022, 3035169, 2022.
39. Scaffaro, R., Citarrella, M.C., Gulino, E.F., Morreale, M., Hedysarum coronarium-based green composites prepared by compression molding and fused deposition modeling. *Materials*, 15, 2, 465, 2022.
40. Atmakuri, A., Palevicius, A., Janusas, G., Eimontas, J., Investigation of hemp and flax fiber-reinforced ecopoxy matrix biocomposites: Morphological, mechanical, and hydrophilic properties. *Polymers*, 14, 21, 4530, 2022.
41. Sumesh, K.R., Kavimani, V., Rajeshkumar, G., Indran, S., Khan, A., Mechanical, water absorption and wear characteristics of novel polymeric composites: Impact of hybrid natural fibers and oil cake filler addition. *J. Ind. Text.*, 51, 4, 5910S–5937S, 2022.
42. Stalin, B. and Athijayamani, A., The performance of bio waste fibres reinforced polymer hybrid composite. *Int. J. Mater. Eng. Innov.*, 7, 1, 15–25, 2016.
43. Gokulkumar, S., Thyla, P.R., Prabhu, L., Sathish, S., Characterization and comparative analysis on mechanical and acoustical properties of camellia sinensis/ananas comosus/glass fiber hybrid polymer composites. *J. Nat. Fibers*, 18, 7, 978–994, 2021.
44. Athijayamani, A., Stalin, B., Sidhardhan, S., Boopathi, C., Parametric analysis of mechanical properties of bagasse fiber-reinforced vinyl ester composites. *J. Compos. Mater.*, 50, 4, 481–493, 2016.
45. Karthi, N., Kumaresan, K., Sathish, S., Prabhu, L., Gokulkumar, S., Balaji, D., Vigneshkumar, N., Rohinth, S., Rafiq, S., Muniyaraj, S., Pavithran, S., Effect of weight fraction on the mechanical properties of flax and jute fibers reinforced epoxy hybrid composites. *Mater. Today Proc.*, 45, 8006–8010, 2021.
46. Fazli, A., Stevanovic, T., Rodrigue, D., Recycled HDPE/natural fiber composites modified with waste tire rubber: A comparison between injection and compression molding. *Polymers*, 14, 15, 3197, 2022.
47. Kavimani, V., Gopal, P.M., Sumesh, K.R., Elanchezhian, R., Improvement on mechanical and flame retardancy behaviour of bio-exfoliated graphene-filled epoxy/glass fibre composites using compression moulding approach. *Polym. Bull.*, 79, 8, 6289–6307, 2022.
48. Sumesh, K.R. and Kanthavel, K., Optimizing various parameters influencing mechanical properties of banana/coir natural fiber composites using grey relational analysis and artificial neural network models. *J. Ind. Text.*, 51, 4, 6705S–6727S, 2022.
49. Alam, M.J., Dewan, M.W., Paul, S.K., Sharmin, K., Investigation of jute and glass fibre reinforced hybrid composites manufactured through compression molding process. *Int. J. Eng. Mater. Manuf.*, 7, 1, 35–46, 2022.
50. Pulikkalparambil, H., Nandi, D., Rangappa, S.M., Prasanth, S., Siengchin, S., Polymer composites from natural fibers and recycled waste surgical masks during COVID-19 pandemic. *Polym. Compos.*, 43, 6, 3944–3950, 2022.
51. Prabhu, L., Krishnaraj, V., Sathish, S., Gokulkumar, S., Karthi, N., Study of mechanical and morphological properties of jute-tea leaf fiber reinforced

hybrid composites: Effect of glass fiber hybridization. *Mater. Today Proc.*, 27, 2372–2375, 2019.
52. Sathish, S., Kumaresan, K., Prabhu, L., Gokulkumar, S., Karthi, N., Vigneshkumar, N., Experimental investigation of mechanical and morphological properties of flax fiber reinforced epoxy composites incorporating SiC and Al2O3. *Mater. Today Proc.*, 27, 2249–2253, 2019.
53. Awais, H., Nawab, Y., Amjad, A., Anjang, A., Akil, H.M., Zainol Abidin, M.S., Effect of comingling techniques on mechanical properties of natural fibre reinforced cross-ply thermoplastic composites. *Compos. B Eng.*, 177, 107279, 2019.
54. Adekunle, K., Åkesson, D., Skrifvars, M., Biobased composites prepared by compression molding with a novel thermoset resin from soybean oil and a natural-fiber reinforcement. *J. Appl. Polym. Sci.*, 116, 3, 1759–1765, 2010.
55. Singleton, A.C.N., Baillie, C.A., Beaumont, P.W.R., Peijs, T., On the mechanical properties, deformation and fracture of a natural fibre/recycled polymer composite. *Compos. B. Eng.*, 34, 6, 519–526, 2003.
56. Mulinari, D.R., Baptista, C.A.R.P., Souza, J.V.C., Voorwald, H.J.C., Mechanical properties of coconut fibers reinforced polyester composites. *Procedia Eng.*, 10, 2074–2079, 2011.
57. Shibata, S., Cao, Y., Fukumoto, I., Press forming of short natural fiber-reinforced biodegradable resin: Effects of fiber volume and length on flexural properties. *Polym. Test.*, 24, 8, 1005–1011, 2005.
58. Prasad, N., Agarwal, V.K., Sinha, S., Banana fiber reinforced low-density polyethylene composites: Effect of chemical treatment and compatibilizer addition. *Iran. Polym. J. (English Edition)*, 25, 3, 229–241, 2016.
59. Hu, R. and Lim, J.K., Fabrication and mechanical properties of completely biodegradable hemp fiber reinforced polylactic acid composites. *J. Compos. Mater.*, 41, 13, 1655–1669, 2007.
60. Khondker, O.A., Ishiaku, U.S., Nakai, A., Hamada, H., Fabrication mechanical properties of unidirectional jute/PP composites using jute yarns by film stacking method. *J. Polym. Environ.*, 13, 2, 115–126, 2005.
61. Zampaloni, M., Pourboghrat, F., Yankovich, S.A., Rodgers, B.N., Moore, J., Drzal, L.T., Mohanty, A.K., Misra, M., Kenaf natural fiber reinforced polypropylene composites: A discussion on manufacturing problems and solutions. *Compos. Part A Appl. Sci. Manuf.*, 38, 6, 1569–1580, 2007.
62. Liu, W., Drzal, L.T., Mohanty, A.K., Misra, M., Influence of processing methods and fiber length on physical properties of kenaf fiber reinforced soy based biocomposites. *Compos. B Eng.*, *38*, 3, 352–359, 2007.
63. Bledzki, A.K. and Faruk, O., Wood fiber reinforced polypropylene composites: Compression and injection molding process. *Polym. Plast. Technol. Eng.*, 43, 3, 871–888, 2004.
64. Yu, T., Ren, J., Li, S., Yuan, H., Li, Y., Effect of fiber surface-treatments on the properties of poly(lactic acid)/ramie composites. *Compos. Part A Appl. Sci. Manuf.*, 41, 4, 499–505, 2010.

65. Nayak, S., Khuntia, S.K., Mohanty, S.D., Mohapatra, J., Mall, T.K., An experimental study of physical, mechanical and morphological properties of alkali treated moringa/areca based natural fiber hybrid composites. *J. Nat. Fibers*, 19, 2, 630–641, 2022.
66. Werlen, V., Rytka, C., Wegmann, S., Philipp, H., Khalaf, Y., Michaud, V., Brauner, C., Dransfeld, C., Novel tooling for direct melt impregnation of textile with variotherm injection moulding: Methodology and proof of concept. *J. Compos. Mater.*, 56, 28, 4245–4257, 2022.
67. Siva, R., Nemali, S.S.R., Kunchapu, S.K., Gokul, K., Kumar, T.A., Comparison of mechanical properties and water absorption test on injection molding and extrusion-injection molding thermoplastic hemp fiber composite. *Mater. Today Proc.*, 47, 4382–4386, 2021.
68. Billah, M.M., Rabbi, M.S., Hasan, A., A review on developments in manufacturing process and mechanical properties of natural fiber composites. *JEA*, 2, 01, 13–23, 2021.
69. Seifu, B., Singh, B., Gutu, J.M., Legesse, D., Mechanical behaviours of hybrid ensete/sisal fiber, reinforced polyethylene composite materials for injection moulding. *SN Appl. Sci.*, 2, 5, 1–25, 2020.
70. Jeffrey, J.A., Sivakumar, A., Kumar, R.N., Anbazhagan, A., Manojkumar, G., Al Obaid, S., Alfarraj, S., Sivakumar, S., Rajkumar, S., Influence of flax fibre hybridization on mechanical behaviour of sisal fibre-polypropylene composites prepared with an injection moulding machine. *Adv. Mater. Sci. Eng.*, 2021, 1–7, 2021.
71. Dolza, C., Fages, E., Gonga, E., Gomez-Caturla, J., Balart, R., Quiles-Carrillo, L., Development and characterization of environmentally friendly wood plastic composites from biobased polyethylene and short natural fibers processed by injection moulding. *Polymers*, 13, 11, 1692, 2021.
72. Nyström, B., Joffe, R., Långström, R., Microstructure and strength of injection molded natural fiber composites. *J. Reinf. Plast. Compos.*, 26, 6, 579–599, 2007.
73. Anuar, H., Zuraida, A., Kovacs, J.G., Tabi, T., Improvement of mechanical properties of injection-molded polylactic acid-kenaf fiber biocomposite. *J. Thermoplast. Compos. Mater.*, 25, 2, 153–164, 2012.
74. Barkoula, N.M., Garkhail, S.K., Peijs, T., Effect of compounding and injection molding on the mechanical properties of flax fiber polypropylene composites. *J. Reinf. Plast. Compos.*, 29, 9, 1366–1385, 2010.
75. Tajvidi, M., Falk, R.H., Hermanson, J.C., Effect of natural fibers on thermal and mechanical properties of natural fiber polypropylene composites studied by dynamic mechanical analysis. *J. Appl. Polym. Sci.*, 101, 6, 4341–4349, 2006.
76. Subasinghe, A.D.L., Das, R., Bhattacharyya, D., Fiber dispersion during compounding/injection molding of PP/kenaf composites: Flammability and mechanical properties. *Mater. Des.*, 86, 500–507, 2015.

77. Sain, M., Suhara, P., Law, S., Bouilloux, A., Interface modification and mechanical properties of natural fiber-polyolefin composite products. *J. Reinf. Plast. Compos.*, 24, 2, 121–130, 2005.
78. Panthapulakkal, S. and Sain, M., Injection-molded short hemp fiber/glass fiber-reinforced polypropylene hybrid composites-mechanical, water absorption and thermal properties. *J. Appl. Polym. Sci.*, 103, 4, 2432–2441, 2007.
79. Aridi, N.A.M., Sapuan, S.M., Zainudin, E.S., AL-Oqla, F.M., Mechanical and morphological properties of injection-molded rice husk polypropylene composites. *Int. J. Polym. Anal. Charact.*, 21, 4, 305–313, 2016.
80. Billah, M.M., Rabbi, M.S., Hasan, A., A review on developments in manufacturing process and mechanical properties of natural fiber composites. *JEA*, 2, 01, 13–23, 2021.
81. Yang, Y., Ota, T., Morii, T., Hamada, H., Mechanical property and hydrothermal aging of injection molded jute/polypropylene composites. *J. Mater. Sci.*, 46, 8, 2678–2684, 2011.
82. Faruk, O., Bledzki, A.K., Fink, H.P., Sain, M., Progress report on natural fiber reinforced composites, in: *Macromolecular Materials and Engineering*, vol. 299, pp. 9–26, Wiley-VCH Verlag, 299, 1, 9–26, 2014.
83. Feldmann, M., The effects of the injection moulding temperature on the mechanical properties and morphology of polypropylene man-made cellulose fibre composites. *Compos. Part A Appl. Sci. Manuf.*, 87, 146–152, 2016.
84. Madsen, B. and Lilholt, H., Physical and mechanical properties of unidirectional plant fibre composites-an evaluation of the influence of porosity. *Compos. Sci. Technol.*, 63, 9, 1265–1272, 2003.
85. Zhang, W., Wang, C., Gu, S., Yu, H., Cheng, H., Wang, G., Physical-mechanical properties of bamboo fiber composites using filament winding. *Polymers*, 13, 17, 2913, 2021.
86. Shrigandhi, G.D. and Kothavale, B.S., Biodegradable composites for filament winding process. *Mater. Today Proc.*, 42, 2762–2768, 2021.
87. Alshgari, R.A., Hemalatha, N., Suryavanshi, A., Prasad, D. V. S. S. S. V., Subalakshmi, R., Abirami, M., Amudha, M.J.R., Wabaidur, S.M., Islam, M.A., Christopher, D., Investigation on physical and mechanical properties of abaca fiber composites using filament winding. *Adv. Polym. Technol.*, 2022, 2022.
88. Ornaghi, H.L., Neves, R.M., Monticeli, F.M., Almeida, J.H.S., Viscoelastic characteristics of carbon fiber-reinforced epoxy filament wound laminates. *Compos. Commun.*, 21, 100418, 2020.
89. Zhang, Q., Wu, J., Gao, L., Liu, T., Zhong, W., Sui, G., Yang, X., Influence of a liquid-like MWCNT reinforcement on interfacial and mechanical properties of carbon fiber filament winding composites. *Polymer*, 90, 193–203, 2016.
90. Alshgari, R.A., Hemalatha, N., Suryavanshi, A., Prasad, D. V. S. S. S. V., Subalakshmi, R., Abirami, M., Amudha, M.J.R., Wabaidur, S.M., Islam, M.A., Christopher, D., Investigation on physical and mechanical properties

of abaca fiber composites using filament winding. *Adv. Polym. Technol.*, 2022, 2022.

91. Dun, M., Hao, J., Wang, W., Wang, G., Cheng, H., Sisal fiber reinforced high density polyethylene pre-preg for potential application in filament winding. *Compos. B Eng.*, 159, 369–377, 2019.
92. Bakar, M.S.A., Salit, M.S., Yusoff, M.Z.M., Zainudin, E.S., Ya, H.H., The crashworthiness performance of stacking sequence on filament wound hybrid composite energy absorption tube subjected to quasi-static compression load. *J. Mater. Res. Technol.*, 9, 1, 654–666, 2020.
93. Ornaghi, H.L., Neves, R.M., Monticeli, F.M., Almeida, J.H.S., Viscoelastic characteristics of carbon fiber-reinforced epoxy filament wound laminates. *Compos. Commun.*, 21, 100418, 2020.
94. Misri, S., Ishak, M.R., Sapuan, S.M., Leman, Z., Filament winding process for kenaf fibre reinforced polymer composites, in: *Manufacturing of Natural Fibre Reinforced Polymer Composites*, pp. 369–383, Springer International Publishing, 369–383, 2015.
95. Mindermann, P., Pérez, M.G., Knippers, J., Gresser, G.T., Investigation of the fabrication suitability, structural performance, and sustainability of natural fibers in coreless filament winding. *Materials*, 15, 9, 3260, 2022.
96. Ansari, S.M., Mohd, C., Ghazali, R., Othman, N.S., The effect of winding speed on the mechanical properties of kenaf fiber reinforced geopolymer composites via filament winding technique. *J. Advanced Res. Eng. Knowledge J. Homepage*, 6, 35–41, 2019.
97. Naseva, S., Srebrenkoska, V., Risteska, S., Stefanovska, M., Srebrenkoska, S., Mechanical properties of filament wound pipes: effects of winding angles. *Qual. Life (Banja Luka)-APEIRON*, 11, 1–2, 10-15, 2015.
98. Rodríguez, E., Petrucci, R., Puglia, D., Kenny, J.M., Vázquez, A., Characterization of composites based on natural and glass fibers obtained by vacuum infusion. *J. Compos. Mater.*, 39, 3, 265–282, 2005.
99. Oksman, K., Wallström, L., Berglund, L.A., Filho, R.D.T., Morphology and mechanical properties of unidirectional sisal-epoxy composites. *J. Appl. Polym. Sci.*, 84, 13, 2358–2365, 2002.
100. Sevkat, E. and Brahimi, M., The bearing strength of pin loaded woven composites manufactured by vacuum assisted resin transfer moulding and hand lay-up techniques. *Procedia Eng.*, 10, 153–158, 2011.
101. Kong, C., Park, H., Lee, J., Study on structural design and analysis of flax natural fiber composite tank manufactured by vacuum assisted resin transfer molding. *Mater. Lett.*, 130, 21–25, 2014.
102. Gu, Y., Tan, X., Yang, Z., li, M., Zhang, Z., Hot compaction and mechanical properties of ramie fabric/epoxy composite fabricated using vacuum assisted resin infusion molding. *Mater. Des.*, 56, 852–861, 2014.
103. Kim, H., Okubo, K., Fujii, T., Takemura, K., Influence of fiber extraction and surface modification on mechanical properties of green composites with bamboo fiber. *J. Adhes. Sci. Technol.*, 27, 12, 1348–1358, 2013.

104. Goutianos, S., Peijs, T., Nystrom, B., Skrifvars, M., Development of flax fibre based textile reinforcements for composite applications. *Appl. Compos. Mater.*, 13, 4, 199–215, 2006.
105. Scalici, T., Pitarresi, G., Badagliacco, D., Fiore, V., Valenza, A., Mechanical properties of basalt fiber reinforced composites manufactured with different vacuum assisted impregnation techniques. *Compos. B Eng.*, 104, 35–43, 2016.
106. Zhang, H., Liu, Y., Kuwata, M., Bilotti, E., Peijs, T., Improved fracture toughness and integrated damage sensing capability by spray coated CNTs on carbon fibre prepreg. *Compos. Part A Appl. Sci. Manuf.*, 70, 102–110, 2015.
107. Kikuchi, T., Tani, Y., Takai, Y., Goto, A., Hamada, H., Mechanical properties of jute composite by spray up fabrication method. *Energy Procedia*, 56, C, 289–297, 2014.
108. Zin, M.H., Abdan, K., Mazlan, N., Zainudin, E.S., Liew, K.E., Norizan, M.N., Automated spray up process for pineapple leaf fibre hybrid biocomposites. *Compos. B Eng.*, 177, 107306, 2019.

4

Electrical Discharge Machining of Al-B_4C Composite for Biomedical Applications

S. Suresh Kumar[1], S. Thirumalai Kumaran[2*], G. Kalusuraman[1] and G. S. Samy[3]

[1]Faculty of Mechanical Engineering, Kalasalingam Academy of Research and Education, Krishnankoil, Tamil Nadu, India
[2]Department of Mechanical Engineering, PSG Institute of Technology and Applied Research, Coimbatore, Tamil Nadu, India
[3]Centre for Research in Advanced Materials and Manufacturing (RAMAM), Department of Mechanical Engineering, Academy of Maritime Education and Training, Chennai, Tamil Nadu, India

Abstract

Aluminum-based metal matrix composites offer a wide range of uses in biomedical applications. In this present study, a detailed examination is done to ascertain the impact of process parameters of electrical discharge machining (EDM) such as current (I), pulse on time (Ton), and voltage (V) during machining of boron carbide (B_4C)-reinforced aluminum AA(6351) composite. The output characteristics such as surface roughness (SR), overcut (OC), and material removal rate (MRR) are investigated after the EDM process on the work material. The experimental results show that if the I and Ton are increased, a poor surface finish is achieved. OC is increased eventually for all of the improvement in the process parameters. In the case of MRR, it increases if I and V are increased. However, a negative trend is observed while increasing Ton. Finally, the machined surface characteristics reported the craters and lay formation.

Keywords: EDM, composite, surface roughness, overcut, material removal rate

**Corresponding author*: thirumalaikumaran@yahoo.com

S. Thirumalai Kumaran and Tae Jo Ko (eds.) Sustainable Machining and Green Manufacturing, (65–78) © 2024 Scrivener Publishing LLC

4.1 Introduction

Electrical discharge machining (EDM) is a novel material removal process that uses electricity to create an electric spark between the electrode and the work material. The spark acts as thermal energy and is primarily responsible for the material removal. As a result, waveforms of current and voltage are employed to monitor the EDM process in real time [1]. The electrodes and the workpiece are submerged in the dielectric medium. The dielectric medium flushes out the eroded materials during machining. However, debris is deposited on the machined surface during the pulse off time [2]. The narrowest spacing between two electrodes determines the position of a spark produced between them. The frequency of sparking might reach thousands of sparks per second that is responsible for producing the craters' shape profile [3]. Harder materials are machined using EDM. EDM studies were conducted on mild steel utilizing various forms of electrodes and the circular one is preferred due to its lower wear and good surface finish. EDM is also used to machine the soft and light materials like aluminum [4]. Machining parameters such as pulse current (I), duty factor and duration/pulse on time (Ton), and voltage (V) have a greater impact on response variables such as the material removal rate (MRR), tool electrode wear, overcut (OC), and surface roughness (SR). The influences of machining parameters on performance response variables such as redial OC, surface quality, and electrode wear have been investigated by gray relational analysis. The key parameters for assessing the process performance are the wear ratio of electrode, quality of surface, and consumption of power. Additionally, the machining performance associated with I, duty factor and Ton, and V [5].

Metal matrix composites (MMCs) are formed by fusing two or more materials with distinct physical and chemical properties. The stir cast aluminum matrix composite has good mechanical properties, allowing it to be used in a variety of applications such as brake discs and automotive drive shafts. The process was investigated using gray relational analysis. Among the several nonconventional machining processes, EDM is appropriate for machining particle-reinforced MMCs [6]. Magnesium, aluminum, and titanium alloys are commonly based on MMCs [7]. Nontraditional machining processes, such as EDM, are used in a variety of such applications to facilitate the machining of aluminum-based MMCs that are well known for its properties and are employed in numerous applications in the engineering industry. The preparation of composite materials involved several manufacturing processes. Stir casting is among the most used methods for producing aluminum-based MMCs. Machining ceramic-reinforced MMCs is much important, and choosing the right cutting parameters is crucial for enhancing the product quality [8, 9]. Because of their light weight, high strength, stiffness, and high temperature resistance, aluminum MMCs have been utilized successfully in the military, automotive, and aerospace industries for decades. Furthermore, aluminum MMCs have superior physical and mechanical qualities when compared to other materials [10]. The most essential component influencing SR is I, and studies have discovered that a lower I results in a superior surface finish [11]. In EDM machining, the control of recast layer formation is a prime key factor to improve the quality of the surface profile. The recast layer is primarily influenced by the parameters such as I, V, Ton, and pulse off time. The recast layer of the machined surface is majorly influenced by an increase in I associated with other parameters [12]. The machining efficiency associated with the MRR, electrode wear, and microstructure. It has been significantly influenced by the Ton and I. The dielectric medium also plays a reasonable contribution to improve the MRR and electrode wear rate along with the input parameters [13]. The Ton highly contributed to improving the MRR and tool life. On the other hand, I also improved the MRR but significantly reduced the tool life rather than the V for machining the ANSI D6 [14]. The attainment of accurate geometrical shape is another major issue in EDM, and this can be achieved by minimizing the OC of the profile. I and Ton are major substantial parameters to control the OC in EDM [15–17]. In EDM, the machining performance is influenced by different input parameters concerning their operating range value, electrode, and work material. Different optimization techniques are used to optimize the suitable input parameters with respect to the output responses, electrode, and work material [18–20]. The magnetic field-assisted EDM (MF-EDM) process improves the MRR and also reduces electrode wear compared to conventional EDM. It is found that the MRR increased by 21.9% and the material erosion and energy utilization efficiency increased by 22.6% and 15.2%, respectively [21]. The MF-EDM technique also improves the stability of machining by flushing out the debris from the machining zone and maintaining the stability of spark discharge during machining [22]. Furthermore, the MF-EDM accelerates multiple spark discharge points that improves the efficiency of energy utilization, MRR, and power consumption and reduces carbon emission. The presence of a magnetic field in the machining zone creates a Lorentz force that helps to flush out the existence of debris between the electrode and workpiece surface [23].

An experimental examination is performed in this research to investigate the machinability performance of boron carbide (B_4C)-reinforced aluminum AA(6351) composite to discover how cutting factors impact machinability. I, Ton, and V are the primary machining parameters used in the EDM process. These processes' primary goal is to explore the impact on the SR, OC, and MRR. Finally, with the help of microscopic examination, the machining quality was assessed through surface morphology.

4.2 Materials and Methods

The AA6351-based composite material is fabricated by 10 wt. % of B_4C that has an average particle size of 100 μm as shown in Figure 4.1. The B_4C is mainly used for absorbing and screening of neutrons in nuclear industries due to its high neutron-absorbing characteristic. Recently, the B_4C plays a vital role in modern medical application as a carrier in Boron Neutron Capture Therapy (BNCT). The researcher found that a low volume of tumor is noticed for B_4C-mediated BNCT compared to neutron and control [24, 25]. The EDS spectrum has been used to measure the particle size and composition of boron and carbide. The analyzed EDS spectrum results depict approximately 73% of boron and 27% carbide in the composition as shown in Figure 4.2.

The sliced AA6351 is heated above the recrystallization temperature to melt the metal, and preheated B_4C particles are mixed with molten aluminum by external stirring action to accomplish the uniform mixing. The B_4C-incorporated molten metal is poured into the circular die to fabricate the cylindrical disc-shaped composite. The irregular peripheral surface of the fabricated composite is removed by fine turning to get a smooth surface. The ASTM B557M-10 test standard is followed to investigate the mechanical properties, and results are shown in Figure 4.3. The composite material

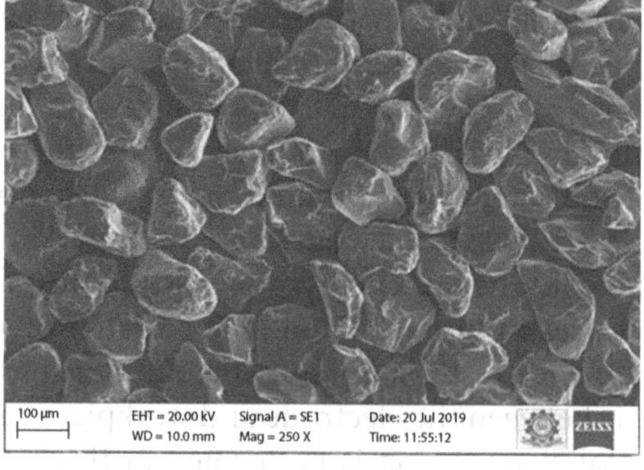

Figure 4.1 Morphology of B_4C particles.

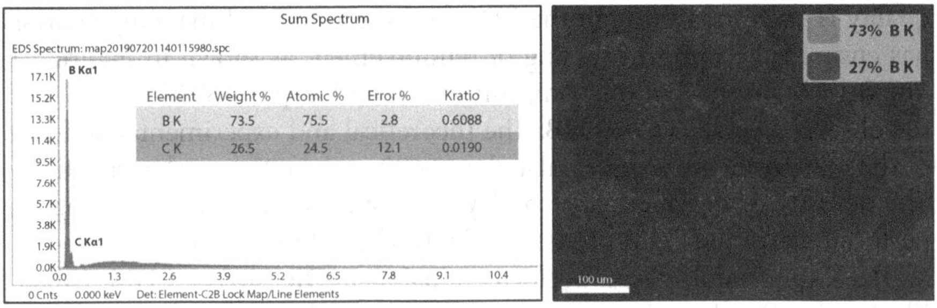

Figure 4.2 EDS spectrum of B_4C particles.

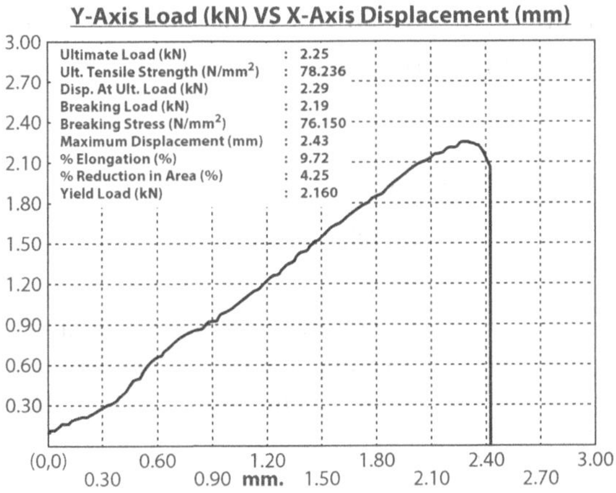

Figure 4.3 Mechanical property of MMC.

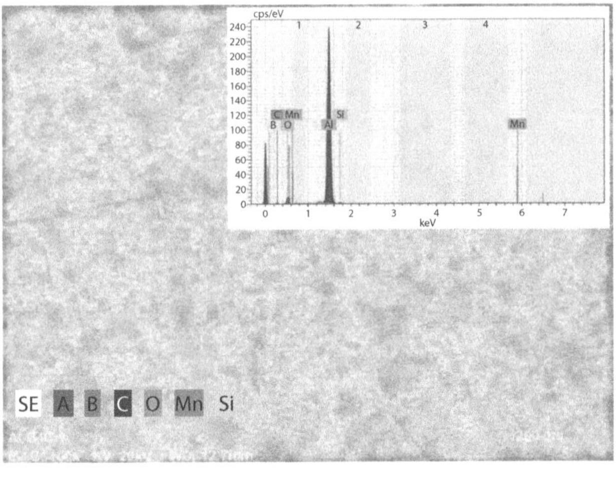

Figure 4.4 EDS elemental mapping of MMC.

quality is examined through the EDS, and the elemental mapping analysis has ensured the uniform mixing of reinforcement as shown in Figure 4.4. The Brinell hardness of the composite has tested as per ASTM E10-15, and the observed value is 69.33 HB. The theoretical and experimental densities of the composite are also calculated as 2,689 and 2,601 kg/m^3, respectively. The porosity of the material also plays an important role in improving the performance of the EDM process, and it has been found as 3.29%.

The experiments were performed using a Die sinker EDM (Make: Electronica Pvt. Ltd., Model: Smart ZNC). All of the trials for machining the samples used a depth of 2 mm. Table 4.1 displays the experimental conditions and process details.

The experiments were carried out by varying the input parameters. The maximum working current of the EDM parameter is 50A. Tool acts

Table 4.1 Experimental particulars.

Equipment	Specification
EDM	Tool: Copper circular electrode (diameter -16 mm and height – 70 mm) Flushing Pressure: 1.5 kg/cm^2 Dielectric medium: Electrol EDM oil
Surface Roughness Tester	Mitutoyo – Surftest SJ-301
Overcut	CMM (Make – Tesa Micro-Hite 3D)

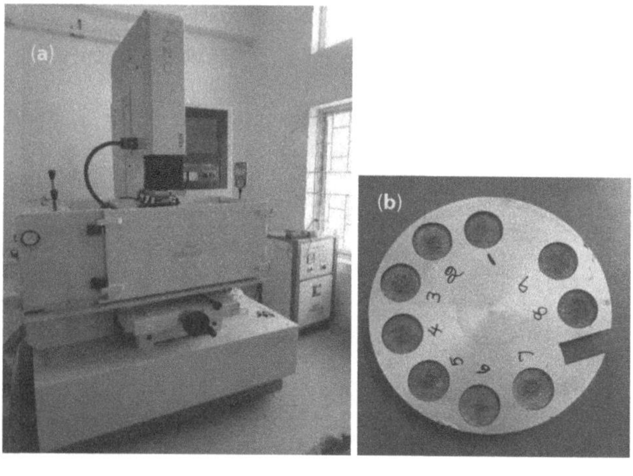

Figure 4.5 (a) EDM facility, (b) machined specimen.

Table 4.2 Experimental results.

S. no	I (amp)	Ton (μs)	V (V)	SR (μm)	OC (mm)	MRR (mm^3/ sec)
1.	10	100	40	11.129	0.771	1.169
2.	10	200	45	13.676	1.046	1.696
3.	10	300	50	15.216	1.096	1.888
4.	15	100	45	10.627	0.907	1.549
5.	15	200	50	13.148	1.004	1.752
6.	15	300	40	15.503	1.010	1.612
7.	20	100	50	13.11	0.98	1.17
8.	20	200	40	14.164	1.165	1.673
9.	20	300	45	13.808	1.402	1.807

as a cathode and workpiece acts as an anode in all of the experimental work. Figure 4.5 (a) depicts the EDM facility, and Figure 4.5 (b) shows the machined specimen. The response characteristics are SR, MRR, and OC. Roughness tester is used to determine the SR of the machined surface. Coordinate measuring machine is used for measuring the OC of the machined section. The excessively machined regions are indicated as OC that are calculated by measuring the deviation between the actual diameter of the hole made and the diameter of the electrode. The quantity of material removed from the sample throughout the machining time for the particular cutting condition is used to calculate the MRR. The experimental results obtained for the varying process conditions are presented in Table 4.2.

4.3 Results and Discussion

4.3.1 Surface Roughness and Overcut

As the input current increases, the more energy is discharged between the electrodes, resulting in a high amount of bombardment impulses emerging on the work material surfaces [1]. These more quantum of bombarding

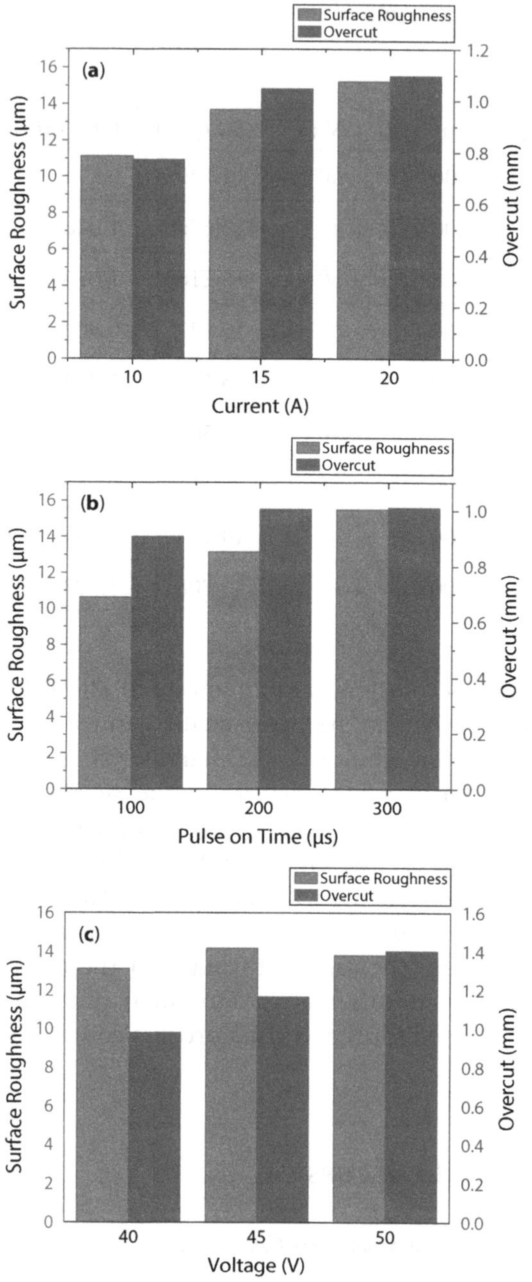

Figure 4.6 (a) Effect of I (V = 45 V, Ton = 200 μs); (b) Effect of Ton (V = 45 V, I = 15 amps); (c) Effect of V (Ton = 200 μs, I = 15 amps).

force develops the cavernous and broader craters on the work material surfaces results in increases in the surface roughness. Also discharge of more spark energy on the work material surface for a prolonged period creates the craters of longer depth that in turn increases the surface roughness. Figure 4.6 depicts the influence of each input factor on SR and OC formed on the work material.

From Figure 4.6 (a), it is clear that the SR is swayed by I. If I is continuously increased, the SR tends to increase. It is because if more pressure outlet from spark is being transferred on the work material through the anode, there is a possibility of overlapping craters on the machined surface. The OC of the work material rises as I increases. The high heat energy produced by the increased I also increases MRR, resulting in a worn surface along the machined corner areas. Therefore, increased I produces a greater amount of dimensional inaccuracy when the scenarios are compared to lower I.

Figure 4.6 (b) shows that SR is gradually increased with an increment in Ton. It may be due to the long duration of I by which the time of spark present in the electrode is high. As a result, more thermal energy is delivered to the work material, resulting in a greater number of melt expulsions on the machined surfaces. It also causes the product SR to rise. However, an improved surface finish can be attained by using a lower I and a shorter Ton. The increased heat energy created in the cutting zone produces corner wear and also prevents the spark energy from being distributed evenly across the machined surface. This causes larger diameter holes to form on the material, resulting in a larger OC. Furthermore, the larger spark created between the electrodes expands the size of the cavity formed. Similar results were obtained while cutting tool steel with a copper electrode [2].

From Figure 4.6 (c), it is understood that SR is gradually increased up to 45 V and then gradually decreased. The higher MRR is due to the more spark energy at the maximal Ton concentrated on the work material. The extremely higher energy produced by the increased V promotes erosion from the material periphery, resulting in an increase in OC.

4.3.2 Material Removal Rate

The EDM input factors and their impacts on the MRR are exemplified in Figure 4.7. Figure 4.7 (a) depicts the changes in MRR while varying the V in the EDM process. MRR drastically increases with an increase in I. This may be due to the increment in I that results in increased intensity of electric discharge spark. This high amount of discharge spark strikes the work material and results in kinetic energy of pulse converted into heat energy.

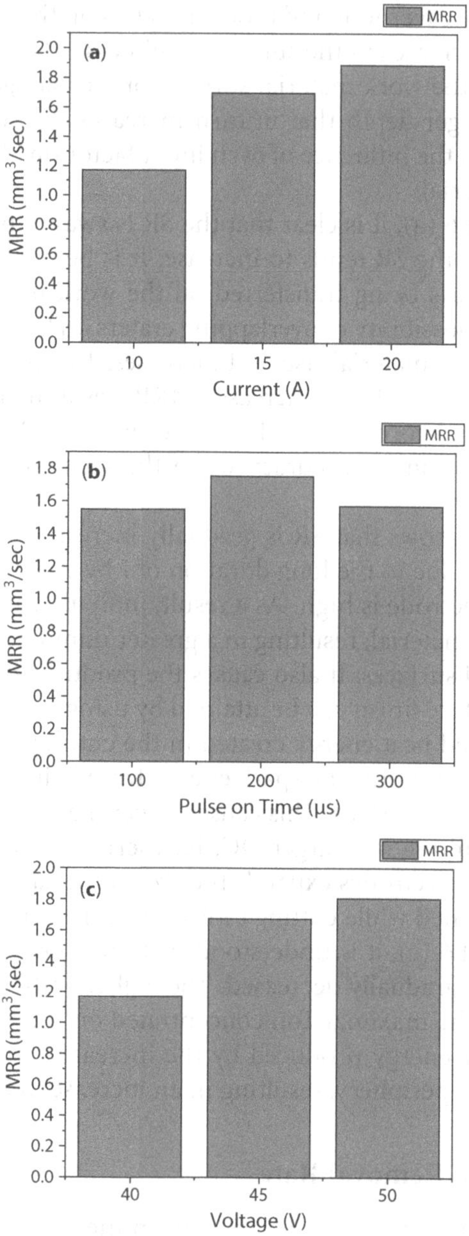

Figure 4.7 (a) Effect of I (V = 45 V, Ton = 200 µs); (b) Effect of Ton (V = 45 V, I = 15 amps); (c) Effect of V (Ton = 200 µs, I = 15 amps).

Due to this high heat energy, the interface between tool and work material temperature rises that causes erosion in the work material and results in the increased MRR.

Figure 4.7 (b) shows that MRR is slightly increased by increasing Ton. Higher Ton increases the amount of heat energy delivered to the workpiece surface. Hence, more material is melted that results in increased MRR. This effect is prevailing to some extent and after that MRR starts decreasing due to debris formed in the spark gap. This debris hinders the contact between the work material and electrode. It also absorbs heat energy and decreases MRR.

According to Figure 4.7 (c), increasing V significantly increases the MRR. It is due to the ions that collide with the work material. Due to the increase in V between the electrode and work material, electron and ion velocities are increased. As a result of the kinetic energy of the electrons, the MRR increases consecutively.

4.3.3 Surface Morphology

The electrical discharge machined surface was examined through a scanning electron microscope (SEM), and the micrographs are presented in Figures 4.8 (a) and (b). During machining, the input energy typically produces a massive fraction of thermal energy on the substance. This causes the material to melt, causing a portion of the work to be vaporized and evacuated by the flushing mechanism.

Furthermore, the discharged molten material generates ridge-rich textures and crater on the workpiece surface. It then approaches the solidification temperature due to the presence of dielectric fluid. Meanwhile, the mass loss creates a plethora of microscopic voids on the workpiece surface, which aggregate to form craters. Higher I and Ton have an effect on

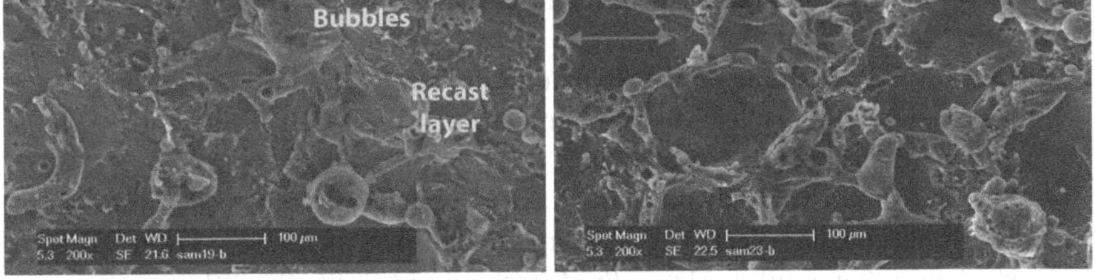

Figure 4.8 (a) I = 15 amps, Ton = 50 μs, V = 40 V; (b) I = 15 amps, Ton = 75 μs, V = 50 V.

size and shape of crater formation. In machined regions, it also generates deeper crater and pits with ridge-rich surfaces.

4.4 Conclusion

In this work, aluminum 6351 reinforced with 5 wt. % B_4C was fabricated using the stir casting process. The EDM was carried out with different process parameters, and the following conclusions were made.

- When the input current is increased, the surface finish of the work material decreases, and the OC increases dramatically as a result of high discharge during machining.
- The surface roughness increases linearly as the pulse on time increases. However, the overcut changes slightly as a result of the spark and removes more metal from the material edges.
- The OC increases as the voltage increases; however, the surface roughness increases up to 45 V and then declines.
- The material's machined face is analyzed, and the creation of craters and layers is recorded.
- The mass loss in the specimen creates an excess number of microscopic voids on the workpiece surface, which aggregate to form the craters of varying sizes and shapes. Similarly, the higher pulse current and time also have an effect on size and shape of crater formation.

References

1. Dwaraka, R. and Arunachalam, N., Investigation on non-invasive process monitoring of die sinking EDM using acoustic emission signals. *Procedia Manuf.*, 26, 1471–1482, 2018.
2. Sidhu, S.S. and Bains, P.S., Study of the recast layer of particulate reinforced metal matrix composites machined by EDM. *Mater. Today Proc.*, 4, 2, 3243–3251, 2017.
3. Prasanna, P., Sashank, T.V., Manikanta, B., Aluri, P., Optimizing the process parameters of electrical discharge machining on AA7075-SiC alloys. *Mater. Today Proc.*, 4, 8, 8517–8527, 2017.
4. Suresh Kumar, S., Uthayakumar, M., Thirumalai Kumaran, S., Parameswaran, P., Mohandas, E., Electrical discharge machining of Al (6351) alloy: Role of electrode shape. *Int. J. Mater. Prod. Technol.*, 53, 1, 86–97, 2016.

5. Suresh Kumar, S., Uthayakumar, M., Thirumalai Kumaran, S., Parameswaran, P., Mohandas, E., Electrical discharge machining of Al (6351)-5% SiC-10% B_4C hybrid composite: A grey relational approach. *Model. Simul. Eng.*, 2014, 7, 2014.
6. Suresh Kumar, S., Uthayakumar, M., Thirumalai Kumaran, S., Parameswaran, P., Electrical discharge machining of Al (6351)–SiC–B_4C hybrid composite. *Mater. Manuf. Process.*, 29, 11-12, 1395–1400, 2014.
7. Singh, M. and Maharana, S., Investigating the EDM parameter effects on aluminium based metal matrix composite for high MRR. *Mater. Today Proc.*, 33, 3858–3863, 2020.
8. Samy, G.S. and Thirumalai Kumaran, S., Measurement and analysis of temperature, thrust force and surface roughness in drilling of AA (6351)-B_4C composite. *Measurement*, 103, 1–9, 2017.
9. Satpathy, A., Tripathy, S., Senapati, N.P., Brahma, M.K., Optimization of EDM process parameters for AlSiC-20% SiC reinforced metal matrix composite with multi response using TOPSIS. *Mater. Today Proc.*, 4, 2, 3043–3052, 2017.
10. Shyn, S., Rajesh, R., Anand, M.D., Modeling and prediction of die sinking EDM process parameters for A6061/6% B_4C metal matrix composite material. *Mater. Today Proc.*, 42, 677–685, 2021.
11. Rizvi, S.A.H., Agarwal, S., Bharti, P.K., Modeling of surface roughness of AISI 4340 using copper-tungsten tool in die sinking EDM. *Mater. Today Proc.*, 22, 2334–2340, 2020.
12. Arooj, S., Shah, M., Sadiq, S., Jaffery, S.H., II, Khushnood, S., Effect of Current in the EDM Machining of Aluminum 6061 T6 and its Effect on the Surface Morphology. *Arab. J. Sci. Eng.*, 39, 4187–4199, 2014.
13. Niamat, M., Sarfraz, S., Shehab, E., Ismail, S.O., Khalid, Q.S., Experimental characterization of electrical discharge machining of aluminum 6061 T6 alloy using different dielectrics. *Arab. J. Sci. Eng.*, 44, 8043–8052, 2019.
14. Barenji, R.V., Pourasl, H.H., Khojastehnezhad, V.M., Electrical discharge machining of the AISI D6 tool steel: Prediction and modeling of the material removal rate and tool wear ratio. *Precis. Eng.*, 45, 435–444, 2016.
15. Dhar, S., Purohit, R., Saini, N., Sharma, A., Kumar, G.H., Mathematical modeling of electric discharge machining of cast Al–4Cu–6Si alloy–10wt.% SiCP composites. *J. Mater. Process. Technol.*, 194, 1–3, 24–29, 2007.
16. Lin, M.Y., Tsao, C.C., Huang, H.H., Wu, C.Y., Hsu, C.Y., Use of the grey-Taguchi method to optimise the micro-electrical discharge machining (micro-EDM) of Ti-6Al-4V alloy. *Int. J. Comput. Integr. Manuf.*, 28, 6, 569–576, 2015.
17. Dewangan, S., Gangopadhyay, S., Biswas, C.K., Study of surface integrity and dimensional accuracy in EDM using Fuzzy TOPSIS and sensitivity analysis. *Measurement*, 63, 364–376, 2015.
18. Singh, H. and Shukla, D.K., Optimizing electric discharge machining parameters for tungsten-carbide utilizing thermo-mathematical modelling. *Int. J. Therm. Sci.*, 59, 161–175, 2012.

19. Işık, A.T., Çakıroğlu, R., Günay, M., Multiresponse optimization of performance indicators through Taguchi-grey relational analysis in EDM of cemented carbide. *CIRP J. Manuf. Sci. Technol.*, 41, 490–500, 2023.
20. Sahoo, S.K., Thirupathi, N., Saraswathamma, K., Experimental Investigation and multi-objective optimization of Die sink EDM process parameters on inconel-625 alloy by using utility function approach. *Mater. Today Proc.*, 24, 2, 995–1005, 2020.
21. Ming, W., Zhang, Z., Wang, S., Zhang, Y., Shen, F., Zhang, G., Comparative study of energy efficiency and environmental impact in magnetic field assisted and conventional electrical discharge machining. *J. Clean. Prod.*, 214, 12–28, 2019.
22. Lin, Y.C. and Lee, H.S., Machining characteristics of magnetic force-assisted EDM. *Int. J. Mach. Tools Manuf.*, 48, 11, 1179–1186, 2008.
23. Ming, W., Zhang, Z., Wang, S., Zhang, Y., Shen, F., Zhang, G., Comparative study of energy efficiency and environmental impact in magnetic field assisted and conventional electrical discharge machining. *J. Clean. Prod.*, 214, 12–28, 2019.
24. Kozień, D., Olearnik, B.S., Rapak, A., Szczygieł, A., Góra, N.A., Boratyński, J., Piasecka, E.P., Bućko, M.M., Pędzich, Z., Boron-rich boron carbide nanoparticles as a carrier in boron neutron capture therapy: Their influence on tumor and immune phagocytic Cells. *Materials (Basel)*, 14, 11, 3010, 2021.
25. Iwagami, T., Ishikawa, Y., Koshizaki, N., Yamamoto, N., Tanaka, H., Masunaga, S., Sakurai, Y., Kato, I., Iwai, S., Suzuki, M., Yura, Y., Boron carbide particle as a boron compound for boron neutron capture therapy. *J. Nuclear Med. Radiat. Ther.*, 5, 5, 2014.

5

Green Manufacturing of Natural Fiber Composite

Meenal Batra and Alka Bali*

University Institute of Pharmaceutical Sciences, Panjab University, Chandigarh, India

Abstract

Natural fiber composites are also termed as green composites, biocomposites, or eco-composites. These are made up of a polymer matrix along with a reinforcing material that can be in the form of either particles or fibers. The demand for natural fiber-reinforced composites has been heightening day by day, and this is attributed to their several advantages like biodegradability, cost-effectiveness, and eco-friendly nature. Hence, these are now being widely employed as an alternative to synthetic fibers, including glass fiber, carbon fiber, etc. This chapter presents a comprehensive overview of various types of natural fibers and their properties and applications coupled with various green manufacturing techniques employed for fabricating the natural fiber composites including resin transfer molding, pultrusion, compression molding, hand lay process, spray up technique, filament winding technique, resin transfer molding with vacuum assistance, vacuum bagging, and resin injection molding. Further various additive manufacturing techniques used for the processing of natural fiber-reinforced composites are described, including fused filament fabrication (FFF), selective laser sintering (SLS), binder jetting (BJ) technique, direct ink writing (DIW), stereolithography (SLA), direct light processing (DLP), laminated object manufacturing (LOM), and electron beam melting (EBM).

Keywords: Natural fiber composites, green manufacturing, additive manufacturing, biodegradability, sustainability

**Corresponding author*: alka.bali@rediffmail.com

S. Thirumalai Kumaran and Tae Jo Ko (eds.) Sustainable Machining and Green Manufacturing, (79–114) © 2024 Scrivener Publishing LLC

5.1 Introduction

Natural fiber composites (NFCs) are referred to as the composite materials in which the natural fibers are embedded in thermoplastic or thermosetting polymer to fabricate the composite [1]. These are made up of a polymer matrix coupled with a reinforcement material as a fiber or particle [2]. Polymer matrix is of two types, i.e., synthetic matrix and natural matrix (also known as biodegradable matrix). The natural fibers employed in composite include plant fibers such as sisal, cotton, agave, flax, jute, hemp, ramie, and coir; animal fibers like mulberry silk, angora, and coarse silk; and mineral fibers including asbestos, ceramic, etc. [3].

Depending upon the proportion of the natural materials that are employed as the reinforcing material, NFCs may be categorized as green composites, semi-green composites, and hybrid composites [4, 5]. In green composite, both polymer matrix and natural fibers are obtained from renewable resources [6]. Hybrid polymer consists of a single polymer matrix and two or more man-made fibers. Semi-green composite consists of both synthetic and natural fibers having natural materials in high proportion [4, 5, 7]. Being eco-friendly, light weight, renewable, cheap, biodegradable, and sustainable materials, currently, there is a huge demand for natural fiber composites with respect to the synthetic fibers [8, 9]. These are widely used in aerospace, civil construction, automobiles, medical sector, chemical industry, construction materials, etc. [10–12]. Natural fiber composites are processed through methods like compression molding, pultrusion, resin transfer molding, etc. [13]. Apart from these, these composites can also be manufactured through additive manufacturing techniques [three-dimensional (3D)]. These techniques involve fabrication through layer by layer building methodology in order to constitute a 3D component by Computer Aided Design (CAD). This process can fabricate materials with complex geometry in lesser amount of time, with almost no wastage of materials, and hence, it is considered as a sustainable manufacturing approach. There are several additive manufacturing techniques put to use for the manufacture of natural fiber composites. Selection of the technique mainly depends on geometric ramification or complexity of the part sought, materials and natural fiber type in use. Most commonly used techniques are fused filament fabrication (FFF), selective laser sintering (SLS), binder jet manufacturing, stereolithography (SLA), direct light processing (DLP), etc. [14].

This chapter gives a comprehensive review and compares various green manufacturing techniques used in the production of natural fiber composites along with their potential application in various sectors.

5.2 Characteristics of Natural Fibers

Natural fibers consist of crystalline cellulose microfibrils embedded inside an amorphous matrix. Fibers are generally light in weight, having a density between 1.1 and 1.6 g/cm^3 [15]. Characteristics of these fibers usually depend on their composition, internal structure, and microfibril angle [16]. Cellulose is the major component present in various types of natural fibers, having a stiffness around 138 GPa [17]. Properties of all natural fibers are not the same owing to the differences in their fiber diameter, fiber length, and fiber bundle width. Mechanical properties of the composite also get affected due to variations in microfibril angle and fiber density of natural fibers [18]. High-density fibers are strong and stiff as compared to low-density fibers. Conversely, fibers having a smaller microfibril contact angle are stiff as compared to fibers with a high microfibril contact angle [19, 20]. Fibers with a low microfibril contact angle show brittle behavior, and those with a high contact angle show large plastic deformation. Bast fibers are generally used to produce composite materials of high mechanical strength [15, 21]. Besides being cheap, eco-friendly, renewable, and biodegradable in nature, the property of being lightweight makes natural fiber composites attractive materials for use in the automotive and aerospace sectors [22, 23]. Furthermore, the low abrasiveness of the natural fibers makes their processing in polymer industry much easier and sustainable compared to synthetic fibers. Apart from several advantages, natural fiber-reinforced composites have their share of disadvantages as well. The physical characteristics of the natural fibers can vary with geographical and environmental factors such as harvesting season, harvesting region, as well as the process employed for harvesting or manufacturing of the fiber. Materials used for manufacturing certain natural fibers can be expensive. Durability can be a cause of concern with natural fiber-reinforced composites, when these are utilized as building material, as defects like concrete cracks (on account of swelling and volume changes) have been noted in several cases. Their high moisture absorption (resulting in lowered processing temperatures) also limits the range of applications for the composites. Furthermore, natural fiber composites are less stiff as compared to other composite materials.

Moreover, their irregular shapes along the length of the fiber make it very difficult to predict their mechanical properties. Another disadvantage of natural fiber composites is their lower thermal stability, as these are stable only up to temperatures of about 2,000 degrees Celsius and get degraded at higher temperatures [24]. Moreover, there is a requirement of preprocessing these fibers on account of their poor adhesion characteristics or incompatibility with polymer matrices.

5.3 Classes of Natural Fibers

Natural fibers may be broadly classified based on their origin, i.e., plant fibers, animal fibers, and mineral fibers (Figure 5.1). Animal fibers include wool, horse hair, goat hair, silk, alpaca hair, angora, etc. Mineral fibers include asbestos, ceramics, inorganic whiskers, etc. Plant fibers are majorly cellulosic fibers [24–26]. Plant fibers commonly used in natural fiber composites include coir [27], jute [28, 29], kenaf [30, 31], flax [32], ramie [33], and sisal [34, 35]. Based on their applications and usage, natural fibers are also classed as primary and secondary fibers. Primary fibers are the ones that are cultivated mainly for harvesting their fibers, whereas secondary fibers are usually plant by-products (including coir fibers, pineapple fibers, agro-residues, etc.). Further classification may be done based on the plant part from which the fiber is obtained, e.g., bast fibers like jute, flax, etc.; seed fibers (coir, cotton); leaf fibers (agave, sisal, banana); core fibers (kenaf, hemp); grass and reed fibers (bagasse, corn, bamboo, etc.).

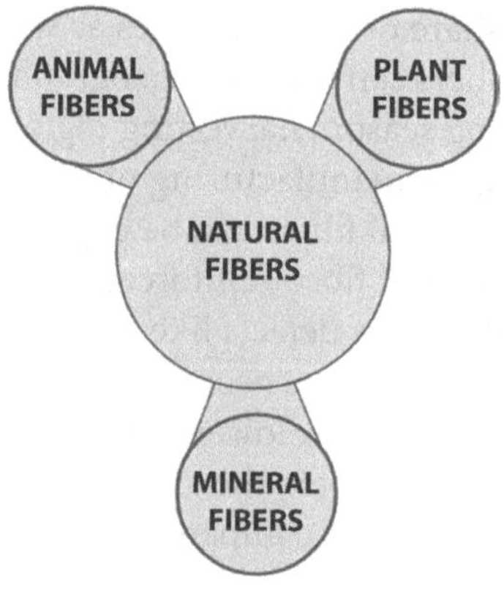

Figure 5.1 Classification of natural fibers.

The source of coir fiber is the husk fiber of coconut. The fabrication of coir fiber-reinforced composites is carried out by using thermoplastic as well as thermosetting resins [24]. Coir fiber-reinforced composite is used in packaging materials, roofing sheets, storage tanks, helmets, post boxes, voltage stabilizer, etc. [36, 37]. These coir-based composites have a high moisture content that can be reduced through chemical treatments like alkalinization, acetylation, etc. Such pretreated coir-based composites possess higher mechanical strength compared to their untreated counterparts [24, 38]. Jute-coir hybrid composites have been used for making doors, building interiors, windows, etc. [39]. Coir-based composites are mainly manufactured through extrusion and compression molding technique [40, 41].

Kenaf fiber is obtained from the outer (bast) fibers and the inner (core) fibers of the kenaf plant. This fiber is mostly found in Asia and Africa. Within 150 days, a kenaf plant can attain a height of 2.4–6 meters. The United States grow a wide variety of kenaf plants for commercial purposes, whose cultivation is entirely dependent on the weather. Several variables, including photosensitivity, plant population, season of growing, day of planting, cultivar, and plant matureness, have a direct impact on the kenaf fiber [24]. Its fiber characteristics are influenced by the sources, age, and separating method [38]. Kenaf fiber has been employed for a wide range of applications, including the fabrication of composites, the production of twin, rope, sackcloth, paper production, animal food production, and the production of oil, medicine, and food additives [24].

Hemp also belongs to the category of bast fibers. Hemp fiber was first used in Asia. Currently, it is being used by the European Union for non-food agriculture. Hemp fiber outperforms cotton fibers in terms of its high tensile strength, absorbency, resistance, high luster, and durability. It is used in construction products and the manufacture of bank notes, pipe, paper, furniture, packaging materials, and composites and for blocking harmful radiation from sun [24]. According to a recent research, hemp fiber accounts for one-third of China's entire production [42]. Hemp fiber-reinforced composite can be fabricated through the processes of extrusion, injection molding, and resin transfer molding [43, 44].

Jute, another type of bast fiber, is among the most important natural fibers, after cotton [45]. It is mainly cultivated in Asian subcontinents, particularly in countries like Bangladesh, Taiwan, Thailand, Nepal, China, Brazil, Cambodia, India, China, Myammar, and some other countries. It offers several advantages such as good insulation properties, good mechanical properties, high tensile strength, thermostability, biodegradability, good water resistance, and greater strength/weight ratio [24, 46].

Flax is another type of bast fiber, mainly grown in temperate regions. Flax fiber is biodegradable, lustrous, and flexible and has good abrasion resistance and strength [47]. As flax fiber has 2–3 times greater strength than cotton fiber, the importance of flax-based composites in the textile industry has progressively risen over the years. It has been used as a raw material in the paper industry, in the production of ropes and twines, for canvas and webbing apparatus, as well as in tablecloths, clothing, sewing thread, ornamental fabrics, and other products [24]. Flax fiber-reinforced composites are produced through the process of vacuum infusion [48].

5.4 Polymer Matrix

Biocomposites are fabricated using natural fibers coupled with a polymer matrix (Figure 5.2), and therefore, it is crucial that the most appropriate polymer matrices be selected for their production. Polymers are macromolecular entities made up of single or several recurring units. A crystalline polymer structure is formed by a unidirectional arrangement of the polymeric chains in the matrix, whereas a random arrangement results in the production of an amorphous structure. The thermoplastic and thermosetting polymers are the two main categories of polymers. When weak as well as strong mechanical qualities are needed for composite materials,

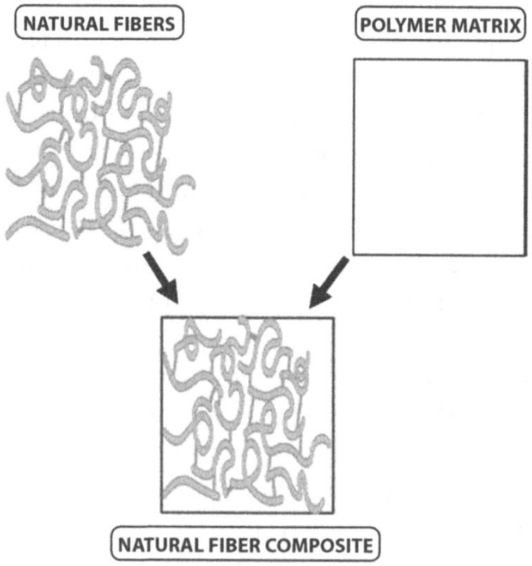

Figure 5.2 Composition of natural fiber composites.

thermoplastic polymers (such as polypropylene, polycarbonate, polyimides, and polyethylene terephthalate) are frequently employed. These polymers exhibit exceptional toughness and wear-and-tear resistance. Conversely, the thermosetting polymers (such as epoxy, silicone, and polyurethane) are tough because they are cross-linked and molded to form 3D networks having strong covalent bonds. These polymers are easier to manufacture and have good wetting qualities, temperature stability, creep resistance, and chemical resistance [3, 49].

5.5 Applications of Natural Fiber Composites

The natural fiber composites find a diverse range of applications on account of their reduced expense, durability, and integrity in the automotive industry, aerospace, medical sector, packaging materials, electric field, sports, chemical industry, etc. [6, 50].

5.5.1 Applications in Automotive and Aerospace Industry

Natural fiber composites have been employed to manufacture lightweight, easily recyclable, safe, and fuel-efficient vehicles [51]. Interior door panels, dashboards, hatch, mats, etc. of cars are fabricated from natural fibers. In the year 2005, around 45,000 tons of natural fiber were expended in automotive manufacturing [52]. Natural fibers are used by German automobile companies (BMW, Audi group, Ford, etc.) to manufacture door panels, headliner panels, noise insulation panel for BMW, to make back seats, side door and back door panels for Audi, and for making door panels, B-pillars for Ford. Companies like Cambridge company and Proton company make use of natural fibers to make different component of cars, majorly employing polyester or propylene along with fibers such as hemp and jute. The global demand for natural fiber-reinforced composites is progressively escalating in the automotive industry. Natural fiber composites also find a variety of aerospace applications amounting to as high as 50% of total aerospace applications. These composites are being employed in the manufacture of helicopters, cockpit satellites, launch vehicles, military fighter aircraft, missiles, etc. This wide usage of the composites in aircrafts is explained by the fact that these scale down the weight of the aircrafts to a considerable extent. Rudders, main wings and wing ribs, and several interior components of aircrafts are all made from natural fibers [11, 12, 16, 24].

5.5.2 Medical Applications

Natural fiber composites are increasingly finding applications in the medical industry today. These composite materials are being employed in dentistry for the restoration of anterior or posterior teeth. The impaired teeth are replaced with composite material, a kind of artificial permanent implant. Composite materials are also employed in orthopedic applications such as bone grafts, hip joint replacement, and bone fixation plates. In addition to these, amalgams that have worn out can be replaced with composites [24].

5.5.3 Construction

Composite materials command a significant place in the construction field. These are used to manufacture civil engineering constructions like door, panels, furniture, fiber boards, windows, etc. [53]. Composite fabricated by combining coir fibers with polymers can be used to manufacture roofing boards, false ceilings, wall panels, etc. With the aid of synthetic resin-based composites, rice husk, paper, straw, etc., it is also possible to make wall panels, fake ceilings, flooring, roofing, lightweight partition walls, cladding, doors, windows, and shutters. As a result, composite materials are frequently utilized in high-rise structures, doors, windows, furniture, bridge components, complete bridge systems, etc. [11, 24].

5.5.4 Sports and Chemical Industry

Composite materials are frequently utilized to manufacture sports equipment because of their lightweight property, toughness, abrasion resistance, high strength, and thermostability. Natural composites are now used to make a variety of sports equipment, including tennis racquets and boards, windsurfing boards, fishing rods, shafts, golf clubs, climbing ropes, slats, skis, bats and swords, and many more. In comparison to other sports materials, equipment made of natural fiber composite has good mechanical properties and is light in weight. In the chemical industry, composites have been employed for manufacturing drive shafts, ducts, casings, below-ground stowage tanks, sheaths and coverings, reactors, composite vessel columns, ducts, pipes, tubes, etc. These are used in the chemical industry due to their fire-resistant properties, moldability, lightweight characteristics, etc. [11, 12, 24, 54].

5.6 Preprocessing of Natural Fiber Composites

Most natural fibers do not adhere to polymer matrixes properly because these fibers possess hydrophilic characteristics that do not blend well with the hydrophobic polymer matrix. Hence, preprocessing of the fibers is done before incorporating them into the polymer matrix with the aim of improving the bonding of the fiber with the matrix [55]. Preprocessing is generally carried out by altering either the fiber surfaces (surface modification) or the polymer matrix with the aid of some compatibilizing agents [15].

5.6.1 Surface Modification

Modification of the surface of the fibers can be executed through both chemical and physical means. Method selection relies on the type of matrix as well as the fiber combination. Chemical methods prevent degradation that occurs due to moisture absorption. During chemical treatment, the hydrophilicity of natural fibers is decreased by substituting hydroxyl groups with appropriate moieties employing suitable chemical agents. Important chemical methods include alkalinization, acetylation, benzoylation, etc. The most commonly employed technique for carrying out surface modification of natural fibers is alkalinization, also known as mercerization [15, 56]. In this process, constituents of fibers like pectin, hemicellulose, waxes, oil, and lignin are removed from the surface by immersing the fibers in a concentrated solution of caustic soda (sodium hydroxide; NaOH) [57]. Subsequent to this surface modification, the binding between fiber and matrix gets significantly improved. Benzoylation utilizes benzoyl chloride to incorporate benzoyl groups into the fiber structure in order to decrease the hydrophilicity of natural fibers and improve the interfacial binding of the natural fiber with the polymer matrix. This process also improves the thermal stability of fibers [58, 59]. Acetylation is another popular surface modification technique where the hydroxyl groups of natural fibers are acetylated with acetic anhydride or acetyl chloride. Similar to benzoylation, this process also reduces the hydrophilic characteristics of the natural fibers, thus improving the binding between the natural fibers and the polymer matrix [60].

Physical methods employed for preprocessing of natural fibers include autoclave treatment, electron beam discharge, plasma treatment, UV treatment, gamma-ray irradiation, etc. [15]. UV treatment or gamma-ray irradiation methods enhance the thermal stability of natural fibers. The latter property is very important during the processing stage of natural fiber-reinforced composites [61]. Similarly, free radicals produced during the process of electron beam discharge can also contribute toward strengthening the bond between the polymer matrix [62]. Heat treatment is the simplest physical method used for upgrading natural fibers in which the latter are subjected to a suitably elevated temperature in order to bring about a reduction in moisture levels, do alteration in their chemistry, polymorphism or crystalline state of cellulose [15, 63]. Plasma treatment, another effective method used for surface modification, increases the hydrophobicity and contact angle with respect to water and decreases the evenness of the surface, resulting in enhanced fiber-matrix adhesion [64].

5.6.2 Modifying the Polymer Matrix with Compatibilizing Agents

The second method used for improving the interfacial binding of the polymer matrix to the natural fiber is by making use of compatibilizing agents. As an example, the chemical modification of the matrix is done by maleation with maleic anhydride (MA) followed by mixing of the maleated polymer with the matrix. In both methods, a strong interaction of the fibers with the polymer matrix due to which the hydrophilicity of natural fibers gets reduced and bonding improves between fiber and matrix [15].

5.7 Fabrication of Natural Fiber Composites

Numerous techniques are there to fabricate natural fiber composites for varying application conditions (Figure 5.3). Based on application conditions [3] and the type of material employed for reinforcement and as polymer matrix, techniques for the fabrication of natural fiber composite can be categorized as open and closed molding techniques [65, 66].

5.7.1 Open Molding Techniques

In open molding techniques, resins are exposed to the atmosphere when they harden or cure [67]. It is less expensive and a relatively simple technique but it requires highly trained operators in order to ensure laminate quality with respect to the fraction of fiber volume and void content in composite laminate [3]. Open molding techniques commonly used to produce composites are hand lay technique, spray up technique, etc. [68, 69].

5.7.1.1 *Hand Lay Technique*

Hand lay is a type of open molding technique used for producing natural fiber composites. This technique involves simple processing steps. First of all, a release gel is sprayed superficially onto the mold in order to ensure that the polymer does not stick to the mold. Following this, plastic sheets with very low thickness are placed on the top side and the bottom of the molding plate in order to obtain the desirable surface texture in the product. After every sheet, reinforcement material is placed as woven roving or chopped strand mats (CSMs). The thermosetting polymer is then poured over the surface of the mat that has been earlier placed in the mold after being completely mixed in the proper proportion with a curing or hardening agent. The polymer is homogeneously distributed with the aid of a brush. Following this, the next layer of the mat is applied to the polymer surface, and the layer of mat and polymer is gently rolled together to eliminate any trapped air and extra polymer. This process is replicated with all subsequent layers of the polymer and the mat until the desired layer number is achieved in the final stack. The release gel is applied as a spray on the inner side of the top mold plate after placement of the plastic sheet. Then, holding the top mold plate on the arranged layers, application of pressure is done. The generated composite part is withdrawn from the mold to be processed further after hardening at any specific temperature, generally at room temperature. The hand layup composite has a density of 1.58 g/cm^3, a mass fraction of 61% fiber, a volume fraction of 46% fiber, and a thickness of 2.96 mm [70, 71].

5.7.1.2 *Spray Up Technique*

The spray up technique is an automatized variant of the hand lay method in which polymer matrices and the chopped strands of reinforcement fibers are sprayed onto an open mold with the help of a sprayer until the required thickness is achieved. Spray up technique is used to manufacture automotive parts, large container tanks, turbine blades, etc. [3].

5.7.2 Closed Molding Techniques

In closed molding technique, resins are not subjected to air exposure when they are being cured or hardened [65]. This technique is used to make high-quality three-dimensional composite parts. Closed molding process generates less waste and has greater productivity as compared to the open molding technique [72, 73]. Resin transfer molding, compression molding, injection molding [74, 75], and resin injection molding are examples of closed molding techniques [3].

5.7.2.1 Resin Transfer Molding

It is a type of closed molding process in which a preformed stack of fibers is impregnated within the mold cavity, which is then sealed. Firstly, reinforcement fibers are placed within the mold cavity, and the resin premixed with the hardener is interjected into the mold at high pressure with the help of a mixing head. This causes the resin to spread and impregnate the fibers. Impregnation is always done in vertical direction. Several factors including the pressure during the injection, temperature employed for preheating of the resin, and mold configuration affect this process. This process has various advantages like good surface finish, low tooling cost, and less void content. RTM is used in manufacturing sports equipment, aerospace industry, automotive industry, etc. [13]. Thermoplastic resins like epoxy resins, polyester, etc. are commonly used in this process because they have high impact strength and better fiber impregnation as compared to thermosetting resins [76].

5.7.2.2 Resin Injection Molding

Resin injection molding is a type of closed molding technique in which composite materials are produced by injecting material into the mold. Nowadays, mostly screw-type injection machines are used. Injection molding causes melting of the material by making use of a screw and a heating device followed by its injection into the mold cavity, subsequently forming the product on cooling of the mold. The injection machine is composed of a system for injecting, a mold system, electric control, hydraulic system, and heating and cooling parts. Broadly, it may be subdivided into three components meant for plasticizing, injection, and pressure driving. The plasticizing part melts the material and plasticizes it uniformly. The injection part injects the formed plasticized material into the mold cavity at a particular speed and pressure. The pressure driving part maintains the pressure until the material gets solidified and also prevents backflow of the material. The clamping system supports the forming mold and prevents the damage of mold from impact. It also facilitates smooth ejection of products. The clamping device includes a draw rod, a power supply, and a fixed template. The hydraulic system helps in determining the technical performance of the injection molding machine. Injection molding technique is of various types such as water-assisted molding, gas-assisted molding, low-pressure molding, microinjection molding, and compression molding. It is a preferred process to produce products with complex shapes and sizes [77]. This technique helps in uniform distribution of resin and provides extra durability to the product. In order to get right finished product through this process, parameters like melt temperature and pressure must be controlled [13].

5.7.2.3 *Compression Molding Technique*

This is a type of closed molding technique where the preheated molding material is laid on the cavity of the lower end of the heated mold, which is then compressed by the upper end so that the molding material fills the mold completely. The compression pressure (generally, 2,000 psi) is maintained until the material gets solidified [13]. The molding material is usually preheated for 15–20 minutes before molding [78]. Composites produced through this technique possess excellent mechanical strength. However, this technique also has several disadvantages like long processing time and restricted size of the mold cavity, which limits the flow of the material in the interior of the cavity and also interferes with the removal of air traps during the production of complex parts. Several factors like heating time, cooling time, melting temperature of the plastic material must be taken into consideration before starting the compression molding process [13].

5.7.2.4 *Vacuum Bagging Process*

It is a type of closed molding technique that is used to produce high-performance glass fiber-reinforced polymer products on a large scale at a relatively cheap cost of tooling. Vacuum bagging was initially evolved for producing large composite parts with superior quality. It makes use of pressure in order to achieve the unique requirements. It has a flexible mold tooling design and can produce complex composite parts with good quality. It bears similarity to the hand lay process and is effortlessly modified to produce different part geometries [71].

5.7.2.5 *Vacuum-Assisted Resin Transfer Molding*

This is among the majorly employed closed molding techniques to manufacture natural fiber composites in which the reinforcing fibers are filled

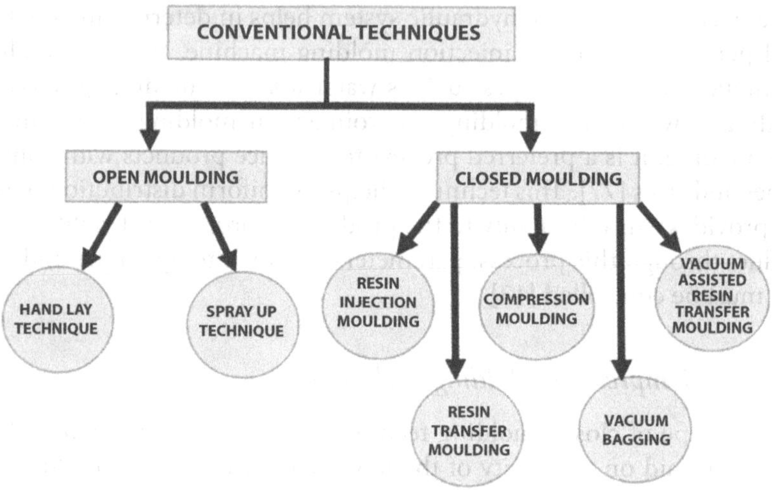

Figure 5.3 Classification of conventional techniques.

into the mold, which is then shut, while allowing the flow of the polymer matrix in and out to saturate the fibers at the same time [79]. The reinforcement fibers are inserted into a mold whose geometry matches the dimensions and surface area required for the composite material [80]. Less void portions and good surface finishes are found in the composite parts produced using this process [81, 82]. In order to create more complicated composite parts with the appropriate finishes at a low price, a variety of modifications are now being made to this fabrication technology [80].

5.8 Additive Manufacturing

Additive manufacturing (Figure 5.4) is another popular technique employed for fabricating natural fiber-reinforced composites. This is also named as 3D printing or additive layered manufacturing, and it makes use of computer-aided designing for building a 3D component by fabricating it on layer-by-layer basis. According to the concept of additive manufacturing, the computer-aided design stage always precedes the additive process. Using this program, models with precise geometric proportions can be created. After sending the file to the device, the process settings and layer division can be made in the machine software or in CAD software (specific software given by the device maker). The developed model is stored with a .stl extension, the latter being supported by the majority of equipment used in additive manufacturing. After that, the manufacturing process of adding of layers starts (with the exception of the CAL method). Depending on the approach, the process can be done either up or down. The supports are removed after the procedure is complete. The CAD software provides for the choice of process parameters that enable the manufacturing of the product having certain attributes, including specialized equipment that implements additive processes. This software can permit the design of a completely new model or it may serve to just alter an existing one [83]. There are various types of additive manufacturing techniques like fused filament fabrication (FFF), SLS, binder jetting (BJ) technique, direct ink writing, etc. [14]. Additive manufacturing has several benefits like high efficiency, high flexibility, and high complexity over traditional or conventional techniques. This technique has a broad range of applications in aerospace or automotive industry, medical industry, electronics, food industry, etc. [84].

Thermoplastics are commonly used in the additive manufacturing process due to their several advantages like low cost, infinite shelf life, good flexibility, and simple processing technique [85–87]. The thermoplastics that are suitable for process include acrylonitrile butadiene styrene (ABS), polyethylene (PE), polylactic acid (PLA), polypropylene (PP), etc. ABS is commonly used in additive manufacturing process like FFF and injection molding technique [87–89]. PLA is another popular thermoplastic polymer employed in additive manufacturing process. PLA has high strength and modulus [90–92]. It does not cause emission of any undesirable gases or give out any type of foul odor through the entire process [93]. It is incorporated into various reinforcing agents with an objective to compound their properties [94]. Apart from these, high-density and low-density polyethylene (HDPE and LDPE, respectively) can also be used. HDPE is preferred over LDPE because it is stiffer in comparison, and its tensile strength is higher compared to that of LDPE. Moreover, it is easy to transform high-density polyethylene into other materials [87, 95].

Additive manufacturing technique can be classified either on the basis of the input material or based on the process type carried out. Input materials commonly used in additive manufacturing technique include solid, liquid, and powder types. For liquid materials, photocuring process is used (use of light for polymerization). Techniques like SLA and direct light processing use liquid as input material. Solid materials are frequently processed through melting. Solid is utilized as a feedstock material in processes like FFF, laminating object manufacturing (LOM), etc. However, sintering techniques are used when the material is powder. Powder is used as a feedstock material in processes like SLS, BJ, electron beam melting (EBM) [83, 96–98], etc.

5.8.1 Stereolithography

Stereolithography is among the most former additive manufacturing methods used for the manufacture of natural fiber composites. On the work surface, a coating of photosensitive resin or photopolymer is placed. A modest-powered laser beam polymerizes the contours of the constructed part, and subsequently, there is creation of cross-reinforcement inside of this contour. The interior of the contour is completely polymerized in the following stage. In this process, light of a particular wavelength is utilized to initiate the polymerization process. The succeeding layer of the photosensitive resin or photo-polymer is then posited after the lowering of the worktable (SL method) by a single-layer thickness. The procedure is repeated until the full portion is produced [83].

5.8.2 Direct Light Processing

DLP is an additive manufacturing technique in which photosensitive polymers or resins are subjected to polymerization. This technique is comparable with the SL/SLA approach, the primary distinction being the light source. In stereolithography, laser is used, whereas in direct light processing, the light source employed is a projector. During projection, the size of a pixel is crucial. Geometrical accuracy of the prepared part increases with the decrease in dimensions of one pixel. The separation distance of the projector from the worktable increases with increasing size of the worktable. Consequently, pixels are increased and there is a decline in the quality of the manufactured parts as the projector's distance from the worktable increases. The smaller the size of the working platform, the higher is the accuracy of the fabricated part. One limitation of this technique is that the parts manufactured by this method could have a lower mechanical strength than those made through other techniques, particularly selective laser sintering [83, 99].

5.8.3 Fused Filament Fabrication

Fused filament fabrication, also called fused deposition modeling, is among the widely used additive manufacturing techniques for fabricating natural fiber-reinforced composites [100, 101]. Feedstock material used in this technique is in the form of a filament made up of thermoplastic materials like polylactic acid, high-impact polystyrene, polyethylene terephthalate glycol (PTEG), thermoplastic polyurethane, polyamides, etc. Here, the filament is introduced into the 3D nozzle where it melts, and then, this melted filament is squeezed out from the extrusion head onto the building platform in order to create objects layer by layer as specified by the CAD model [87]. The quality of the product manufactured through this technique depends on several factors such as deposition speed, nozzle diameter, deposition head temperature, filament feed rate, etc. [15]. It is a simple, easy-to-use, and eco-friendly technique that is much cheaper in comparison to other techniques. Some drawbacks of this technique include wavy finishing on the surface, anisotropy, and shrinking of thermoplastic resulting in parts with reduced accuracies [102, 103].

5.8.4 Laminated Object Manufacturing

In this widely used additive manufacturing technique, a foil having a laminate (heat-activated adhesive) over it is put onto the surface of the worktable. Subsequently, a heated roller is used to bond or weld the foil with the worktable. The correct shape of the layer (contour) is cut out using a laser, and the remaining foil (the outside part of the layer or contour) is then divided into little squares so that extra material may be easily removed after the procedure. The worktable is again lowered for application of the subsequent foil layer. Until the procedure is finished, this cycle gets repeated numerous times, so that each part gets manufactured layer by layer [104–106].

5.8.5 Direct Ink Writing

Direct ink writing or 3D plotting is another additive manufacturing technique for the production of natural fiber-reinforced composites. It is almost similar to fused filament fabrication. Here, a viscous solution is dispensed out from a small nozzle followed by layer-by-layer deposition in a definite path as per the model of CAD to form the 3D structures [15]. Post fabrication is done after the deposition with UV light, heating, or by using reactive feedstock in order to harden the deposited part [107]. Extrusion pressure used in this process is low as compared to fused filament fabrication and no solidification or heating is required in this process. This technique may be employed to fabricate parts with functional and compositional gradients [108]. Feedstock material used in this process must be devoid of any air entrapment or aggregation. Printed material produced through this process must possess adequate stiffness so that it becomes self-supporting after printing [15].

5.8.6 Selective Laser Sintering

SLS is another technique of additive manufacturing, which employs a high-powered laser for fusion of small particles forming a mass possessing the desired 3D shape. Powder present on the bed surface is selectively fused by a laser beam through scanning of cross-sections fashioned from 3D digitized depiction of the part. Once the scan of each cross-section gets completed, it is followed by lowering of the powder bed by another single-layer thickness so that a fresh layer can be deposited. Repetition of this process is done until the complete part is formed. During the processing of polymers by SLS technique, enhancement of melt viscosity occurs during building of the parts because of high temperature of the powder bed, which causes impairment of the surface, commonly referred to as "orange-peel effect". Likewise, usage of recycled powder increases its molecular weight due to which bonding between the layers is enhanced with concomitant improvement in mechanical properties [15]. In order to produce high-grade polymer products, superior quality feedstock material is required to be used [109]. Polymers like acrylonitrile butadiene styrene, polyamide, polycarbonate can be used in this process [110–113].

5.8.7 Binder Jetting

Binder jetting technique is almost similar to SLS; however, here, a binder is used to effect the fusion of the powder instead of laser beam. The technique involves printing of this binder onto the powder bed using a printer concord with the CAD-CAM model and instantly heated using a heat source for drying the binder. After solidification of layers, the excess powder is removed. All types of polymers available in powder form can be used in the binder jetting technique. A wide range of products can be fabricated with this technique, which is simple, eco-friendly, and economical and does not require high temperatures [15]. The quality and properties of the product depend on several factors like powder size, powder shape, powder distribution, speed of print head, number of printing layers, binder viscosity, etc. [114, 115].

5.8.8 Electron Beam Melting

Electron beam melting process is conducted *in vacuo*, in which, an electron beam is emitted from an electron emitter (or gun) with a fiber cathode

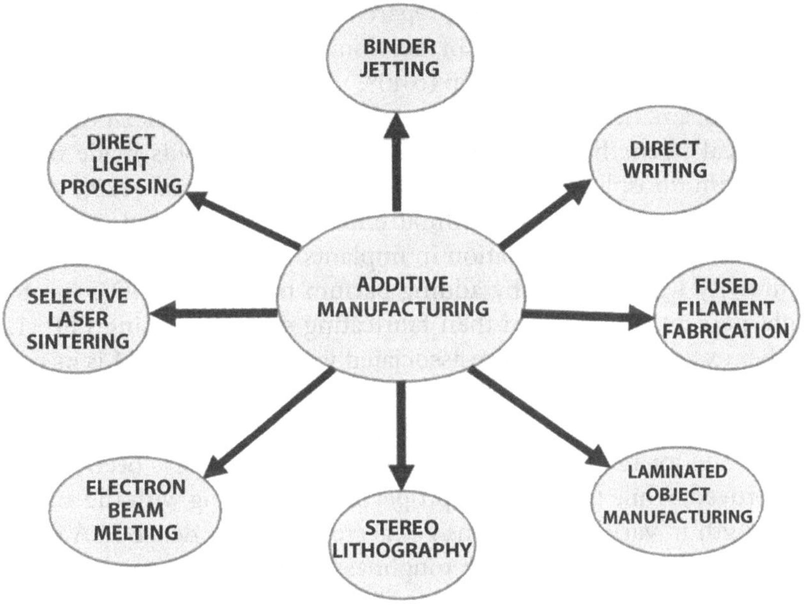

Figure 5.4 Classification of additive manufacturing techniques.

made of tungsten, which is subsequently accelerated in the direction of the worktable. These accelerated electrons have a high kinetic energy, and these are made to collide with the metal powder at high velocities. The powder melts due to conversion of kinetic energy into thermal energy. There are no mechanical moving parts in the electron gun to cause deflection of the electron beam. The beam is manipulated and operated with the help of focusing and deflecting coils [116, 117].

5.9 Additive Manufacturing of Different Composites

Using the abovementioned AM technologies, composites of polymer matrices reinforced with various materials such as micro-form or nano-form particulate matter (microscale and nanoscale) and fibers (continuous or short-fibers) have been created. Particle reinforcements are less expensive, simpler to work with, and more readily incorporated into the matrix as compared to the other fiber reinforcements. As a result, it is also rather simple to prepare feedstock for SLS/BJ and FFF in the form of powder or filament. Among various AM technologies, majority of the reported studies relate to the processing of polymer composites utilizing FFF. Diverse reinforced filaments have been prepared and treated [118] using metals and ceramics. Reinforcement of acrylonitrile butadiene styrene (ABS) with iron and copper powder (up to 40% v/v) using FFF yields ABS-based composites, and the composites show a considerable enhancement in their mechanical strength and thermostability [119]. Scaffolds made of PLA-Tri Calcium Phosphate were examined by Drummer *et al.* [120], utilizing FFF, and these were found to exhibit impressive mechanical and biological characteristics for application in implants. A 240% increase in relative permittivity was achieved by adding barium titanate ($BaTiO_3$) (up to 70 weight percent) to ABS and then fabricating structures using FFF [121]. Another exceptional advantage associated with the use of AM is its capacity for the production of novel periodical structures with desirable compositional diversity. By combining the elastomer of FFF composite (titanium dioxide, jute for reinforcement) with ABS, the surface has been shown to be improved, while their ductile properties [122] along with the anisotropic strength in various directions get decreased. Fully dense PLA samples made using FFF exhibit higher toughness than samples made using IM processing (5 MPa-$m^{1/2}$ *vs* 3 MPa-$m^{1/2}$) [123]. Nylon-6-Al_2O3 composite filament was created by Singh *et al.* [124], and components manufactured with FFF displayed noticeably improved tribological properties [15, 125].

Recent times have seen a significant increase in the use of nanomaterials as reinforcement for fabrication of polymer composites because these offer several benefits like higher mechanical strength, better electrical and thermal conductivity, increased resistance to fire and heat, exceptional strength-weight ratios, along with added functional properties, when equated with their corresponding micromaterial counterparts [126, 127]. The creation and development of these nanoparticle-reinforced polymer composites thus offer new opportunities, but these are not bereft of newer challenges. For example, a composite filament composed of ABS and graphene nanoplatelets (xGnP) developed and manufactured *via* FFF showed improved tensile modulus and decreased creep compliance of ABS regardless of the direction with respect to the compression molding technique [128]. Additional characteristics, like tensile strength and elongation, were shown to be direction-dependent. Increasing the elastic modulus of ABS-GnP composites using FFF is seen as advantageous, although this method has no influential effects on their elongation. In another report, FFF was used for aligning graphite flakes alongside the components' through-plane directions, producing conductive parts [129]. According to Shofner *et al.* [130], by optimizing the mixing, extrusion, and FFF processes, it is possible to uniformly distribute the vapor-grown carbon fibers (VGCFs) into the ABS matrix having the least amount of porosity. As the VGCFs get aligned along the direction of extrusion, the composites also show an augmented tensile strength and modulus (39% and 60%). With these composites, a significant loss in elongation is observed, as adding on of VGCFs to ABS changes their originally ductile fracture mode to brittle. Electric conductivity of ABS gets improved nearly four times in graphene-reinforced ABS composites made with FFF [131]. Usage of nano-clay particles of the natural clay Montmorillonite as reinforcement material in ABS composites has demonstrated an enhanced tensile strength of 15% in the composite with respect to the injection molding (IM) composite [132]. Composites created with FFF likewise show enhanced thermostability. Numerous carbon-based nanomaterials, including nanotubes and nanofibers, have been the subject of extensive research [15, 131], which show considerable improvement in mechanical, electrical, and thermal characteristics. The construction of nanoparticle-reinforced polymer composites has been described by several researchers employing photocurable resins [133]. However, high concentrations of nanoparticles have been found to significantly shield and scatter laser light [134]. In order to increase the stability of nanofillers (silica, montmorillonite, and palygorskite) in SLA resin, Weng *et al.* [135] have changed the surfaces of these materials. Composites containing 5% w/w SiO_2 show an improvement in tensile strength of around 21% coupled with 65% enhancement in modulus. The surface of nano TiO_2 particles has been altered using a silane-based coupling reagent followed by evaluation of the effect of titanium dioxide concentration on cure depth and resin thickness with SLA processing [136]. With the application of improved resin, a notable improvement with respect to the tensile strength, ductility, and thermostability of the composites has been demonstrated. Additionally, some researchers have reported that silane and other couplants can improve the characteristics of the composite by transferring the stress perpetrated on the reinforcement toward the matrix [137]. PS-Al_2O_3 composites have been made utilizing SLS, and similarly, surface modification of nano-Al_2O_3 particles have been carried out with polystyrene (PS). Impact (up 50%) and tensile strength (up 300%) were increased due to improved laser absorption by coated particles. Therefore, the use of AM methods to create fully dense nanocomposites requires proper surface treatment of nanoparticles [138].

Another method of producing polymer composites is by using fiber as reinforcing materials with resultant improvement in their properties [139]. FFF is the most well-liked and extensively researched AM method for fabricating fiber-reinforced polymer composites. Even while there are several reports of application of additional AM techniques including DR, SLS, and SLA, these are not as commonly used as FFF. The usage of short fibers is the main focus of research on FFF. The alignment of fibers along the direction of extrusion or along the deposition of the material is the key feature of FFF method of fiber-based polymer composites. According to Ferreira *et al.* [140], polylactic acid-carbon fiber composites made by FFF have improved mechanical properties because of the favorable alignment of the short carbon fibers (up to 15 wt%). Similarly, the tensile strength and modulus of short carbon fiber (CF)-reinforced ABS composites made with FFF have been shown to be raised by 115% and 700%, respectively [141]. FFF produced preferential orientation of the fibers (around 92%) in the direction of deposition of the material, counteracting the negative effects of porosity at these concentrations of composites, which results from the elevated thermal conductivity. Ning *et al.* [142] have also shown meliorations in the tensile and bending characteristics of CF-reinforced thermoplastics, where longer CFs were found to lower the ductility and toughness of the composites while increasing the tensile strength and modulus. Ning *et al.* [143] have reported the effects of various FFF process parameters on the mechanical characteristics of CF-reinforced plastics. An alternative study has employed short glass fibers (GF) for strengthening of the ABS filament, but this had the unintended consequence of reducing filament flexibility and processibility. This was primarily because swelling was reduced and the tape on the print head stiffened up with FFF, which had an adverse effect on flexibility [144]. However, it was discovered that adding a small quantity of plasticizers and compatibilizers to the ABS-glass fiber composite filament made it easier to process using FFF. Carneiro *et al.* [145] have investigated the effect of FFF process factors including layer thickness, orientation of the filament, and infill degree on the manufacture of polypropylene (PP)-GF composite filaments. The authors come to the conclusion that because FFF composite parts have better mechanical qualities than compression molded parts, these can be used right away as functioning parts. The matrix and the reinforcement fibers in the polymer composites are premixed to construct composite filaments. Yet, employing such composites to create composite powder with desirable free-flow properties for SLS is a huge challenge. Moreover, the SLS component quality is heavily influenced by powder properties and consistency of the powder bed. Therefore, processing of polymer composites reinforced with short fibers is a daunting task using this technique [146]. Fabricating continuous fiber-reinforced polymer composites utilizing additive manufacturing technology is another crucial and difficult challenge due to the fact that these composites have superior mechanical properties in comparison with the composites reinforced with short fibers [102]. A novel FFF approach has recently been devised that involves feeding continuous fiber and polymer filament toward the print head separately aimed at in-nozzle saturation proximately ahead of getting printed [147]. Jute fibers, along with continuous carbon fibers (CCFs), have been employed as reinforcement materials in this study, with PLA serving as the matrix. The tensile strength and modulus of the PLA-CCF composites were 185.2 MPa and 19.5 GPa, respectively, which measured around six to four times more compared to those of lone PLA [15].

Using a marketed printer having a dual print head, meant for the fiber as well as the matrix, analogous composites fabricated from CCF inserted within nylon are reported [148]. Substantial voids have been noted with elevated CCFs, and the composites have a reduced strength. Mechanical characteristics of these composites have been shown to be strongly affected with various process variables including print head temperature, thickness of layers, and speed of printing, just like with normal FFF [149]. Other researchers [150] have also reported a range of findings linked with the manufacture of continuous fiber-reinforced composites utilizing FFF. It was also successful to fabricate newer composites with cross lap-based and panel-core lap-based designing that had good compressive strength [151]. With a 5-fold increase in fiber concentration to 10%, the specific strength of the composites enhanced significantly (18 kN m/kg *vs* 28 kN m/kg). Additionally, the corrugated-core construction with 11.5 vol% fiber demonstrated a good compressive strength, comparing well with the corresponding parts manufactured by employing other methods. Utilizing AM also has the advantage of producing complicated structures. Recently, the effect of type and/or orientation of the fibers and their volume fraction has been studied on the mechanical characteristics of composites made by reinforcing nylon with CCFs, Kevlar, and glass using a commercial-grade FFF machine. The method involves stacking the reinforcing fiber bundles within the matrix deposit. Although carbon fiber reinforcement produced the best strength, it was discovered that increasing the fiber concentration caused a significant quantity of voids. Glass fibers seem to offer composites free of voids with an increase in concentration (up to 22.5%) among all fibers [152]. In a related investigation, interlaminar shear strength was found to be the highest for carbon fibers and lowest for Kevlar fibers, which have poor wettability in nylon matrix [153]. The glass fibers were discovered to be the best for high-impact strength, while carbon and Kevlar fiber performance is comparable. Li *et al.* [150] created PLA-CCF composites using FFF and showed how PLA-methylene dichloride surface modification of carbon fibers can significantly increase fiber-matrix adhesion. Tensile and flexion strengths of the improved fiber composites were enhanced by 14% and 164%, respectively, and further their storage modulus was also increased by fiber surface treatment. In a different study [154], CCFs were treated with polyamide (PA6) to increase adherence, which increased interlaminar shear strength by around 42% in comparison to untreated fibers [15].

5.10 Critical Issues During Processing of Natural Fiber Composites

5.10.1 Thermal Stability

Natural fibers get decomposed beyond a particular temperature (2,000 degrees Celsius); therefore, thermal stability is regarded as a significant factor during manufacturing of composites. Changes like discoloration, dehydration, depolymerization, decarboxylation, recrystallization occur when natural fibers are subjected to elevated temperatures [155]. These changes can affect mechanical properties of composites produced with these fibers [156, 157]. Thus, thermal stability has an important role to play in the development of natural fiber-reinforced composites [13, 15, 158].

5.10.2 Hydrophilic Nature of Natural Fibers

Natural fibers are highly hydrophilic (presence of OH groups); hence, their bonding with hydrophobic polymer is not strong, with resultant poor mechanical properties of composites. Microbial degradation can also take place due to high hydrophilicity of natural fibers. Therefore, before processing, these fibers must be dried. Furthermore, after drying, these fibers must be stored and handled with care in order to prevent moisture absorption. Excessive moisture can cause pore formation in the composites reinforced with natural fibers while these are being processed. Hydrophilicity of natural fibers can be improved by various fiber treatments like alkali treatment, steam explosion, acetylation, electron beam irradiation, silanization, graft copolymerization, gamma-ray irradiation, grinding, benzoylation, plasma treatment, etc. [15, 159].

5.10.3 Fiber Breakage During Processing

Fiber breakage is another important issue that can arise during processing of natural fiber-reinforced composite, and this can cause degradation of composite materials. Fibers can break due to multiple reasons. During initial stages, breakage may occur due to concurrent action of pressure and temperature. Sometimes, fibers get ruptured due to collision between fiber and mold or between fibers. Breakage mainly occurs due to longer mixing time, high processing temperature, or due to high rotor speed. So, in order to prevent breakage, these processing parameters must be carefully controlled [15].

5.10.4 Moisture Absorption and Distribution of Fiber Inside the Matrix

The presence of high moisture reduces interfacial joining of the fiber to the matrix because of which the mechanical strength of composites is decreased [160]. Thus, to obtain composite materials of high strength and high quality, moisture content should be reduced below 3% weight. Moisture content can be reduced by chemically treating the natural fibers through a variety of procedures [161]. Mechanical characteristics of fibers can also be impacted by poor fiber distribution within the matrix. This distribution can be ameliorated by employing various physical and chemical treatment methods [15].

5.11 Conclusion

Composite materials based on natural fibers have assumed considerable significance in recent times, finding a broad array of applications in various sectors by modifying existing materials and technologies toward greener approaches. Several types of naturally obtained fibers are obtainable from plant, animal, or mineral sources that have found use for fabrication of natural fiber composites. The composites created with natural fibers offer several advantages over synthetic fiber-based materials in being economical, eco-friendly, lightweight, renewable, sustainable, and biodegradable materials. Several environmental issues are encountered when dealing with conventional materials that have actuated interest in various manufacturers who are developing modified technologies to adopt these composite materials for a variety of applications. These natural fiber-reinforced composites are finding a diverse array of applications in the automotive, industrial, medical, and aerospace sectors. Natural fiber composites can be manufactured using numerous techniques. The most widely employed techniques include compression molding, resin transfer molding, spray up technique, hand lay technique, etc. Apart from these, conventional methods, additive manufacturing techniques are also being employed for the manufacture of NFCs. It can manufacture composites with site-specific performance and functionality. Direct ink writing, stereolithography, binder jetting, selective laser sintering, direct light processing, etc., are some of the commonly used additive manufacturing techniques for producing composites. Despite their numerous advantages, further research is required in this area to tide over the drawbacks of natural fiber-based composites, including their moisture-absorbing properties, poor adhesiveness, flawed prediction of mechanical properties, etc.

References

1. Ramamoorthy, S.K., Skrifvars, M., Persson, A., A review of natural fibers used in biocomposites: Plant, animal and regenerated cellulose fibers. *Polym. Rev.*, 55, 107, 2015.
2. Mitra, B.C., Environment friendly composite materials: Biocomposites and green composites. *Def. Sci. J.*, 64, 244, 2014.
3. Syduzzaman, M.D., Abdullah, M.D., Faruque, A.L., Bilisik, K., Naebe, M., Plant-based natural fibre reinforced composites: A review on fabrication, properties and applications. *Coatings*, 10, 973, 2020.
4. Maghsoudi-Ganjeh, M., Lin, L., Wang, X., Zeng, X., Bioinspired design of hybrid composite materials. *Int. J. Smart Nano Mater.*, 10, 90, 2018.
5. Subramani, N., Murali, J.G., Suresh, P., Sankar, V.V.A., Review on hybrid composite materials and its applications. *Int. Res. J. Eng. Technol.*, 4, 1921, 2017.
6. Peças, P., Carvalho, H., Salman, H., Leite, M., Natural fibre composites and their applications: A review. *J. Compos. Sci.*, 2, 51, 2018.
7. Kanitkar, Y.M., Kulkarni, A.P., Wangikar, K.S., Characterization of Glass Hybrid composite: A Review. *Mater. Today Proc.*, 4, 9627, 2017.
8. Safri, S.N.A., Sultan, M.T.H., Jawaid, M., Jayakrishna, K., Impact behaviour of hybrid composites for structural applications: *A review. Compos. B Eng.*, 133, 112, 2018.
9. Satyanarayana, K.G., Arizaga, G.G.C., Wypych, F., Biodegradable composites based lignocellulosic fibers-an overview. *Prog. Polym. Sci.*, 34, 982, 2009.
10. Ramesh, M., Palanikumar, K., Reddy, K.H., Plant fibre- based bio-composites: Sustainable and renewable green materials. *Renew. Sustain. Energy. Rev.*, 79, 558, 2017.
11. Faruk, O., Bledzki, A.K., Fink, H.P., Sain, M., Progress report on natural fiber reinforced composites. *Macromol. Mater. Eng.*, 299, 9, 2014.
12. Holbery, J. and Houston, D., Natural-fiber-reinforced polymer composites in automotive applications. *JOM*, 58, 80, 2006.

13. Masri, M.N., Mohd, S.H., Abu Bakar, M.B., Rusli, N.W., Ismail, M.F., Amini, M.H.M., Al-rashdi, A.A., Review of manufacturing process of natural fiber reinforced polymer composites. *Int. J. Integr. Eng.*, 13, 172, 2021.
14. Ganguly, A., Shankar, S., Das, A., Shukla, M., Swaroop, C., Bhardwaj, T., Natural fiber reinforced composites: A review based on additive manufacturing routes and biodegradability perspective. *Mater. Today Proc.*, 62, 131, 2022.
15. Balla, V.K., Kunal, H., Kate, K.H., Satyavolu, J., Singh, P., Tadimeti, J.G.D., Additive manufacturing of natural fiber reinforced polymer composites: Processing and prospects. *Compos. B Eng.*, 174, 106956, 2019.
16. Ahmad, F., Choi, H.S., Park, M.K., A review: Natural fiber composites selection in view of mechanical, light weight, and economic properties. *Macromol. Mater. Eng.*, 300, 10, 2015.
17. Shah, D.U., Developing plant fiber composites for structural applications by optimising composite parameters: A critical review. *J. Mater. Sci.*, 48, 6083, 2013.
18. Burgert, I., Exploring the micromechanical design of plant cell walls. *Am. J. Bot.*, 93, 1391, 2006.
19. Martinschitz, K.J., Boesecke, P., Garvey, C.J., Gindl, W., Keckes, J., Changes in microfibril angle in cyclically deformed dry coir fibers studied by in-situ synchrotron X-ray diffraction. *J. Mater. Sci.*, 43, 350, 2008.
20. Keckes, J., Burgert, I., Fruhmann, K., Müller, M., Kolln, K., Hamilton, M. *et al.*, Cell-wall recovery after irreversible deformation of wood. *Nat. Mater.*, 2, 810, 2003.
21. Anandjiwala, R.D. and Blouw, S., Composites from bast fibres-prospects and potential in the changing market environment. *J. Nat. Fibers*, 4, 91, 2017.
22. Nguong, C.W., Lee, S.N.B., Sujan, D., A review on natural fibre reinforced polymer composites. *Int. J. Miner. Metall. Mater.*, 7, 52, 2013.
23. O'Donnell, A., Dweib, A., Wool, R., Natural fiber composites with plant oil-based resin. *Compos. Sci. Technol.*, 64, 1135, 2004.
24. Keya, K.N., Kona, N.A., Koly, F.A. *et al.*, Natural fiber reinforced polymer composites -history, types, advantages and applications. *Mater. Eng. Res.*, 1, 69, 2019.
25. Bongarde, U.S. and Shinde, V.D., Review on natural fiber reinforcement polymer composites. *Int. J. Eng. Innov. Technol.*, 3, 431, 2014.
26. Vijay, K.T., Manju, K.T., Raju, K.G., Review: Raw natural fiber-based polymer composites. *Int. J. Polym. Anal. Charact.*, 19, 256, 2014.
27. Siakeng, R., Jawaid, M., Asim, M., Saba, N., Sanjay, M.R., Siengchin, S., Fouad, H., Alkali treated coir/pineapple leaf fibers reinforced PLA hybrid composites: Evaluation of mechanical, morphological, thermal and physical properties. *Polym. Lett.*, 14, 717, 2020.
28. Fatima, S. and Mohanty, A.R., Acoustical and fire-retardant properties of jute composite materials. *Appl. Acoust.*, 72, 108, 2011.

29. Liu, L., Yu, J., Cheng, L., Yang, X., Biodegradability of poly (butylene succinate) (PBS) composite reinforced with jute fiber. *Polym. Degrad. Stab.*, 94, 90, 2009.
30. Jaafar, C.A., Zainol, I., Ishak, N., Ilyas, R., Sapuan, S., Effects of the liquid natural rubber (LNR) on mechanical properties and microstructure of epoxy/silica/kenaf hybrid composite for potential automotive applications. *J. Mater. Res. Technol.*, 12, 1026, 2021.
31. Sabaruddin, F.A., Tahir, P.M., Sapuan, S.M., Ilyas, R.A., Lee, S.H., Abdan, K., Mazlan, N., Roseley, A.S.M., Hps, A.K., The effects of unbleached and bleached nanocellulose on the thermal and flammability of polypropylene-reinforced kenaf core hybrid polymer bionanocomposites. *Polym. J.*, 13, 116, 2020.
32. Nassiopoulos, E. and Njuguna, J., Thermo-mechanical performance of poly (lactic acid)/flax fibre-reinforced biocomposites. *Mater. Des.*, 66, 473, 2015.
33. Syafri, E., Kasim, A., Abral, H., Asben, A., Cellulose nanofibers isolation and characterization from ramie using a chemical-ultrasonic treatment. *J. Nat. Fibers*, 16, 1145, 2018.
34. Battegazzore, D., Noori, A., Frache, A., Hemp hurd and alfalfa as particle filler to improve the thermo-mechanical and fire- retardant properties of poly (3-hydroxybutyrate-co-3-hydroxyhexanoate). *Polym. Compos.*, 40, 3429, 2019.
35. Gupta, M. and Singh, R., PLA-coated sisal fibre-reinforced polyester composite: Water absorption, static and dynamic mechanical properties. *J. Compos. Mater.*, 53, 65, 2019.
37. Verma, D., Gope, P.C., Shandilya, A., Maheshwari, M.K., Coir fiber reinforcement and application in polymer composites: A review. *J. Mater. Environ. Sci.*, 4, 263, 2013.
36. Ayrilmis, N., Jarusombuti, S., Fueangvivat, V., Bauchongkol., P., White, R.H., Coir fiber reinforced polypropylene composite panel for automotive interior applications. *Fibers Polym.*, 12, 919–926, 2011.
38. Dash, D., Samanta., S., Gautum, S.S. *et al.*, Mechanical characterization of natural fiber reinforced composite material. *Adv. Mater. Manuf. Charact.*, 3, 275, 2013.
39. Charoenvaisarocha, A.C., Jongjit, H., Joseph, K., Materials and mechanical properties of pretreated coir-based green composites. *Compos. B. Eng.*, 40, 633, 2009.
40. Harish, S., Michael, D.P., Bensely, A., Lal, D.M., Rajadurai, A., Mechanical property evaluation of natural fiber coir composite. *Mater. Charact.*, 60, 44, 2009.
41. Brahmakumar, M., Pavithran, C., Pillai, R.M., Coconut-fibre reinforced Polyethylene composites: Effect of natural waxy surface layer of the fibre on fibre/matrix interfacial bonding and strength of composites. *Compos. Sci. Technol.*, 65, 563, 2005.

42. Shahzad, A., A study in physical and mechanical properties of hemp fibres. *Adv. Mater. Sci. Eng.*, 8, 1, 2013.
43. Beckermann, G.W. and Pickering, K.L., Engineering and evaluation of hemp fibre reinforced polypropylene composites: Fibre treatment and matrix modification. *Compos. Part A Appl. Sci. Manuf.*, 39, 979, 2008.
44. Rouison, D., Sain, M., Couturier, M., Resin-transfer molding of hemp-fibre composite: Optimization of the process and mechanicalproperties of the materials. *Compos. Sci. Technol.*, 66, 895, 2006.
45. Ali, M., Natural fibres as construction materials. *J. Civ. Eng. Constr. Technol.*, 3, 80, 2012.
46. Rohit, K. and Dixit, S., A review—future aspect of natural fiber reinforced composite. *Polym. Renew. Resour.*, 7, 43, 2016.
47. Zhu, J.C., Zhu, H.J., Njuguna, J. *et al.*, Recent development of flax fibres and their reinforced composites based on different polymeric matrices. *Mater. J.*, 6, 5171, 2013.
48. Rodríguez, E., Petrucci, R., Puglia, D., Kenny, J.M., Vázquez, A., Characterization of composites based on natural and glass fibers obtained by vacuum infusion. *Compos. Mater.*, 39, 265, 2005.
49. Yao, S.S., Jin, F.L., Rhee, K.Y., Hui, D., Park, S.J., Recent advances in carbon-fibre-reinforced thermoplastic composites: A review. *Compos. B Eng.*, 142, 240, 2018.
50. May-Pat, A., Valadez-Gonzlez, A.P., Herrera-Franco, J., Effect of fiber surface treatments on the essential work of fracture of HDPE-continuous henequen fiber-reinforced composites. *Polym. Test.*, 32, 1114, 2013.
51. Koronis, G., Silva, A., Fontul, M., Green composite: A review of adequate materials for automotive application. *Compos. B Eng.*, 44, 120, 2013.
52. Pickering, K.L., Efendy, M.G.A., Le, T.M., A review of recent developments in natural fibre composites and their mechanical performance. *Compos. Part A Appl. Sci. Manuf.*, 83, 98, 2006.
53. Lila, M.K., Kumar, F., Sharma, S., Composites from waste for civil engineering applications. *I-manager's J. Mater. Sci.*, 1, 1, 2013.
54. Wang, J.L., Application of composite materials on sports equipments. *Appl. Mech. Mater.*, 155, 903, 2022.
55. George, J., Sreekala, M.S., Thomas, S., A review on interface modification and characterization of natural fiber reinforced plastic composites. *Polym. Eng. Sci.*, 41, 1471, 2001.
56. Ray, D., Sarkar, B.K., Rana, A.K., Bose, N.R., Effect of alkali treated jute fibres on composite properties. *Bull. Mater. Sci.*, 24, 129, 2004.
57. Arrakhiz, F.Z., El Achaby, M., Malha, M., Bensalah, M.O., Fassi-Fehri, O., Bouhfid, R., Mechanical and thermal properties of natural fibers reinforced polymer composites: Doum/low density polyethylene. *Mater. Des.*, 43, 200, 2013.

58. Manikandan, N.K.C., Thomas, S., Groeninckx, G., Thermal and dynamic mechanical analysis of polystyrene composites reinforced with short sisal fibres. *Compos. Sci. Technol.*, 61, 2519, 2001.
59. Xue, L., Lope, G.T., Satyanarayan, P., Chemical treatment of natural fibre for use in natural fibre-reinforced composites: A review. *J. Polym. Environ.*, 15, 25, 2007.
60. Sreekala, M.S., Kumaran, M.G., Joseph, S., Jacob, M., Oil palm fibers reinforced phenol formaldehyde composites: Influence of fibers surface modifications on the mechanical performance. *Appl. Compos. Mater.*, 7, 295, 2000.
61. Gassan, J. and Gutowski, V.S., Effects of corona discharge and UV treatment on the properties of jute-fibre epoxy composites. *Compos. Sci. Technol.*, 60, 2857, 2000.
62. Huber, T., Biedermann, U., Müssig, J., Enhancing the fibre matrix adhesion of natural fibre reinforced polypropylene by electron radiation analyzed with the single fibre fragmentation test. *Compos. Interfac.*, 17, 371, 2010.
63. Rong, M.Z., Zhang, M.Q., Liu, Y., Yang, G.C., Zeng, H.M., The effect of fiber treatment on the mechanical properties of unidirectional sisal-reinforced epoxy composites. *Compos. Sci. Technol.*, 61, 1437, 2001.
64. Sinha, E. and Panigrahi, S., Effect of plasma treatment on structure, wettability of jute fiber and flexural strength of its composite. *J. Compos. Mater.*, 43, 1791, 2009.
65. Rajak, D.K., Pagar, D.D., Menezes, P.L., Linul, E., Fiber-reinforced polymer composites: Manufacturing, properties, and applications. *Polym. J.*, 11, 1667, 2019.
66. Crutchlow, R., Changing from open to closed moulding. *Reinf. Plast.*, 48, 40, 2004.
67. Idicula, M., Boudenne, A., Umadevi, L., Ibos, L., Candau, Y., Thomas, S., Thermophysical properties of natural fibre reinforced polyester composites. *Compos. Sci. Technol.*, 66, 2719, 2006.
68. Gunge, A., Koppad, P.G., Nagamadhu, M., Kivade, S., Murthy, K.S., Study on mechanical properties of alkali treated plain woven banana fabric reinforced biodegradable composites. *Compos. Commun.*, 13, 47, 2019.
69. Elkington, M., Bloom, D., Ward, C., Chatzimichali, A., Potter, K., Hand layup: Understanding the manual process. *Adv. Manuf. Polym. Compos. Sci.*, 1, 138, 2015.
70. Cucinotta, F., Guglielmino, E., Sfravara, F., Life cycle assessment in yacht industry: A case study of comparison between hand lay-up and vacuum infusion. *J. Clean. Prod.*, 142, 3822, 2017.
71. Abdurohman, K., Satrio, T., Muzayadah, N.L., Teten, A comparison process between hand lay-up, vacuum infusion and vacuum bagging method toward e-glass EW 185/lycal composites. *J. Phys. Conf. Series.*, 1130, 1742, 2018.
72. Brouwer, W., Van Herpt, E., Labordus, M., Vacuum injection moulding for large structural applications. *Compos. Part A Appl. Sci. Manuf.*, 34, 551, 2003.

73. Loff, A., Choosing the right closed moulding process. *Reinf. Plast.*, 48, 46, 2004.
74. Saleem, Z., Rennebaum, H., Pudel, F., Grimm, E., Treating bast fibers with pectinase improves mechanical characteristics of reinforced thermoplastic composites. *Compos. Sci.Technol.*, 68, 471, 2008.
75. Thomason, J.L., Interfacial strength in thermoplastic composites-At last an industry friendly measurement method? *Compos. Part A Appl. Sci. Manuf.*, 33, 1283, 2002.
76. De Farias, M.A., Amico, S.C., Coelho, L.A.F., Pezzin, S.H., Multi-component nanocomposites of epoxy/ silsesquioxane reinforced with carbon fibers and carbon nanotubes processed by resin transfer molding. *Polym.-Plast. Technol. Mater.*, 59, 517, 2020.
77. Fu, H., Xu, H., Liu, Y., Yang, Z., Kormakov, S., Wu, D., Sun, J., Overview of injection molding technology for processing polymers and their composites. *ES Mater. Manuf.*, 8, 3, 2002.
78. Jaafar, J., Siregar, J.P., Tezara, C., Hamdan, M.H.M., Rihayat, T., A review of important considerations in the compression molding process of short natural fiber composites. *Int. J. Adv. Manuf. Technol.*, 105, 3437, 2019.
79. Bender, D., Schuster, J., Heider, D., Flow rate control during vacuum-assisted resin transfer molding (VARTM) processing. *Compos. Sci. Technol.*, 66, 2265, 2006.
80. Yoon, M.K., Baidoo, J., Gillespie, J.W., Heider, D.J., Hsiao, K.T., Vacuum g (VARTM) process incorporating gravitational effects: A closed-form solution. *J. Compos. Mater.*, 39, 2227, 2005.
81. Kedari, V.R., Farah, B.I., Hsiao, K.T., Effects of vacuum pressure, inlet pressure, and mold temperature on the void content, volume fraction of polyesterne-glass fiber composites manufactured with VARTM process. *J. Compos. Mater.*, 45, 2727, 2011.
82. Yalcinkaya, M.A., Sozer, E.M., Altan, M.C., Fabrication of high- quality composite laminates by pressurized and heated-VARTM. *Compos. Part A Appl. Sci. Manuf.*, 102, 336, 2017.
83. Gawel, T.G., Review of additive manufacturing method. *Solid State Phenom.*, 308, 1, 2020.
84. Parupelli, S.K. and Desai, S., A comprehensive review of additive manufacturing (3D Printing) processes, applications and future potential. *Am. J. Appl. Sci.*, 16, 244, 2019.
85. Rosa, S.M.L., Santos, E.F., Ferreira, C.A., Nachtigall, S.M.B., Studies on the properties of rice-husk-filled-PP composites: Effect of maleated PP. *Mater. Res.*, 12, 333, 2009.
86. Avinc, O. and Khoddami, A., Overview of poly (lactic acid) (PLA) fibre. *Fibre Chem.*, 42, 68, 2010.
87. Royan, R., Leong, J.S., Chan, W.N., Tan, J.R., Shamsuddin, Z.S.B., Current State and challenges of natural fibre-reinforced polymer composites as feeder in FDM-based 3D printing. *Polymers*, 13, 2289, 2021.

88. Mazzanti, V., Malagutti, L., Mollica, F., FDM 3D printing of polymers containing natural fillers: A review of their mechanical properties. *Polymers*, 11, 1094, 2019.
89. Osman, M.A. and Atia, M.R., Investigation of ABS-rice straw composite feedstock filament for FDM. *Rapid. Prototyp. J.*, 24, 1067, 2018.
90. Zhang, K., Mohanty, A.K., Misra, M., Fully biodegradable and bio-renewable ternary blends from polylactide, poly (3-hydroxybutyrate-co-hydroxyvalerate) and poly (butylene succinate) with balanced properties. *ACS. Appl. Mater. Interfaces*, 4, 3091, 2012.
91. Stoof, D., Pickering, K., Zhang, Y., Fused deposition modelling of natural fibre/polylactic acid composites. *J. Compos. Sci.*, 1, 8, 2017.
92. Dong, Y., Milentis, J., Pramanik, A., Additive manufacturing of mechanical testing samples based on virgin poly (lactic acid) (PLA) and PLA/wood fibre composites. *Adv. Manuf.*, 6, 71, 2018.
93. Gkartzou, E., Koumoulos, E.P., Charitidis, C.A., Production and 3D printing processing of bio-based thermoplastic filament. *Manuf. Rev.*, 4, 1, 2017.
94. Nyambo, C., Mohanty, A.K., Misra, M., Effect of maleated compatibilizer on performance of PLA/wheat straw-based green composites. *Macromol. Mater. Eng.*, 296, 710, 2011.
95. Fernández, C. and Puig, C.C., Changes in melting behavior and morphology on annealing of linear and branched polyethylene. *J. Macromol. Sci. B*, 41, 991, 2002.
96. Ngo, T.D., Kashani, A., Imbalzano, G., Nguyen, K.T.Q., Hui, D., Additive manufacturing (3D printing): A review of materials, methods, applications and challenges. *Compos. B Eng.*, 143, 172, 2018.
97. Leon, A.C.D., Chen, Q., Palaganas, N.B., Palaganas, J.O., Manapat, J., Advincula, R.C., High performance polymer nanocomposites for additive manufacturing applications. *React. Funct. Polym.*, 103, 141, 2016.
98. Dobrzański, L.A., Achtelik-Franczak, A., Król, M., Computer aided design in selective laser sintering (SLS)–application in medicine. *J. Achiev. Mater. Manuf. Eng.*, 60, 66, 2013.
99. Kadry, H., Wadnap, S., Xu, C., Ahsan, F., Digital light processing (DLP) 3D-printing technology and photoreactive polymers in fabrication of modified-release tablets. *Eur. J. Pharm. Sci.*, 135, 60, 2019.
100. Anitha, R., Arunachalam, S., Radhakrishnan, P., Critical parameters influencing the quality of prototypes in fused deposition modelling. *J. Mater. Process. Technol.*, 118, 385, 2001.
101. Mohan, N., Senthil, P., Vinodh, S., Jayanth, N., A review on composite materials and process parameters optimisation for the fused deposition modelling process. *Virtual Phys. Prototyp.*, 12, 47, 2017.
102. Parandoush, P. and Lin, D., A review on additive manufacturing of polymer-fiber composites. *Compos. Struct.*, 182, 36, 2007.
103. Wang, X., Jiang, M., Zhou, Z., Gou, J., Hui, D., 3D printing of polymer matrix composites: A review and prospective. *Compos. B Eng.*, 110, 442, 2017.

104. Sridharan, A.K. and Joshi, S., An octree-based algorithm for the optimization of extraneous material removal in laminated object manufacturing (LOM). *J. Manuf. Syst.*, 19, 355, 2001.
105. Park, J., Tari, M.J., Hahn, H.T., Characterization of the laminated object manufacturing (LOM) process. *Rapid. Prototyp. J.*, 6, 36, 2000.
106. Liao, Y.S., Chiu, L.C., Chiu, Y.Y., A new approach of online waste removal process for laminated object manufacturing (LOM). *J. Mater. Process. Technol.*, 140, 136, 2003.
107. Billiet, T., Vandenhaute, M., Schelfhout, J., Van Vlierberghe, S., Dubruel, P., A review of trends and limitations in hydrogel-rapid prototyping for tissue engineering. *Biomaterials*, 33, 6020, 2012.
108. Hardin, J.O., Ober, T.J., Valentine, A.D., Lewis, J.A., Microfluidic printheads for multimaterial 3D printing of viscoelastic inks. *Adv. Mater.*, 27, 3279, 2015.
109. Schmid., M., Amado, A., Wegener, K., Materials perspective of polymers for additive manufacturing with selective laser sintering. *J. Mater. Res.*, 29, 1824, 2014.
110. Williams, J.M., Adewunmi, A., Schek, R.M., Flanagan, C.L., Krebsbach, P.H., Feinberg, S.E. *et al.*, Bone tissue engineering using polycaprolactone scaffolds fabricated via selective laser sintering. *Biomaterials*, 26, 4817, 2005.
111. Ghita, O.R., James, E., Trimble, R., Evans, K.E., Physico-chemical behaviour of poly (ether ketone) (PEK) in high temperature laser sintering (HT-LS). *J. Mater. Process. Technol.*, 214, 969, 2014.
112. Bai, J., Goodridge, R.D., Hague, R.J.M., Song, M., Improving the mechanical properties of laser-sintered polyamide 12 through incorporation of carbon nanotubes. *Polym. Eng. Sci.*, 53, 1937, 2013.
113. Salmoria, G.V., Paggi, R.A., Lago, A., Beal, V.E., Microstructural and mechanical characterization of PA12/MWCNTs nanocomposite manufactured by selective laser sintering. *Polym. Test.*, 30, 611, 2011.
114. Utela, B., Storti, D., Anderson, R., Ganter, M., A review of process development steps for new material systems in three- dimensional printing (3DP). *J. Manuf. Process.*, 10, 96, 2008.
115. Tarafder, S., Balla, V.K., Davies, N.M., Bandyopadhyay, A., Bose, S., Microwave-sintered 3D printed tricalcium phosphate scaffolds for bone tissue engineering. *J. Tissue. Eng. Regenerat. Med.*, 7, 631, 2013.
116. Parthasarathy, J., Starly, B., Raman, S., Christense, A., Mechanical evaluation of porous titanium (Ti6Al4V) structures with electron beam melting (EBM). *J. Mech. Behav. Biomed. Mater.*, 3, 249, 2010.
117. Murr, L.E., Gaytan, S.M., Ramirez, D.A., Martinez, E., Hernandez, J., Amato, K.N., Shindo, P.W., Medina, F.R., Wicker, R.B., Metal fabrication by additive manufacturing using laser and electron beam melting technologies. *J. Mater. Sci. Technol.*, 28, 1, 2012.
118. Novakova-Marcincinova, L., Application of fused deposition modeling technology in 3D printing rapid prototyping area. *Manuf. Ind. Eng.*, 11, 35, 2012.

119. Nikzad, M., Masood, S.H., Sbarski, I., Thermo-mechanical properties of a highly filled polymeric composites for fused deposition modeling. *Mater. Des.*, 32, 3448, 2011.
120. Drummer, D., Cifuentes-Cuellar, S., Rietzel, D., Suitability of PLA/TCP for fused deposition modeling. *Rapid. Prototyp. J.*, 18, 500, 2012.
121. Castles, F., Isakov, D., Lui, A., Lei, Q., Dancer, C.E.J., Wang, Y. *et al.*, Microwave dielectric characterisation of 3D-printed BaTiO3/ABS polymer composites. *Sci. Rep.*, 6, 22714, 2016.
122. Torrado Perez, A.R., Roberson, D.A., Wicker, R.B., Fracture surface analysis of 3D-printed tensile specimens of novel ABS-based materials. *J. Fail. Anal. Prev.*, 14, 343, 2014.
123. Song, Y., Li, Y., Song, W., Yee, K., Lee, K.Y., Tagarielli, V.L., Measurements of the mechanical response of unidirectional 3D-printed PLA. *Mater. Des.*, 123, 154, 2017.
124. Singh, R., Singh, S., Fraternali, F., Development of in-house composite wire-based feed stock filaments of fused deposition modelling for wear-resistant materials and structures. *Compos. B Eng.*, 98, 244, 2016.
125. Singh, R., Singh, N., Amendola, A., Fraternali, F., On the wear properties of Nylon-6- SiC-Al_2O_3 based fused deposition modelling feed stock filament. *Compos. B Eng.*, 119, 125, 2017.
126. Chiappone, A., Roppolo, I., Naretto, E., Fantino, E., Calignano, F., Sangermano, M. *et al.*, Study of graphene oxide-based 3D printable composites: Effect of the *in-situ* reduction. *Compos. B Eng.*, 124, 9, 2017.
127. Sun, X., Sun, H., Li, H., Peng, H., Developing polymer composite materials: Carbon nanotubes or graphene? *Adv. Mater.*, 25, 5153, 2013.
128. Dul, S., Fambri, L., Pegoretti, A., Fused deposition modelling with ABS-graphene nanocomposites. *Compos. Part A Appl. Sci. Manuf.*, 85, 181, 2016.
129. Jia, Y., He, H., Geng, Y., Huang, B., Peng, X., High through-plane thermal conductivity of polymer- based product with vertical alignment of graphite achieved *via* 3D printing. *Compos. Sci. Technol.*, 145, 55, 2017.
130. Shofner, M.L., Lozano, K., Rodriguez-Macias, F.J., Barrera, E.V., Nanofiber-reinforced polymers prepared by fused deposition modeling. *J. Appl. Polym. Sci.*, 89, 3081, 2003.
131. Wei, X., Li, D., Jiang, W., Gu, Z., Wang, X., Zhang, Z., Sun, Z., 3D printable graphene composite. *Sci. Rep.*, 5, 11181, 2015.
132. Weng, Z., Wang, J., Senthil, T., Wu, L., Mechanical and thermal properties of ABS/montmorillonite nanocomposites for fused deposition modeling 3D printing. *Mater. Des.*, 102, 276, 2016.
133. Kim, K., Zhu, W., Qu, X., Aaronson, C., McCall, W.R., Chen, S. *et al.*, 3D optical printing of piezoelectric Nanoparticle-Polymer composite materials. *ACS Nano*, 8, 9799, 2014.
134. Azzopardi, N., Moharamzadeh, K., Wood, D.J., Martin, N., van Noort, R., Effect of resin matrix composition on the translucency of experimental dental composite resins. *Dent. Mater.*, 25, 1564, 2009.

135. Weng, Z., Zhou, Y., Lin, W., Senthil, T., Wu, L., Structure-property relationship of nano enhanced stereolithography resin for desktop SLA 3D printer. *Compos. Part A Appl. Sci. Manuf.*, 88, 234, 2016.
136. Yugang, D., Yuan, Z., Yiping, T., Dichen, L., Nano-TiO_2 -modified photosensitive resin for RP. *Rapid. Prototyp. J.*, 17, 247, 2011.
137. Ligon-Auer, S.C., Schwentenwein, M., Gorsche, C., Stampfl, J., Liska, R., Toughening of photo-curable polymer networks: a review. *Polym. Chem.*, 7, 257, 2016.
138. Zheng, H., Zhang, J., Lu, S., Wang, G., Xu, Z., Effect of core-shell composite particles on the sintering behavior and properties of nano-Al_2O_3/polystyrene composite prepared by SLS. *Mater. Lett.*, 60, 1219, 2006.
139. Love, L.J., Kunc, V., Rios, O., Duty, C.E., Elliott, A.M., Post, B., Rachel, J., Smith, J., Blue, C.A., The importance of carbon fiber to polymer additive manufacturing. *J. Mater. Res.*, 29, 1893, 2014.
140. Ferreira, R.T.L., Amatte, I.C., Dutra, T.A., Bürger, D., Experimental characterization and micrography of 3D printed PLA and PLA reinforced with short carbon fibers. *Compos. B Eng.*, 124, 88, 2017.
141. Tekinalp, H.L., Kunc, V., Velez-Garcia, G.M., Duty, C.E., Love, L.J., Naskar, A.K., Blue, C.A., Oxcan, S., Highly-oriented carbon fiber-polymer composites *via* additive manufacturing. *Compos. Sci. Technol.*, 105, 144, 2014.
142. Ning, F., Cong, W., Qiu, J., Wei, J., Wang, S., Additive manufacturing of carbon fiber reinforced thermoplastic composites using fused deposition modeling. *Compos. B Eng.*, 80, 369, 2015.
143. Ning, F., Cong, W., Hu, Y., Wang, H., Additive manufacturing of carbon fiber reinforced plastic composites using fused deposition modeling: Effects of process parameters on tensile properties. *J. Compos. Mater.*, 51, 451, 2017.
144. Zhong, W., Li, F., Zhang, Z., Song, L., Li, Z., Short fiber reinforced composites for fused deposition modeling. *Mater. Sci. Eng. A*, 301, 125, 2001.
145. Carneiro, O.S., Silva, A.F., Gomes, R., Fused deposition modeling with polypropylene. *Mater. Des.*, 83, 768, 2015.
146. Guo, N. and Leu, M.C., Additive manufacturing: Technology, applications and research needs. *Front. Mech. Eng.*, 8, 214, 2013.
147. Matsuzaki, R., Ueda, M., Namiki, M., Jeong, T.K., Asahara, H., Horiguchi, K. *et al.*, Three- dimensional printing of continuous-fiber composites by in-nozzle impregnation. *Sci. Rep.*, 6, 23058, 2016.
148. Der Klift, F., Van Koga, Y., Todoroki, A., Ueda, M., Hirano, Y., Matsuzaki, R., 3D printing of continuous carbon fibre reinforced thermo-plastic (CFRTP) tensile test specimens. *Open J. Compos. Mater.*, 6, 18, 2016.
149. Tian, X., Liu, T., Yang, C., Wang, Q., Li, D., Interface and performance of 3D printed continuous carbon fiber reinforced PLA composites. *Compos. Part A Appl. Sci. Manuf.*, 88, 198, 2016.
150. Li, N., Li, Y., Liu, S., Rapid prototyping of continuous carbon fiber reinforced polylactic acid composites by 3D printing. *J. Mater. Process. Technol.*, 238, 218, 2016.

151. Hou, Z., Tian, X., Zhang, J., Li, D., 3D printed continuous fibre reinforced composite corrugated structure. *Compos. Struct.*, 184, 1005, 2018.
152. Dickson, A.N., Barry, J.N., McDonnell, K.A., Dowling, D.P., Fabrication of continuous carbon, glass and Kevlar fibre reinforced polymer composites using additive manufacturing. *Addit. Manuf.*, 16, 146, 2017.
153. Caminero, M.A., Chacon, J.M., García-Moreno, I., Reverte, J.M., Interlaminar bonding performance of 3D printed continuous fibre reinforced thermoplastic composites using fused deposition modelling. *Polym. Test.*, 68, 415, 2018.
154. Liu, T., Tian, X., Zhang, M., Abliz, D., Li, D., Ziegmann, G., Interfacial performance and fracture patterns of 3D printed continuous carbon fiber with sizing reinforced PA6 composites. *Compos. Part A. Appl. Sci. Manuf.*, 114, 368, 2018.
155. Gassan, J. and Bledzki, A.K., Thermal degradation of flax and jute fibers. *J. Appl. Polym. Sci.*, 82, 1417, 2001.
156. Poletto, M., Zattera, A.J., Santana, R.M.C., Thermal decomposition of wood: kinetics and degradation mechanisms. *Bioresour. Technol.*, 126, 7, 2012.
157. Tserki, V., Matzinos, P., Kokkou, S., Panayiotou, C., Novel biodegradable composites based on treated lignocellulosic waste flour as filler. Part I. Surface chemical modification and characterization of waste flour. *Compos. Part A Appl. Sci. Manuf.*, 6, 965, 2005.
158. Taj, S., Munawar, M.A., Khan, S., Review: Natural fibre reinforced polymer composites. *Proc. Pak. Acad. Sci.*, 44, 129, 2007.
159. John, M.J. and Anandjiwala, R.D., Recent developments in chemical modification and characterization of natural fiber-reinforced composites. *Polym. Compos.*, 29, 187, 2008.
160. Sarani, Z. and Poh, K.L., Polystyrene-benzoylated EFB reinforced composites. *Polym. Plastic. Technol. Eng.*, 41, 951, 2002.
161. Wang, B., Panigrahi, S., Tabil, L., Crerar, W., Pre-treatment of flax fibres for use in rotationally molded biocomposites. *J. Reinf. Plast. Compos.*, 26, 447, 2007.

6

Manufacturing Issues and Process Parameters of Composite Filament for Additive Manufacturing Process

Jafrey Daniel James D.[1], Ramesh M.[1*], Sakthi Balan G.[2] and Aravind Raj S.[2]

[1]SRM Institute of Science and Technology, Ramapuram, Chennai, India
[2]Department of Manufacturing Engineering, School of Mechanical Engineering, Vellore Institute of Technology, Vellore, Tamil Nadu, India

Abstract
Researchers are becoming increasingly interested in polymeric composites and additive manufacturing due to their numerous potential uses. Polymer composite filaments were manufactured using a melt intercalation technique and a twin-screw extruder. In this study, significant complexities of manufacturing filament of polybenzimidazole (PBI) enhanced with high-density polyethylene (HDPE) in a twin-screw extruder due to variations in process parameters (screw speed, temperature profile, etc.) were extensively examined. The manufacturing issues such as porosity, bending, thinning, impurities, and bulges of filament were discussed. The filament was characterized by the tensile test, whereas FE-SEM analysis reveals a homogeneous dispersion of PBI in HDPE. We found that a degassing zone of 40 mm Hg pressure, a cooling distance of 50 cm, and a humped temperature profile attained with a twin-screw rotation of 65 rpm and a screw rotation of 7 rpm for the volumetric feeder (170°C–190°C) provided the optimal filament parameters.

***Keywords*:** HDPE/PBI composite, manufacturing issues, twin-screw extrusion, FE-SEM

**Corresponding author*: sastra.ramesh90@gmail.com

S. Thirumalai Kumaran and Tae Jo Ko (eds.) Sustainable Machining and Green Manufacturing, (115–130) © 2024 Scrivener Publishing LLC

6.1 Introduction

Three-dimensional (3D) printing often known as additive manufacturing is one of the most fascinating developments in the field of material science to take place in the last several decades. In a wide variety of applications, polymers are gradually taking the place of more conventional materials [1]. Nanocomposites of polymers are polymers that have been strengthened with nanoscale fillers (PNC). Extrusion, liquid deposition modeling, and granular material binding are three typical AM techniques for polymeric materials (selective laser sintering). Green manufacturing techniques like 3D printing produce few by-products and need less oversight because of their versatility and adaptability [2]. These advantages make it possible to 3D print highly customizable components at lower manufacturing costs for specific uses [3]. The implementation of these PNC results in considerable alterations to the properties of virgin polymers. Improvements have been made to the mechanical, fireproof, thermal, and permeability properties of these PNC. When it comes to deciding how a PNC will behave, the particle size, the degree to which it adheres to the polymer matrix, and the kind of addition procedure all play key roles [4]. Recent research on PNC has resulted in the production of new structures, such as exfoliated and intercalated structures, that integrate fillers such as carbon nanotubes, nanoclays, graphene, and a great deal more. These structures have been created as a result of the research [5].

There are a variety of distinct approaches that may be taken when manufacturing nanocomposites. Among these approaches, mixing in the solid state, template synthesis, in situ polymerization, mixing in solvents, and melt intercalation are some of the techniques used [6]. The manufacturing process known as melt intercalation is the one that is used the most frequently out of those three. Extruders with either a single screw or twin screws are utilized in the process of melt intercalation [7]. Because of the numerous advantages they offer, twin-screw extruders are used far more frequently than their single-screw counterparts. During every stage of this extrusion process, fillers receive the appropriate amount of polymer reinforcement. The type of extruder and the design of screws that are employed in an extrusion process that uses twin screws have a significant role in determining the level of success that may be achieved [8]. The findings of Ray *et al.* [17] indicate that the melt intercalation technique offers a superior benefit than other ways due to the fact that it is an operation that is less harmful to the environment. There was not much investigation into applications in the assessment, which focused on the difficulties and future potential of processing these materials [9]. Although Yang *et al.* [10] gave a comprehensive evaluation of biocomposite fillers in composites extruded AM specimens, with an emphasis on their usage in biomedical applications, they did not evaluate many additional uses for composite materials. Several authors in the literature utilized different fillers in a high-density polyethylene (HDPE) matrix; however, polybenzimidazole (PBI) powder was not among them. The issues of manufacturing the filament for additive manufacturing process are to be minimized by optimizing the twin-screw extrusion parameters.

6.2 Materials and Properties

The matrix was composed of HDPE from Reliance India, Ltd.; specifically, REPOL HD50MA180 grade. The GAZOLE™ 5000 PBI particles (90–130 nm) were supplied by Merck, India Ltd.

6.2.1 Preparation of HDPE/PBI Filaments

The HDPE/PBI nanocomposites were created utilizing a segmented twin-screw extruder with a 40:1 length-to-diameter ratio through the melt intercalation process. Both the hopper and die regions of the twin-screw extruder were maintained at a constant temperature of 120°C to 190°C, and the screw speed was held constant at 65 revolutions per minute. The long filaments that emerged from the extruder were cooled using water and air. Figures 6.1 and 6.2 show a line representation of the extrusion machine and procedure.

Twin-screw extruder (TSE) with variable pitch. It is separated into input feed, plasticizing, melting, mixing, degassing, and pumping zones. The intake feed screw pitch is 14 mm. When the screw is tightened, this component blocks material from flowing toward the seal. After the pitch

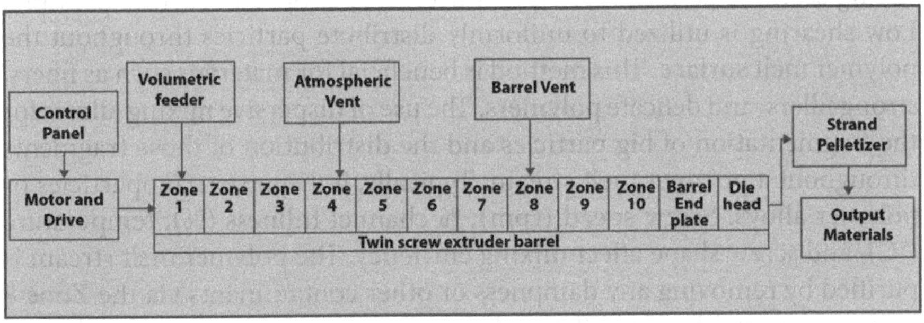

Figure 6.1 Zones in an extrusion machine [4].

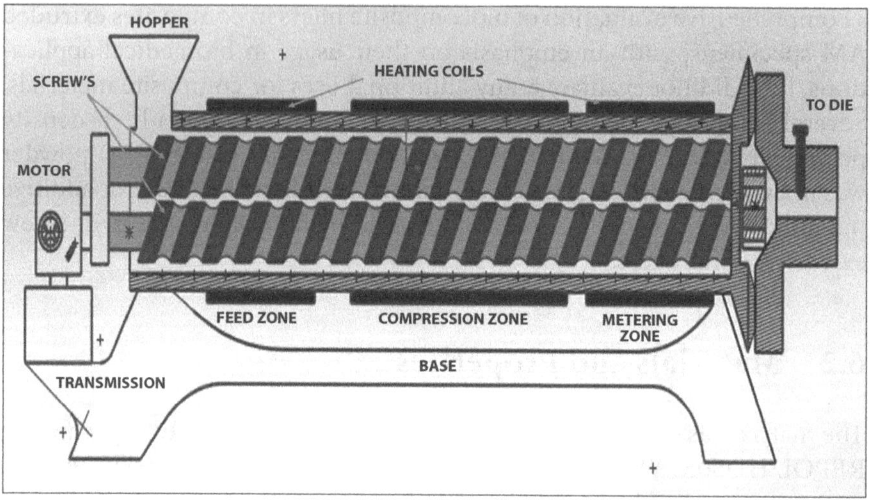

Figure 6.2 Schematic representation of twin-screw extrusion machine.

of 14 mm, the component pitch is adjusted to 21 mm with a notch that permits material to enter the extruder and pushes it to the next zone. Extruder motors provide between 78% and 88% of the heating to plasticize various types of polymers. As polymers are carried, barrel walls heat them. Conductive heat is transferred to polymers. Each TSE screw's pitch depends on the zone's material. Due to pitch changes, the TSE blends the free volume [4].

Before transporting materials from Zone 2 to Zone 3, they are warmed and compressed. Zone 1 is kept cool with water to prevent the intake polymers from overheating. Zone 2 heats materials, while Zone 3 crushes them. Most extruder screws are left-threaded. Polymers melt in Zone 3 of the TSE. The molten material is transferred to the zone for screw transmission (i.e., Zone 4 barrel and Zone 5 barrel). Comparatively to Zone 3, the melt conveying section's 28 mm screw pitch decreases effort and surface pressure. Low shearing is utilized to uniformly distribute particles throughout the polymer melt surface. This method is beneficial for materials such as fibers, strong fillers, and delicate polymers. The use of dispersive mixing allows for the fragmentation of big particles and the distribution of those fragments throughout a polymer melt stream. Typically, it disperses nanoparticles or polymer alloys. Screw speed (rpm), % channel fullness (%), temperature (°C), and screw shape affect mixing efficiency. The polymer melt stream is purified by removing any dampness or other contaminants via the Zone 8 barrel opening. The Zone 8 barrel element screw pitch has been adjusted from 14 mm to 28 mm, reducing the pressure and enabling polymers to move slowly. During the production process of polymer composites, a vacuum is applied to the vent barrel in order to eliminate humidity and any other contaminants. The point at which a material melts is the primary factor in determining the partial vacuum. Zone 8 is cooled below Zone 7 to avoid polymer ejection via the vent barrel during manufacture. With lower temperatures, Zone 8 polymer hardens, preventing particles from exiting the vent barrel. The pumping zone moves material from the previous zones to the die zone. As Zone 9 pitch drops from 28 to 21 mm, Zone 10 pitch drops to 14 mm. These components' smaller pitch pumps polymers into the die zone faster.

During the manufacturing of HDPE/PBI nanocomposites, the use of TSE's numerous barrel zones, which included the transportation zone, melt zone, blending zone, and softening zone, was helpful in ensuring that the PBI was equally distributed throughout the HDPE matrix. The degassing zone pressure is 40 mm Hg, the cooling length is 50 cm, and the humped temperature profile is achieved with a twin-screw speed rotation of 65 rpm, and the speed of the screw is 7 rpm for the volumetric feeder (120°C–190°C). For the purpose of characterizing the processed HDPE/PBI, X-ray diffraction (XRD) was used to examine the intercalation/exfoliation structure formation, and to investigate the dispersion of PBI inside the HDPE matrix, a field emission scanning electron microscope, also known as a FE-SEM, was used [11].

6.2.2 Processing Conditions for HDPE/PBI Nanocomposites

Following is the list of the parameters that were chosen prior to the fabrication of various wt% HDPE/PBI nanocomposites employing a twin-screw extruder:

- The speed of the screw on the volumetric feeder used in the manufacture of HDPE/PBI nanocomposite material
- The temperature gradient for the TSE, as well as the temperature levels of the barrel regions, was chosen in order to produce HDPE/PBI nanocomposite material
- The rotational speed of the TSE screw used in the production of HDPE/PBI nanocomposite material
- The pressure inside the degassing zone during the production of the HDPE/PBI nanocomposite
- The length of the cooling process for the HDPE/PBI nanocomposite strand that was fabricated

6.3 Results and Discussion

6.3.1 Screw Speed of Volumetric Hopper for the Production of the HDPE/PBI Composite

The design of the extruder and the material that is fed into the TSE are the primary factors that determine how polymers are fed into the TSE when the twin-screw extrusion process is being carried out. In most cases, the feeder will be positioned immediately just above the feed orifice of the TSE. This will allow the materials to be fed directly into the TSE. In order to feed the polymer, either a volumetric feeder or a particle acceleration feeder might be used. Since the steady flow of the material input throughout the extrusion process, the volumetric feeder is used in the vast majority of cases because it is more convenient. The narrow and deep screw path of the volumetric feeder makes it easier for materials to flow in the correct direction inside the TSE. The HDPE polymers were fed by a volumetric feeder that was set at different speeds, including 20, 15, and 10. The twin-screw rotation inside the extrusion process became overloaded when an exaggerated quantity of material became deposited between both the two rotary screws, which resulted in the cessation of screw turning even during system of twin-screw extruder. The speeds of the volumetric feeder were set at 20, 15, and 10. After going through a number of iterations of trial and error, the inflow was at last set at 7 rpm.

6.3.2 Temperature Profile Distribution and Zone Barrel Temperatures for the TSE to Fabricate HDPE/PBI Nanocomposites

The temperature profiles that are developed even during the twin-screw extrusion process are determined by a number of factors, including the kind of polymer that is used, the type of filler that is selected, and the throughput rate. The procedure of extrusion is the transmission of rotational power from an electric motor, which sometimes heats the plastic and sometimes cools it, changing it from a solid to a liquid. It is essential that the materials being introduced have a viscosity that allows them to be transported to the subsequent zone of TSE. The friction that occurs within the barrel or the increased chamber heaters that are located on the outside are the two sources of heat. The barrel zones may be cooled by either water or air [12]. If the temperature gradient is set lesser than the speed at which the polymer melts, there is a possibility that the material may not actually melt completely. This could cause the material to stick together between the screws and stop the polymers from moving to the next zone. If the temperature chosen is too significant, it could also cause polymers to break down, which would reduce the characteristics of the manufactured polymer [13].

The temperature is controlled throughout the extrusion process, beginning at the input region (also known as the feed area) and continuing all the way to the die zone [14]. Depending on how the TSE is made, the number of barrels can range from 4 to 12. The humped temperature gradient was chosen most of the time, and it was employed in this investigation. According to Figure 6.3, the temperature of the extruder was adjusted to be somewhere between 130°C and 180°C. The temperature of the intake zone, also known as the feed zone and including Zones 1 and 2, is maintained at a level that is lower than the HDPE matrix's point of melting. It is often done for the purpose of preventing polymers from melting too quickly and adhering to the inflow zone, both of which might result in the polymers being trapped at the inlet zone. The temperature of the barrels was gradually increased from the third barrel zone all the way up to the seventh barrel zone. Because barrel Zone 8 has a vent barrel, the temperature in that zone was dropped in order to facilitate devolatilization, and then it was elevated once again in barrel Zone 9. The temperature had to be brought down from barrel Zone 10 to the die area so that there would be an adequate discharge of HDPE/PBI composites in the die region.

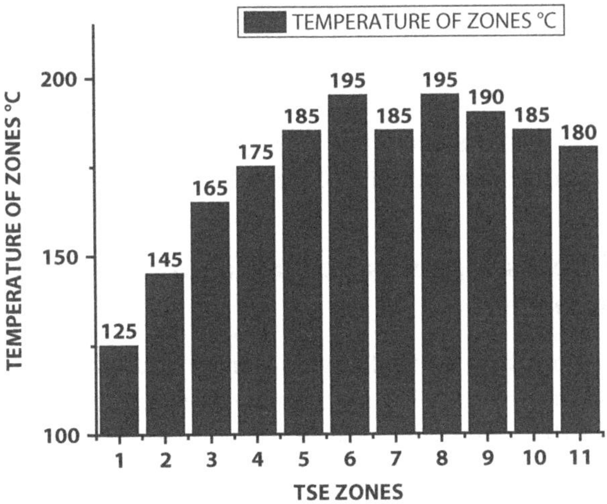

Figure 6.3 Temperatures are maintained at various TSE zones during the production process of HDPE/PBI composites.

The temperature that is maintained in the die zone throughout the twin-screw extruder is the subsequent most critical component. It is essential to choose the appropriate temperature for the die zone in order to get the desired flow of material out of the extruder. The HDPE/PBI nanocomposite was extruded in strand form from the TSE's die, which measures 3 mm in diameter. The HDPE/PBI composite materials from the extrusion die remained liquid and difficult to pull into strands at an initial die zone temperature of 195°C. Because the temperature of the die region was dropped to 185°C, it is difficult to release the strands from the extruder because the HDPE/PBI nanocomposite materials have already hardened in the die zone. This is due to the fact that the die zone temperature was lowered. In the end, the temperature of the die zone was raised to 180°C, which proved to be the perfect setting for drawing the HDPE/PBI composite material strands.

6.3.3 TSE Screw Speed for HDPE/PBI Composite Material Manufacture

Since it is the factor that decides how long polymers remain inside the TSE, the screw speed of the TSE needs to be adjusted before the twin-screw extruder process can begin. The TSE's screw speed was decreased to 50 rpm, which resulted in a number of complications, including an increase in the dwell duration as well as polymer decomposition inside the screw chamber. When the screw speed of the TSE was high (that is, 100 rpm), difficulties such as (i) increased polymer temperature, (ii) drag flow was formed as a result of excessive pressure within the die, and the polymer began to leak out through the degassing vent, and (iii) inappropriate blending of polymers with filler appeared inside the screw compartments. After multiple trials and errors, the screw speed for the production of HDPE/PBI composites was determined to be 65 rpm.

6.3.4 TSE Degassing Pressure for Manufacturing of HDPE/PBI Composites

The removal of moisture as well as gasses from treated materials is the primary function of the degassing unit. This may result in an improvement in the qualities of the treated material. In most cases, degassing is completed with the assistance of a vacuum pump that is capable of producing a peak pressure difference of 650 mm Hg. The melting temperature of the polymer being used and the kind of polymer being used both have a role in determining the potential of the induced vacuum pressure. Initially, the extrusion technique was used to manufacture HDPE/PBI composite materials without degassing, which resulted in the development of water air pockets and cavities at the surfaces of the filaments and the inappropriate dispersion of PBI filler in HDPE matrix. When the same situation was treated using a degassing device, the PBI in the HDPE matrix was properly mixed. As HDPE has a low melting point temperature, a peak pressure (vacuum) of 50 mm Hg was employed during the fabrication of HDPE/PBI composites. When it was applied to more over 50 mm Hg, the HDPE/PBI composite polymers began to leak via the degassing vent. A temperature reduction in Zone 8 allowed for the partial solidification of HDPE/PBI composite material. If the HDPE/PBI composite materials were not consolidated in Zone 7, there is a potential that material will escape through the degassing outlet, since they were in a molten form at that time. In the event that the substance only partly solidifies, there will be a decreased probability of material being expelled via the degassing vent.

6.3.5 TSE Cooling Length for Manufactured HDPE/PBI Composite Strands

When using a twin-screw extruder machine, the pace that the filament product is cooled is critical in defining the optimal qualities of the finished product. The moment that the glass transition temperature (Tg) is attained, the cooling process is connected with the atomic-level chemical bonding. When the polymer reaches a temperature below Tg, it becomes glassy. Polymer is extruded in a set size and may be withdrawn from the end zone of the die extruder by pulling it through the die. Since the outer surface area is less than Tg, the size of the surface stays constant; however, the inner area, which is larger than Tg, induces faster motion of molecules, which causes the surface to shrink as a result of shrinkage. If the cooling is not uniform across the whole strand, there is a chance that the process of cooling will be slower and more uneven in the area that is in the middle of the strand. This might lead to cavities and wetness being trapped inside the filaments.

The strands that are produced by the extruder are first cooled with water in order to facilitate the solidification process. The existing situation calls for the cooling range to be designed at a span of 120 cm. It was revealed that there were water bubbles present in the HDPE/PBI composite filaments as a consequence of the longer duration of time necessary for cooling, and initially, 120 cm of liquid cooling system was given for the strands.

It was revealed that the wetness was still retained in the filament despite that the length had been shortened by the same amount even though it had been dropped to a shorter length of 80 cm throughout the chilling process. The length of the HDPE/PBI composite strands was brought down to 40 cm after further reduction, and further testing confirmed that there was no moisture in the strand. The properties of the materials are what ultimately decide whether or not strand cooling with the use of water will be implemented. After being allowed to cool, the strands were put through a pelletizer in order to be cut into 3-mm pieces.

6.3.6 Extrusion Parameters for Manufacturing of HDPE/PBI Composites

The extrusion parameters for manufacturing of HDPE/PBI composites were selected from trial experimentation and were shown in Table 6.1. The HDPE/PBI composites were manufactured using a twin-screw extruder.

6.3.7 Manufacturing Issues During the Process of Filament Preparation

6.3.7.1 *Thinning of Filament*

The filament that gets extruded from the die will be at semi solid state with the temperature at die. will be at solid state with temperature. The temperature plays a vital role in maintaining the filament diameter. The cooling down of the filament from the die end to the water region had to be maintained with constant speed of extrusion. The sudden variable temperature reduction after extrusion from die end will make the filament go to the thinning process [15].

Table 6.1 Process variables of manufacturing HDPE/PBI composites.

Process variables	Parameter kept
Barrel zone temperatures	120°C–190°C (i.e., inflow zone to end zone)
Screw speed of extruder	65 rpm
Volumetric feeder of hopper	7 rpm
Degassing pressure for moisture removal	50 mm Hg
Cooling distance of filament	40 cm

6.3.7.2 *Bending of Filament*

The bending of the filament occurs due to the temperature difference of the materials and reinforcement (Figure 6.4). The time taken during cooling and the control of screw speed in extrusion of material also play a greater role in preparing the uniform-diameter filament. The temperature difference and accumulation of material flow near the nozzle make the filament bend and the uniformity of extrusion is disturbed.

6.3.7.3 *Impurities and Bulges*

The impurities are the foreign particles that come along the filament extrusion. The plastic flow of the materials takes some particles along with the flow during the screw torque. The foreign particles are like previous extruded material, old pellets, fibers, etc. that are present inside the twin-screw extruder. The foreign particles get accumulated near the nozzle, and this makes those impurities create bulges and get added to the filament.

6.3.7.4 *Porosity in Filament*

The porosities in the filament are the defects caused due to the oxygen liberated inside the filament. The reinforcement particles get mixed with matrix material at the molten state with difference in the melting temperature of two materials causing the air bubbles to get generated during the extrusion process. The air molecules generated come out along with filaments, causing the hollow filament section as shown in Figure 6.5.

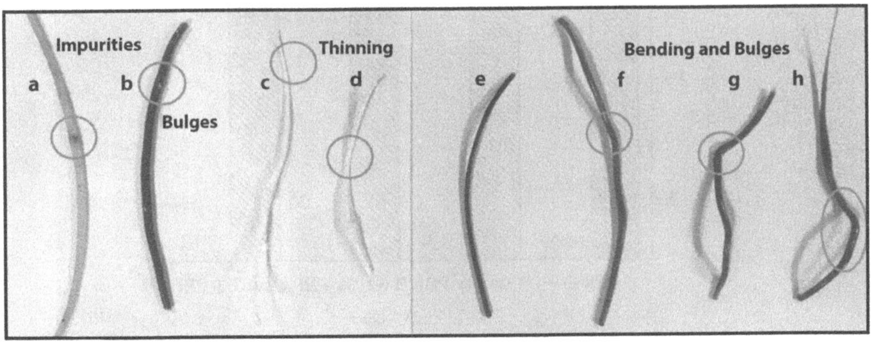

Figure 6.4 Filaments taken during the extrusion of HDPE/PBI composites.

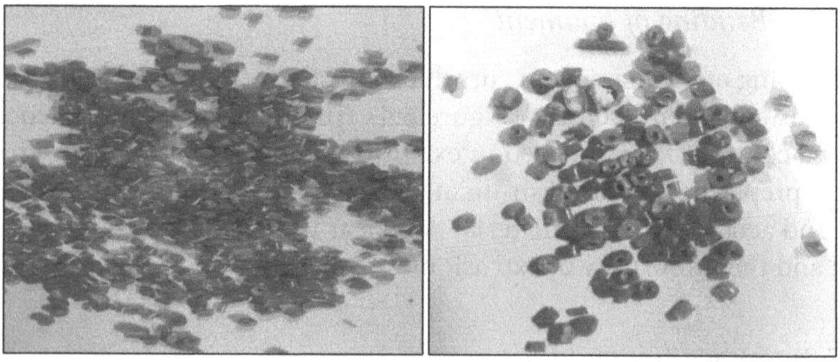

Figure 6.5 Porosity formed on extruded HDPE/PBI filament.

6.3.8 Tensile Test

There are several variables that contribute to the tensile strength of a composite, including the strength of the fiber, the fiber modulus, the infusion size of reinforcement, and the binding between the fiber and the matrix. It should be noted that the tensile strength and tensile modulus were also increased when 3 and 5 wt.% PBI filler was added to the HDPE matrix (Figure 6.6). Addition of 3 wt.% increased the tensile strength to 27.5 MPa and the tensile modulus to 2.18 GPa by a substantial 4.2% compared to addition of 1 wt.%. There was a 2.2% increase in tensile strength with the

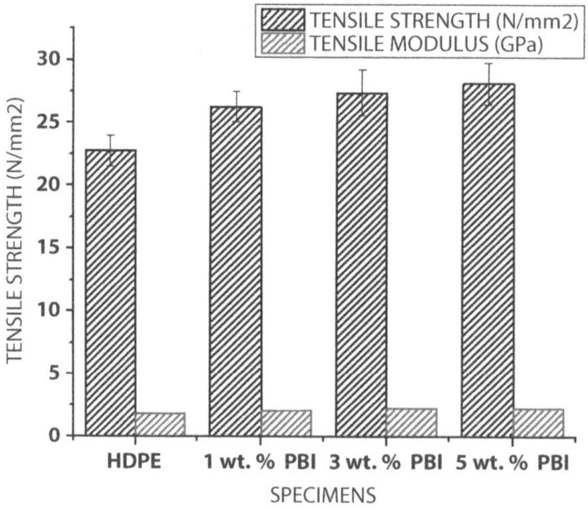

Figure 6.6 Tensile test and tensile modulus graph on HDPE/PBI filament.

5 wt.% PBI reinforcement compared to the 3 wt.% addition and a 22% increase compared to pure HDPE.

HDPE/PBI composites have superior interfacial bonding and stress transmission, as evidenced by a significant improvement in total tensile strength. It is possible that the steadily decreasing proportion of tensile strength rise between the reinforcement material additions is due to the phase's immiscibility, cavities, and aggregation of fillers within the matrix, which generates some poor adhesion of bonds. If so, this would explain why the percentage has been decreasing [16].

6.3.9 FE-SEM

Figure 6.7 (a) shows the pure HDPE where there are no reinforcements present. Figure 6.7 (b) shows the presence of little reinforcement particles. The dispersion of particles was good and uniformly distributed. Figure 6.7 (c) shows some particles not dissolved fully inside HDPE matrix and producing a strong resistance to fracture. Figure 6.7 (d) shows more amount of agglomeration of reinforcement, and distribution is affected by the hard PBI particles.

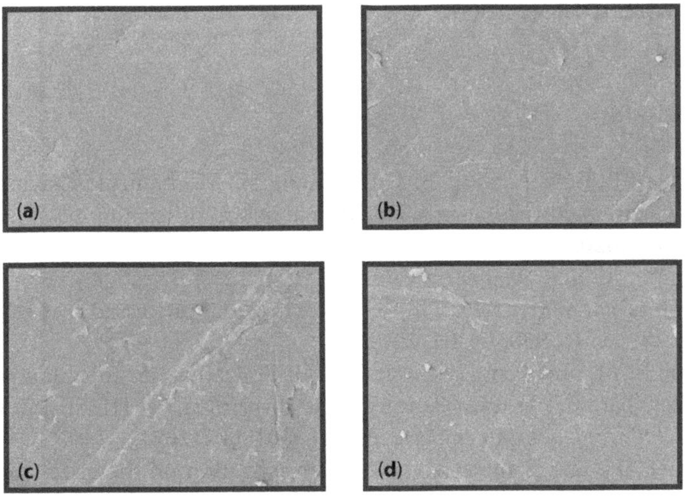

Figure 6.7 (a–d) FE-SEM results of HDPE with 0, 1, 3, and 5 wt.% PBI-reinforced composite.

6.4 Conclusion

The parameters for the twin-screw extrusion process for fabricating HDPE/PBI filaments were discussed as well as the manufacturing issues, and the characterization was done for the HDPE/PBI filaments.

- The preparation of HDPE/PBI filaments through twin-screw extrusion at different zones of temperature was discussed as well as the process parameters to be known in determining the filament extrusion.
- The zones with various temperature profiles of the extrusion process show the issues in manufacturing the filament with uniform distribution and without defects.
- The defects such as porosity, bending, thinning, impurities, and bulges are due to the application of improper extrusion parameters such as temperature profile, screw speed, volumetric feeder, and the cooling length.
- The optimized obtained parameters were used to produce filaments with different PBI reinforcement, and the tensile values were analyzed. The results show that the 5 wt.% PBI has shown higher tensile strength and modulus.
- FE-SEM results show the distribution of PBI particulates in the HDPE matrix and the agglomeration of particles in 5 wt.% reinforcement.

References

1. Beesetty, P., Kale, A., Patil, B., Doddamani, M., Mechanical behavior of additively manufactured nanoclay/HDPE nanocomposites. *Compos. Struct.*, 247, 112442, 2020.
2. Ramesh, M. and Panneerselvam, K., Mechanical investigation and optimization of parameter selection for nylon material processed by FDM. *Mater. Today Proc.*, 46, 9303–9307, 2019.
3. Ramesh, M. and Panneerselvam, K., Optimization for injection moulding process parameters towards warpage and shrinkage of HDPE-PBI composites. *IOP Conf. Ser. Mater. Sci. Eng.*, 1183, 012001, 2021.
4. Daniel, D.J. and Panneerselvam, K., Investigation on the effects of Cloisite 30B and copolymer (ethylene and butyl acrylate) reinforcement with polypropylene thermoplastic by melt intercalation method. *J. Thermoplast. Compos. Mater.*, 31, 1371–1392, 2018.

5. Ravi Kumar, B.N., Venkataramareddy, M., Suresha, B., Two-body abrasive wear behavior of nano-clay filled LDPE/EVA composites. *J. Reinf. Plast. Compos.*, 28, 2999–3007, 2009.
6. Jafrey, D.D. and Panneerselvam, K., Study on tensile strength, impact strength and analytical model for heat generation in friction vibration joining of polymeric nanocomposite joints. *Polym. Eng. Sci.*, 57, 5, 495–504, 2017.
7. Jafrey Daniel, D. and Panneerselvam, K., Modeling of tensile properties, dispersion studies, and hardness evaluation of Cloisite 30B in polypropylene with Elvaloy AC 3427 as compatibilizer. *J. Compos. Mater.*, 50, 3219–3227, 2016.
8. Kumaran, P., Mohanamurugan, S., Madhu, S., Vijay, R., Singaravelu, D.L., Vinod, A., Sanjay, M.R., Siengchin, S., Investigation on thermo-mechanical characteristics of treated/untreated Portunus sanguinolentus shell powder-based jute fabrics reinforced epoxy composites. *J. Ind. Text.*, 50, 427–459, 2020.
9. Madhu, G., Bhunia, H., Bajpai, P.K., Chaudhary, V., Mechanical and morphological properties of high density polyethylene and polylactide blends. *J. Polym. Eng.*, 34, 813–821, 2014.
10. Yang, Z., Peng, H., Wang, W., Liu, T., Crystallization behavior of poly(ε-caprolactone)/layered double hydroxide nanocomposites. *J. Appl. Polym. Sci.*, 116, 2658–2667, 2010.
11. Mohan, R. and Kavan, P., Influence of polybenzimidazole nanoparticle on the thermo-mechanical characteristics of high density polyethylene composite. *Phys. Scr.*, 97, 035706, 2022.
12. Sarabi, M.T., Behravesh, A.H., Shahi, P., Daryabari, Y., Effect of polymeric matrix melt flow index in reprocessing extruded wood-plastic composites. *J. Thermoplast. Compos. Mater.*, 27, 881–894, 2014.
13. Tanoue, S., Hasook, A., Itoh, T., Yanou, M., Iemoto, Y., Unryu, T., Effect of screw rotation speed on the properties of polystyrene/organoclay nanocomposites prepared by a twin-screw extruder. *J. Appl. Polym. Sci.*, 101, 1165–1173, 2006.
14. Daniel, D.J. and Panneerselvam, K., Mechanical and thermal behaviour of polypropylene/cloisite 30B/elvaloy AC 3427 nanocomposites processed by melt intercalation method. *Trans. Indian Inst. Met.*, 70, 1131–1138, 2017.
15. Daniel, D.J. and Panneerselvam, K., Manufacturing issues of polypropylene nanocomposite by melt intercalation process. *Mater. Today Proc.*, 4, 4032–4041, 2017.
16. Dawkins, B.G., Qin, F., Gruender, M., Copeland, G.S., Polybenzimidazole (PBI) high temperature polymers and blends, in: *High Temperature Polymer Blends*, pp. 174–212, 2014.
17. Ray, S.S. and Okamoto, M., Polymer/layered silicate nanocomposites: A review from preparation to processing. *Prog. Polym. Sci.*, 28, 1539–1641, 2003.

7

Material Sustainability During Friction Stir Joining

Raheem Al-Sabur[1]* and M. Serier[2]

[1]*Mechanical Department, Engineering College, University of Basrah, Basrah, Iraq*
[2]*Mechanical Engineering Department, University of Ain Temouchent, Ain Temouchent, Algeria*

Abstract
Sustainable manufacturing aspires to operate available resources with minimum environmental impact by executing eco-friendly strategies. Fusion welding dominates many industrial processes whether simple bonding operations or manufacturing more complex structures. Many manufacturers and environmental protection organizations have been worried about fusion welding problems such as harmful fumes, the high consumption of welding electrode shielded gases, etc. In the last decade, the environmental index granted many gains to the process of friction stir welding (FSW) because of its success in reducing the energy used and solving most fusion welding environmental problems, in addition to its ability to weld asymmetric metals, which made it consider a green technology that gives rise to hope. This chapter deals with the essential aspects of friction stir welding as a sustainable technology, including process history, types, benefits, and limitations compared to other types of welding, as well as recent trends and most frequently used applications.

Keywords: Sustainability, friction stir welding, environmental impact, energy consumption

**Corresponding author*: raheem.musawel@uobasrah.edu.iq

S. Thirumalai Kumaran and Tae Jo Ko (eds.) Sustainable Machining and Green Manufacturing, (131–154) © 2024 Scrivener Publishing LLC

7.1 Introduction

Although the term "sustainability" appeared in literature several centuries ago, there is no consensus on a comprehensive definition of it. Sustainability can be defined as finding a balance between ensuring future generation rights and the present need. During the new millennium, many countries considered sustainability an influencing factor in environmental safety. As a result, many factories have faced harsh legislation, forcing many to close [1]. In addition to confronting global warming, legislation to adopt industrial sustainability has become a prerequisite for all new projects and newly established factories. In manufacturing, studies focused on sustainability in the field of welding. They looked at many things, such as the effect on costs, the effect on the environment, reducing resource use, improving joint quality, and maximizing joint quality [2]. Welding is a very influential factor in the industry renaissance, dominating all jointing operations such as bolting or riveting. Welding processes are divided mainly into fusion welding and solid-state welding, in which the welding temperature does not reach the melting point of the welded metals [3]. Fusion welding is widely used and is characterized by its many types that try to cover most industrial uses. Despite these advantages, workers in the welding field are forced to bear difficulties, especially concerning health aspects such as gases and fumes. For example, shielding gases are necessary to solve the problem of hydrogen embrittlement and impurities [4]. In addition, the emission of infrared, ultraviolet, and even visible radiation is common in arc welding, one of the main types of fusion welding. It is also noted that filler metal is required in many types of fusion welding, and the heat-affected zone is relatively large, affecting the welding quality. Martensite welding and the welding of dissimilar metals are complex processes in fusion welding [5]. In the early 1990s, in one of the most famous welding research and technology organizations (TWI), there were several attempts to invent a new type of welding. Wayne Thomas gave the practical principles of friction stir welding (FSW) [6]. FSW was intended to be a revolutionary leap for light metal joining, such as aluminum and magnesium [7, 8].

Friction stir welding is based on a nonconsumable rotating tool consisting of two connected parts. The upper cylindrical part, called the shoulder, is connected to the welding machine (usually a CNC) and the other end ends in a tapered cylindrical part (usually) called the pin. The pin can be square, triangular, obturator, or cylindrical (not tapered) and can be threaded [9]. In friction stir welding, two plates of similar or dissimilar metals (considering the melting point of each metal) are arranged as a butt

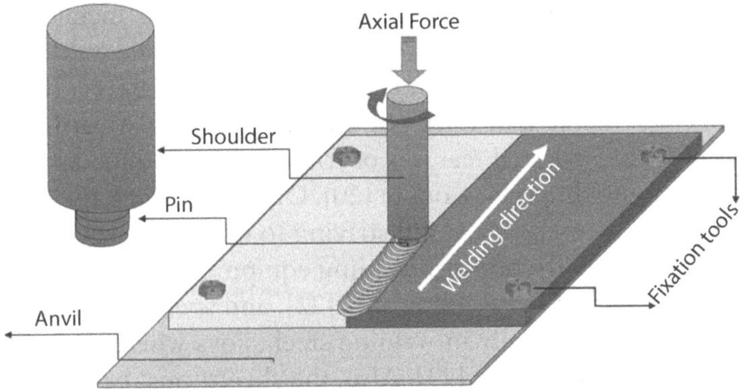

Figure 7.1 Scheme of FSW process and fixation tools.

joint. The two plates are firmly fixed to an anvil to reduce vibration and obtain a homogeneous joint, as shown in Figure 7.1. When the rotational tool reaches the weld seam, the contact area temperature rises, allowing the pin to gradually sink into the two plates until the lower shoulder surface comes into direct contact with the weld seam. In the second stage, the welding tool moves rotationally and linearly toward the weld seam until it reaches the end of the area to be welded. In the third stage, the linear motion stops, and the rotational tool rises, leaving behind welded plates and a hole at the end of the weld [10]. The choice of welding conditions (tool rotation speed, welding speed, and penetration force) depends on the welded materials.

For understanding the metal flow during the process, the spinning tool offers a continuous hot working activity that pushes the material around a pin at temperatures lower than the joining sheets' melting points. This procedure moves metal from the top face of the pin to its trailing edge while plasticizing it in a constrained area. The pin torque delivers the energy to form a viscous texture material, permitting it to flow around the pin as a boundary layer between the pin and the solid parent material driven away or diminished in thickness at the pin's top edge.

The relatively low temperature of friction stir welding has many benefits. In addition to solid-state welding, where there is no melting, the process is considered free from solidification defects such as cracks and porosity [11], and the heat-affected zone becomes better [12]. Moreover, most mechanical properties improve, such as tensile strength [13], fatigue life [14], hardness [15], and fracture toughness [16], which is offset by a decrease in distortion [17] and an ultrafine granular structure [18].

The ability to weld similar and sometimes dissimilar metals and polymers is one of the benefits of friction stir welding [19]. The welding process is automated, and the penetration is high with fewer defects. The welding tool is nonconsumable and does not require post-weld treatment. Safety and health are the essential features of friction stir welding, as there is no spatter, fumes, shielding gases, or arc [20]. On the contrary, there are many limitations that researchers are still striving to overcome, such as the high initial cost of robotic friction stir welding equipment [21], the presence of a keyhole at the end of the welding area [22], and special requirements for the tool materials, especially in welding steel alloys where the tool can be exposed to wear and damage [23]. Also, the sheets must be firmly fixed to the anvil, and the sheets that need to be welded must touch each other.

Sustainability has emerged as an insistent factor, and many restrictions have been placed on factories. On the other hand, countries, especially in Europe, encouraged renewable energies based on sustainability and decarbonization. In production lines, reducing production costs and reducing resource consumption, in addition to preserving the environment and health, are the main factors through which the production line can be considered sustainable.

In this chapter, the concept of sustainability in friction stir welding will be highlighted by reducing three aspects: cost, environmental impact, and energy consumed to maximize the efficiency of the resulting weld joints. It will also refer to the factors affecting FSW as "solid-state welding" and the technical improvements that accompanied the development of FSW tools. In addition, several industrial applications and projects will be mentioned in the context of industrial sustainability.

7.2 FSW Parameters

7.2.1 Rotation Tool Speed and Traverse Velocity

During FSW, speeds are the essential parameters, as there are two types of speed [24]. The first type is the tool rotational speed, which represents the speed at which the pin penetrates the metal, and the shoulder presses the sheet contact region. The second is the velocity of tool travel (traverse) along the required welding line. It can be considered a linear velocity associated with the rotational speed of the tool.

It is difficult to determine the optimal values for both types of speed, as they vary according to the joined metals and other parameters. The force acting on the tool can be on multiple axes, the longitudinal force, the vertical force, and the lateral force for the X, Y, and Z axes. The longitudinal force performs parallel to the tool movement and is positive in the x-direction. Since this force emerges due to the material's resistance to the tool motion, these forces will likely decrease as the material's temperature increases. The vertical or downward force is crucial to keep the tool positioned at or below the material surface. Some FSW equipment works under load control, but the tool's vertical position is often preset so that the load will alter during FSW. As torque is applied to rotate the tool, it depends on the downforce, friction coefficient, and material flow. The downforce, or lateral force, operates perpendicular to the traverse direction of the tool and is described as positive toward the FSW advancing side. When added to the effect of thermal impact, these forces can cause the fixture and the sheets being processed to deform and affect the tool's wear. These effects will make it harder to figure out how residual stresses will form in the welding and how long tools will last.

It is worth noting that controlling the traverse velocity and the rotational speed contributes significantly to sustainability. Reduced speed reduces energy consumption and production costs, significantly contributing to environmental improvement. In general, the forces generated in FSW processes are affected by the amount of rotational speed and travel velocity. The speed increase can help increase the force on the shoulder, but it may be off the pin. Table 7.1 shows the relationship between the rotational speed, the traverse velocity, and the effect of force components.

Table 7.1 Effect of tool rotational speed and travel velocity on force components.

Tool part	Force component	Tool rotational speed*	Tool traverse velocity
Pin	Longitudinal	↓	↑
	Vertical	↑	—
	Lateral	↑	—
Shoulder	Longitudinal	↓	↑
	Vertical	↑	↓
	Lateral	↑	—

*Symbols ↑ (increase), ↓(decrease), and—(no effect).

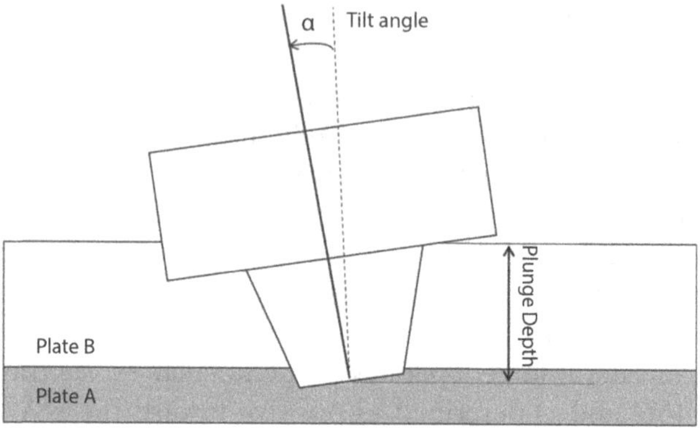

Figure 7.2 Tilted tool and plunge depth of FSW tool (cross-sectional view).

7.2.2 Plunge Depth and Tilt Angle

The plunge depth, a crucial factor in maintaining weld quality, is described as the distance between the shoulder lowest point and the plate surface. The pressure below the tool is increased by lowering the shoulder below the sheet surface, ensuring proper forging of the material at the tool's back. If the plunge depth is too great, the pin could rub against the surface of the backing plate, or there could be a big difference between the weld depth and the base material.

During the FSW, the plunge depth adjustment is necessary to ensure suitable downward pressure and complete penetration of the tool into the welded plate. The tilting of the FSW tool by $\alpha = 2\text{–}4$ degrees is necessary to overcome the incredible tool deflects and reset the plunge depth to the standard setting, as shown in Figure 7.2 [25].

7.3 FSW Sustainability Review

The term sustainability has appeared in many studies of the FSW in recent years. The section reviews only the previous articles in which sustainability appeared in their titles. Bevilacqua *et al.* [26] used life cycle assessment software to analyze the results of AA5754 sheet joining by friction stir welding. They found that it is more sustainable than other welding methods, as it reduces energy consumption and resource depletion. They also indicated a noticeable sustainability in the environmental impact commensurate with the speed of rotation of the tool. Sued *et al.* [27] concluded that the total power used for friction stir welding was at most 130 kW when joining AA3000 aluminum plates. The total energy used is very little compared to fusion welding. They used a travel speed of 155 mm/min, a dwell time of 15 sec, and three rotational speeds of 700, 800, and 900 rpm. Sethi *et al.* [28] reviewed several studies in the FSW of similar and dissimilar metals regarding welding process parameters and their impact on the interlayer, microscopic structure, and the resulting mechanical properties. The researchers did not address cost calculations, energy consumption, or the environmental and health impacts, which are the most studied topics in the subject of sustainability in manufacturing. The same matter also happened in many articles, such as [29–31], where sustainability studies should have been discussed. The study presented by Majeed *et al.* [32] is considered one of the best published in the sustainability studies of FSW, as it included a clear explanation and helpful schematic diagrams about the potential aspects of sustainability in FSW, in addition to a comparison of economic feasibility with other types. The researchers discussed the environmental friendliness criterion and did not overlook the limitations of FSW processes. Al-Wajidi [33] studied the effect of using lubrication on FSW sustainability for joining 6061-T651 and 5052-H32 plates. The study focused on four aspects, cost, energy consumption, environmental impacts, and waste management, and showed that adding lubrication can reduce the required welding forces in addition to improving the mechanical properties and microstructure. Suresh *et al.* [34] suggested an algorithm to predict the best eco-friendly FSW parameters when welding AA6061-T6 alloys. In the study, nanoparticles of Al_2O_3 were put in the keyhole; the tool's rotation speed was 1,387 rpm, and its traverse speed was 7 mm/min. Azeez and Akinlabi [35] conducted an in-depth study on the relationship between the tensile strength and generated heat of FSW joints and the sustainability criteria through economic, environmental, and health aspects. The emissivity, welding energy, and life cycle assessment (LCA) were analyzed. It was found that the lower FSW temperature led to a higher Weibull modulus, which is considered a criterion for better sustainability. To improve sustainability, Sharma *et al.* [36] used the LCA methodology to study the environmental effects associated with friction stir processing (FSP) of alloyed alloys. The study addressed the method of manufacturing the FSP tool and suggested developing it through "zinc-coated cutting wires," which makes it more suitable for environmental considerations.

7.4 FSW Sustainability Aspects

The system is supposed to be unsustainable when resources are consumed, and waste is created faster than nature's ability to cope [37]. Friction stir

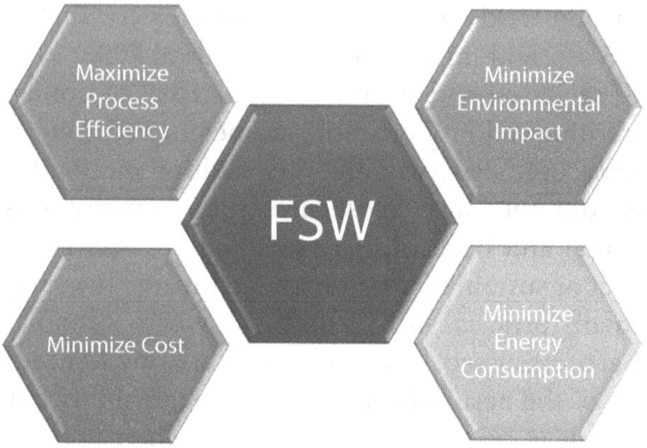

Figure 7.3 FSW sustainability aspects.

welding is considered a successful and sustainable alternative to fusion welding. In addition to its high efficiency, the energy used in it is much less than that in similar types of welding. This section discusses the FSW as a sustainable tool according to six sections, as indicated in Figure 7.3; minimize cost, maximize process efficiency, minimize environmental impact, and minimize energy consumption.

7.4.1 Minimizing the FSW Costs

Reducing the cost of production while ensuring high performance is the main goal in consideration of economic feasibility. Cost reduction is the key to achieving sustainability. Estimating the total cost of production is critical when deciding on welding types. Several studies have tried to estimate and compare the production costs of many welds, including FSW [38]. Boeing showed that the use of FSW reduced the welding cost to about 60% in Delta IV and Delta II, and the welding period was also reduced by 74%, which is a very encouraging factor if the FSW is compared to other types of welding [39]. Boeing reports were an essential and decisive factor in favoring FSW over others. Despite the strict safety requirements in airplanes, the increasing use of this type of welding in the aviation sector indicates that FSW has taken the lead in the welding industry. Regarding FSW cost estimation, the cost of machinery, power, tools, and fixtures, in addition to the labor cost, are the main elements for calculating the total cost, where the machinery and labor costs constitute 95%, while the total costs of tools and power are 5%, as shown in Figure 7.4 [40].

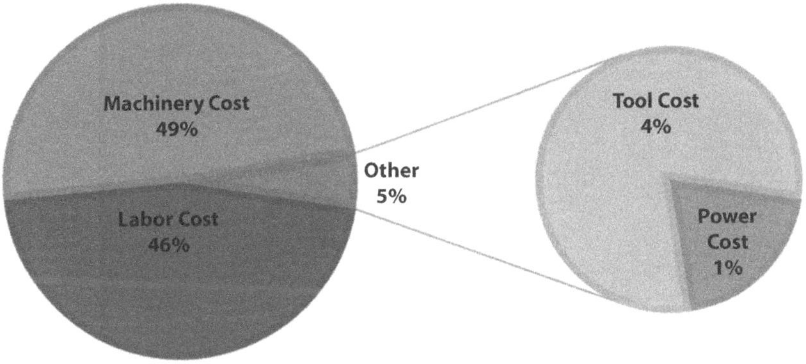

Figure 7.4 Cost distribution in the FSW process.

The machinery cost (Cm) and labor cost (C_{lc}) are calculated for FSW from the following equations [40]:

$$C_{mc} = \frac{C_M \times [(W_T \times \mathrm{n}) + T_s + T_{ch}]}{MR \times 60} \tag{7.1}$$

$$C_{lc} = \left(\frac{W_T \times \mathrm{n}}{OF} + T_W \right) \times \frac{C_L}{60} \tag{7.2}$$

where C_M is machine rate ($/hr), W_T is time to weld (min), n is number of passes, T_S is setup time (min), T_{Ch} is tool change-over time (min), MR is machine reliability (assuming 95%), T_W is weld preparation time (min), C_L = Labor rate ($/hr), and OF = Operating factor. The remaining costs (tools, fixtures, and power) are restricted to 5% of the total cost [33].

Several studies tried to compare the cost of FSW with other processes. When comparing the labor costs of the MIG welding process vs. the FSW, it was discovered to be roughly double, at 78% for MIG vs. 46% for FSW [40]. When comparing FSW with submerged arc welding (SAW), it was found that the direct machine cost is 47% in SAW vs. 95% for FSW while the personal and overhead cost was 30% in SAW vs. only 3% in FSW [41]. In the test evaluation, the FSW production time was 82% of the corresponding MIG [42].

7.4.2 Minimizing the FSW Energy Consumption

FSW is solid-state welding, meaning that the process takes place at temperatures lower than the melting point, and this has contributed to a

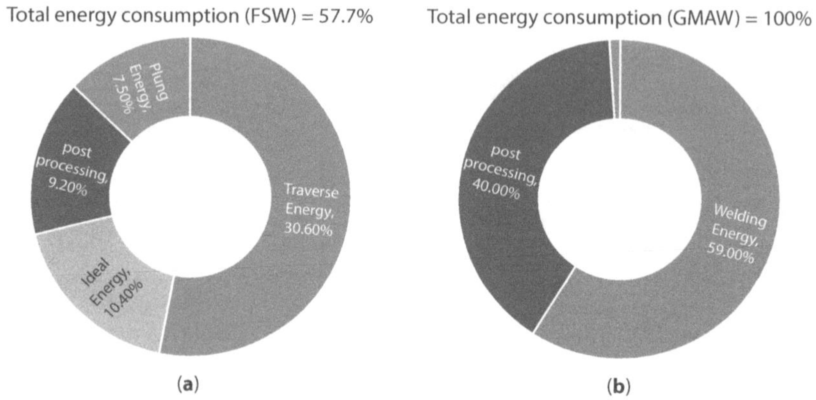

Figure 7.5 Comparison between the total energy consumption of (a) FSW and (b) GMAW.

decrease in the energy consumption used in welding. In addition, the FSW process does not require a post-weld, which is another positive factor in reducing energy consumption. A low energy consumption in FSW is an attractive factor for manufacturers and investors because it supports government requirements toward sustainability and reducing global warming.

Majeed *et al.* [32] summarized a set of comparisons between energy consumption in FSW and other welding processes, especially Gas Metal Arc Welding (GMAW). They concluded that the energy consumption is much lower than that of all arc welding processes. When comparing FSW with GMAW, FSW consumes about 42% of the energy consumed in GMAW [43]. Moreover, Dawood *et al.* [44] confirmed that the energy consumed in FSW consumes a quarter of what is consumed by GMAW. It is also comparable to other types of welding; when compared to laser welding, FSW is superior by less than 2.5% energy consumption [45]. The FSW energy consumption consists of postprocessing (cutting), traverse, plunge, and idle energies, while in the GMAW, it consists of postprocessing (grooving), weld energy, and standby energy. Figure 7.5 compares FSW and GMAW energy consumption using subdividing ratios [46]. The welding energy was reduced from 59% in GMAW to only 7.5% in FSW, which is a clear indication of the sustainability achieved by switching to FSW.

7.4.3 Maximizing the FSW Process Efficiency

The efficiency of a joint, or how much load it can withstand under service conditions before failing, is used to determine its quality [47]. The ratio of the joint tensile strength to the base metal is called joint efficiency [48]. Concerning the welding of similar metals in FSW, the results of joint efficiency in aluminum alloys were surprising and very encouraging. For the AA1100 joints, the ultimate joint efficiency was 114%; for AA6061-T651, it was 69% [48].

Many studies in the field of FSW of dissimilar metals have focused on the joint's efficiency for aluminum and magnesium. A joint efficiency of 61% was achieved for the FSW of AZ31B magnesium alloy with A5052-H aluminum [49]. Masoudian *et al.* [50] noticed that a brittle-mode fracture occurred when the joint efficiency reached 76% for the FSW of AA6061 Al and AZ31 Mg alloys. Malarvizhi and Balasubramanian [51] found that for the same alloys, the joint efficiency could be increased by up to 89% if the shoulder diameter was made to be about 3.5 times the sheet thickness. Mahamud *et al.* [52] studied improving joint efficiency using ER5356 filler to enhance intermetallic compounds for the FSW of AZ31B magnesium and A6061 aluminum alloys. The use of the filler increased joint efficiency by 9%, from 67% to 76%. The post-weld heat treatment can improve the overall joint efficiency and promote irregular grain growth through different characteristics and softening behaviors formed in the resulting joints [53, 54].

The FSW process, on the other hand, improved thermal efficiency, which is the percentage of heat flow into the weldment compared to total heat generated [55, 56]. The excited thermal efficiency of Al 7075 T6 for FSW was 88% [57], 79% for 7039 alloy [58], and it was around 85%–88% for the AA6061 alloy [59, 60]. It was optimized up to 95% for AA2195 aluminum alloy using the M2 steel tool [61]. The thermal efficiency in the magnesium AM60B alloy was achieved at 91% using a hardened steel tool [55]. For steel alloys, the thermal efficiency was 50% for 304L stainless steel [62], while it was 75% for the AISI 1018 steel alloy when using the molybdenum- and tungsten-based FSW tool [63].

In summary, FSW contributes to the improvement of the performance of weld joints, as the ultimate tensile strength and yield strength are increased. Not only that, but the hardness also improved significantly, and the defects were reduced or absent in many cases. Welding has become more uniform in the surface layer and has better penetration into other metal layers. All of these improvements are factors affecting the Industrial Sustainability Index.

7.4.4 Minimizing the Environmental Impact

Environmental impacts are short- or long-term changes that affect the inhabitants of the ecosystem system in the natural or built environment [64].

Regarding welding, the processes are excessively affecting the health of its workers and the environment by generating fumes, dust, greenhouse gas emissions, and process wastes, in addition to the impact on energy and resource consumption [65]. The LCA can be utilized to expect the environmental impact of discrete metal-forming processes [66]. LCA provides a more accurate, comprehensive view of the actual environmental impact of a process over its entire life cycle [67]. In FSW, the environmental impact studies focused on analyzing the influence of FSW parameters using LCA methodology, such as the effect of rotational and welding speeds on pin tool wear.

Bevilacqua *et al.* [65] used the LCA method to compare FSW, gas tungsten arc welding (GTAW), and laser beam welding using the same functional unit. The study proved that FSW had greater environmental sustainability than the GTAW process and little environmental benefits over laser beam welding. For AA5754-H114 aluminum alloys, the rotational speed influences the environmental impact less than the welding velocity. The variation in the welding velocity clearly affects reducing the energy absorbed during the friction stir welding processes of aluminum alloys [26, 65]. For aluminum alloys, the absorbed energy is inversely proportional to the increase in welding speed for all rotational speed ranges, as shown in Figure 7.6.

In summary, the combination of cost reduction and resource consumption with improved weld joint performance points to the fact that the FSW is an influential contributor to environmental impact. This contribution, aided by a significant amount of applied research, was also aided by the rapid adoption of stir welding in many industrial applications because it

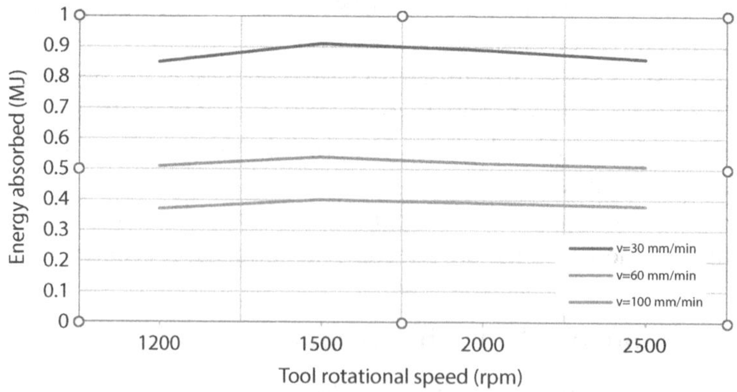

Figure 7.6 Energy absorption of Al alloys during FSW.

is considered an environmentally friendly welding method. Therefore, expectations indicate that environmental safety determinants and global warming requirements will increase the use of environmentally friendly welds such as FSW.

7.5 Recent Modifications in FSW Processes

The continuous development of FSW equipment is evidence that this industry is influential and moving in the right direction. Recently, there have been many developments on the research side as well as on the applied side. In the research aspect, the follower can note the massive articles on FSW that are published annually. Many companies have devoted a special section to the friction welding industry in the industrial field, and the matter has developed further so that we have many companies worldwide specialized in providing friction stir welding services or selling their equipment. This part reviews some research developments in FSW tools.

7.5.1 Double-Sided FSW Tool

In order to obtain more symmetrical welds and reduce the reactive torque and input heat, TWI researchers developed an FSW tool using two opposing tools on both sides of the joined sheets, as shown in Figure 7.7. This technique is known as the double-sided FSW tool [68]. The pins need not connect but should be arranged adequately close together so that the softened material around the two pins overlaps near the pin ends to forge a good FSW process.

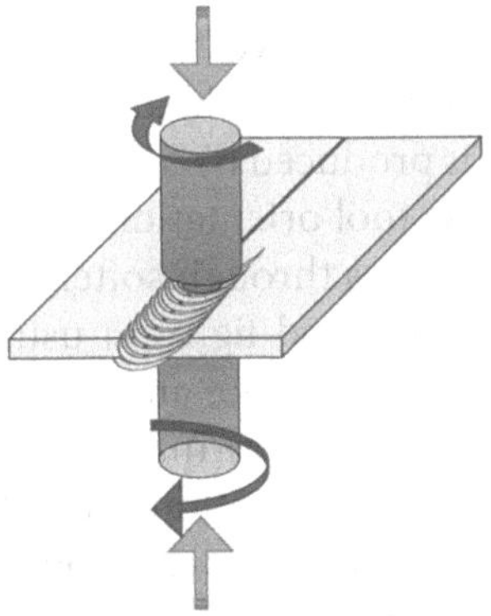

Figure 7.7 Scheme of the double-sided FSW process.

The double-sided tool was later used in many studies, such as studying the influence of tool size on the tensile strength of the Al alloys [69] and using convex and concave double-sided tools to study the plastic deformation and heat generation of the magnesium alloys [70]. It is also extended to investigate the effects of adding nanoparticles to the nugget zone [71] and double-sided friction stir spot welding of aluminum and steel alloy sheets [72]. In conclusion, even though double-sided FSW is being used increasingly, its industrial applications still need to be more ambitious and may require more in-depth scientific research.

7.5.2 Twin-Tool FSW Process

In some cases, the entire FSW process requires more than one pass; additional passes are needed to improve the welding quality and the mechanical properties obtained. However, using more than one pass is not desirable in solid-state welding processes, which initially depend on sustainability, maximizing resources, and minimizing consumption. Therefore, the twin-tool FSW concept was born to overcome the requirements of using more than one pass in the welding process. TWI first developed the idea in 1999 [73] using parallel twin tools, and it was later developed using tandem twin tools [74]. In both types of twin-tool FSW, the welding time will be significantly reduced, the rotating speed of the tools can be reduced, and the travel speed can be increased while maintaining the heat generated by friction. These improvements can be considered positive steps toward achieving sustainability and reducing resource use. In order to distinguish between the two types, in parallel twin-tool FSW, the arrangement of the two tools is according to the welding direction: side by side and transverse, while in tandem twin-tool FSW, they are arranged in line with the welding direction, as shown in Figure 7.8.

The tandem strategy will improve the weld's quality by breaking up any leftover oxide layers in the first weld area. Without losing mechanical properties, a second weld is produced over the previous one in the opposite direction. Similarly, because tool orientation dictates that one tool follows the other, the second tool passes through softened material.

Many studies have tried to shed light on using the twin-tool FSW. A comparative study between a twin tool and a single tool for welding aluminum alloys showed that plastic deformation occurred more when using the twin tool. The increase in deformation is due to the steady rise in heat generated along the joint seam [75]. In addition to obtaining a defect-free joint, an increase in hardness was observed.

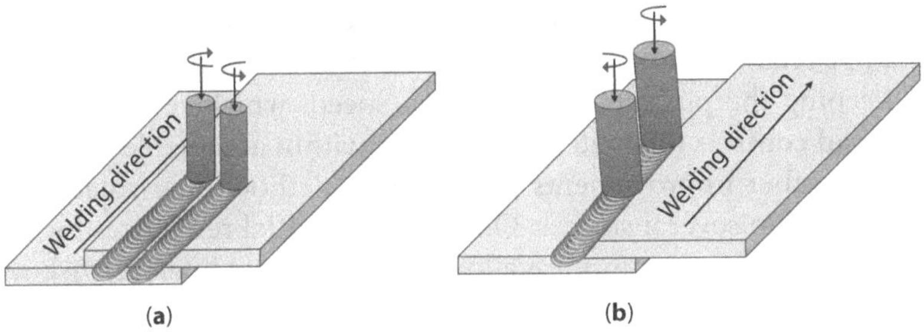

Figure 7.8 Scheme of (a) parallel-twin tool FSW, (b) tandem twin-tool FSW.

7.5.3 Dual-Rotation FSW Process

In traditional FSW, the shoulder and the pin rotate at the same speed, which is often constant throughout the welding process. This is also unaffected by recent advancements such as twin-tool FSW and double-sided FSW. The other interesting point in traditional FSW is that the relative speed starts at zero in the center of the pin and gradually increases until it reaches the maximum in the outer diameter of the shoulder. In the case of dual-rotation FSW, the FSW tool is made of two separate parts, the pin is centered on the shoulder, and both rotate in the same direction, as shown in Figure 7.9.

The developed technology provides many advantages, such as the ability to control the relative speed and rotate the shoulder at a speed different from the speed of the pin. Furthermore, this development allows the shoulder and the pin to rotate in opposite directions. The sheets overheating and uncontrolled material flow rate at high temperatures are determinants of

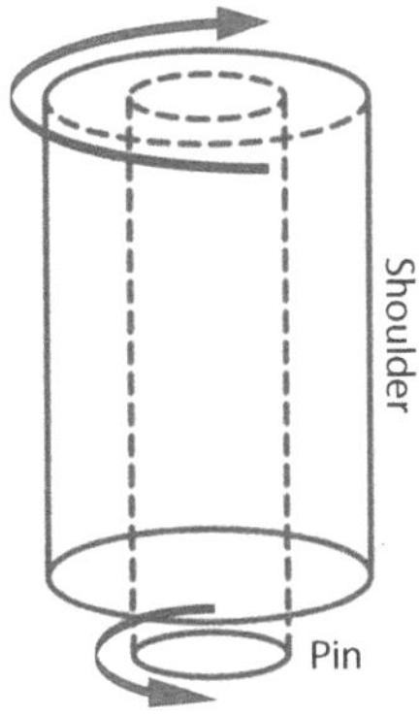

Figure 7.9 Dual-rotation FSW of the shoulder and pin arrangement.

FSW, so dual-rotation FSW can be considered one of the reliable solutions for that. A relatively low rotational speed is often used for the shoulder while keeping the pin rotating at a high speed, which reduces excessive heating and controls the material flow rates within acceptable limits.

As with other improvements in the FSW method, dual-rotation FSW has been used in several studies. One of the essential references for studying this technique on aluminum alloys was made by Liu *et al.* [76], where 5A06 aluminum alloy was used. The study showed that good joint efficiency was obtained using a low rotational speed, not exceeding 200 rpm for the shoulder and 400 rpm for the pin. On the other hand, conventional FSW was compared with dual-rotation FSW, and it was found that the joint results were defect-free when using dual-rotation FSW. Also, the welding process was more stable due to the remarkable improvement in the metal flow rate [77].

7.5.4 Friction Stir Spot Welding

FSSW is a special issue of traditional FSW, including all of its benefits, especially sustainability concerns [78]. The primary distinction between traditional FSW and FSSW is that there is no travel speed or movement in the FSSW, as shown in Figure 7.10, which concentrates the entire process in one point, so this type is classified as spot welding. From a manufacturing sustainability perspective, FSSW is superior to corresponding joining methods such as riveting, resistance, and laser spot welding. FSSW consumes less energy and does not require melting the weld region. It also does not need a professional worker, and no consumer rods need to be replaced periodically. About 90% of the consumed energy can be minimized by replacing resistance spot welding (RSW) with FSSW, and the cost can be reduced to 40% [79].

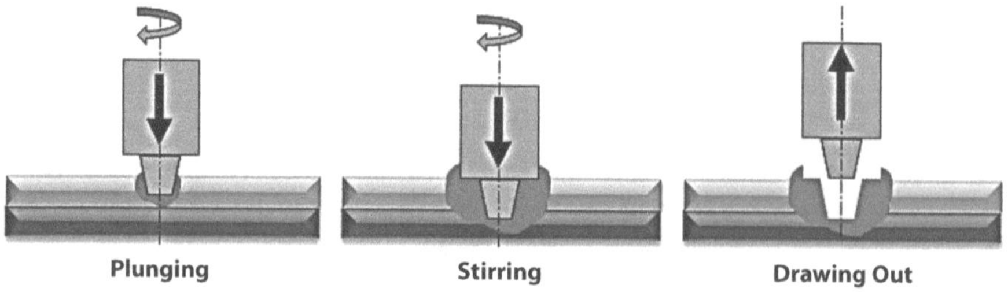

Figure 7.10 Scheme of FSSW stages.

There are several forms of FSSW used in literature for many applications. Traditional FSSW [80], which is widely used, refill FSSW to overcome the keyhole [81], pinless FSSW with only one shoulder [82], and swing FSSW with swinging movement over the surface of the weld or coating surfaces [83].

7.6 Recent Applications of FSW

Friction stir welding has progressed from the realm of research and limited application to active participation in various countries worldwide in the last 10 years. The increasing demand for friction stir welding as a sustainable tool has led to many companies specializing in manufacturing equipment for this type of welding or providing unique services ever are moving toward. FSW can be automated, which improves precision and speeds up production compared to other types of welding. Because of the weight and cost savings, FSW was embraced for bonding fuel tanks on satellite vehicles, commercial aircraft manufacture, alloy rims, engine blocks, and battery boxes. FSW was used in the Eclipse 500 business jet, the Boeing 747 freighter, the Legacy 500 aircraft, and the Embraer aircraft [84]. Several cooperative projects in the field of FSW have recently been completed, including the 6.8-million-euro European Industrialization of Friction Stir Welding (EuroStir®), the 88.0-million-euro Technology Application for the Near-Term Business Goals of the Aerospace Industry (TANGO), and the 5.1-million-euro Welding of Airframes by Friction Stir (WAFS). Furthermore, FSW was used in several rail applications, such as the Commuter EMU Series 20000 and Express EMU Series 885 built by Hitachi and the 700 Series Shinkansen managed on the Tokaido and Sanyo Lines in Japan [85].

7.7 Conclusions

This chapter attempted to shed light on the concept of sustainability and its applications to friction stir welding through four essential aspects: reducing the environmental impact, reducing the energy and resources consumed, reducing the cost of production and labor, and maximizing the welding process efficiency.

The chapter showed that by reviewing several novel studies, friction stir welding succeeded effectively in all four aspects compared to other welding processes. This remarkable success has led to the growing role of this type of welding in the industry in recent years. Moreover, it has become an integral part of many industries, especially transportation, such as planes, trains, and cars, due to its significant contribution to reducing weight and cost. It has also been assigned many recent applications, such as manufacturing batteries. In conclusion, friction stir welding will contribute to the global vision of industrial sustainability in the coming years, and its areas of use will increase steadily.

References

1. McGrath, J.E., Hickner, M.A., Höfer, R., Introduction: Polymers for a sustainable environment and green energy, in: *Polymer Science: A Comprehensive Reference 10*, 1–3, 2012.
2. Jamal, J., Darras, B., Kishawy, H., A study on sustainability assessment of welding processes. *Proc. Inst. Mech. Eng. B J. Eng. Manuf.*, 234, 3, 501–512, 2020.
3. Akca, E. and Gürsel, A., Solid state welding and application in aeronautical industry. *Period. Eng. Nat. Sci. (PEN)*, 4, 1, 1–8, 2016.
4. Mohan, D.G. and Wu, C.S., A review on friction stir welding of steels. *Chin. J. Mech. Eng. (English Edition)*, 34, 1, 2021.
5. Liu, L., Introduction to the welding and joining of magnesium, in: *Welding and Joining of Magnesium Alloys*, L. Liu, (Ed.), pp. 3–8, Woodhead Publishing, Cambridge, 2010.
6. Al-Sabur, R., Jassim, A.K., Messele, E., Real-time monitoring applied to optimize friction stir spot welding joint for AA1230 Al-alloys. *Mater. Today Proc.*, 42, 2018–2024, 2021.
7. Xu, W.F., Wu, X.K., Ma, J., Lu, H.J., Luo, Y.X., Abnormal fracture of 7085 high strength aluminum alloy thick plate joint via friction stir welding. *J. Mater. Res. Technol.*, 8, 6, 6029–6040, 2019.
8. Ma, Z.Y., Feng, A.H., Chen, D.L., Shen, J., Recent advances in friction stir welding/processing of aluminum alloys: Microstructural evolution and mechanical properties. *Crit. Rev. Solid State Mater. Sci.*, 43, 4, 269–333, 2018.
9. Bahedh, A.S., Mishra, A., Al-Sabur, R., Jassim, A.K., Machine learning algorithms for prediction of penetration depth and geometrical analysis of weld in friction stir spot welding process. *Metall. Res. Technol.*, 119, 3, 305, 2022.
10. Asmare, A., Al-Sabur, R., Messele, E., Experimental investigation of friction stir welding on 6061-t6 aluminum alloy using Taguchi-based GRA. *Metals*, 10, 11, 1480, 2020.
11. Jacquin, D. and Guillemot, G., A review of microstructural changes occurring during FSW in aluminium alloys and their modelling. *J. Mater. Process. Technol.*, 288, 116706, 2021.

12. Khalaf, H., II, Al-sabur, R., Abdullah, M.E., Kubit, A., Derazkola, H.A., Effects of underwater friction stir welding heat generation on residual stress of AA6068-t6 aluminum alloy. *Materials*, 15, 6, 2223, 2022.
13. Al-Sabur, R., Tensile strength prediction of aluminium alloys welded by FSW using response surface methodology-comparative review. *Mater. Today Proc.*, 45, 4504–4510, 2021.
14. Silva, A.C.F., Braga, D.F.O., de Figueiredo, M.A.V., Moreira, P.M.G.P., Ultimate tensile strength optimization of different FSW aluminium alloy joints. *Int. J. Adv. Manuf. Technol.*, 79, 5–8, 805–814, 2015.
15. Güleryüz, G., Relationship between FSW parameters and hardness of the ferritic steel joints: Modeling and optimization. *Vacuum*, 178, 109449, 2020.
16. Nishida, H., Ogura, T., Hatano, R., Kurashima, H., Fujimoto, M., Hirose, A., Fracture toughness and fatigue crack behaviour of A3003/SUS304 lap friction stir welded joints. *Weld. Int.*, 31, 4, 3901, 2017.
17. He, W., Li, M., Song, Q., Liu, J., Hu, W., Efficacy of external stationary shoulder for controlling residual stress and distortion in friction stir welding. *Trans. Indian Inst. Met.*, 72, 1349–1359, 2019.
18. Morishige, T., Tsujikawa, M., Oki, S., Kamita, M., Chung, S.W., Higashi, K., Friction stir processing of cast Mg-Y-Zn alloy. *Adv. Mater. Res.*, 15–17, 369–374, 2007.
19. Al-Sabur, R., Khalaf, H., II, Świerczyńska, A., Rogalski, G., Derazkola, H.A., Effects of noncontact shoulder tool velocities on friction stir joining of polyamide 6 (PA6). *Materials*, 15, 12, 4214, 2022.
20. Rangel Pacheco, M., Paul Kabche, J., Thesi, I., Nunes Diesel, F., Temperature distribution analysis in plates joined by friction stir welding. *ASME 2009 International Mechanical Engineering Congress and Exposition, Proc*, vol. 11, 163–169, 2010.
21. Smith, C.B., Robotic friction stir welding using a standard industrial robot. *2nd Friction Stir Welding International Symposium*, 2000.
22. Mehta, K.P. and Patel, R., On FSW keyhole removal to improve volume defect using pin less tool. *Key Eng. Mater.*, 821, 215–221, 2019.
23. Fernandez, G.J. and Murr, L.E., Characterization of tool wear and weld optimization in the friction-stir welding of cast aluminum 359+20% SiC metal-matrix composite. *Mater. Charact.*, 52, 1, 65–75, 2004.
24. Buchibabu, V., Reddy, G.M., Kulkarni, D., De, A., Friction stir welding of a thick Al-Zn-Mg alloy plate. *J. Mater. Eng. Perform.*, 25, 3, 1163–1171, 2016.
25. Yoon, J., Kim, C., Rhee, S., Performance of plunge depth control methods during friction stir welding. *Metals*, 9, 3, 283, 2019.
26. Bevilacqua, M., Ciarapica, F.E., D'Orazio, A., Forcellese, A., Simoncini, M., Sustainability analysis of friction stir welding of AA5754 sheets. *Proc. CIRP*, 62, 529–534, 2017.
27. Sued, M.K., Samsuri, S.S.M., Kassim, M.K.A.M., Nasir, S.N.N.M., Sustainability of welding process through bobbin friction stir welding. *IOP Conf. Ser. Mater. Sci. Eng.*, 318, 1, 012068, 2018.

28. Sethi, S.R., Das, A., Baruah, M., A review on friction stir welding: A sustainable way of manufacturing. *Mater. Today Proc.*, 44, 2685–2688, 2021.
29. Ramesha, K., Sudersanan, P.D., Santosh, N., Jangam, S., Sustainable processsing of dissimilar aluminium alloy joints by friction stir welding. *IOP Conf. Ser. Mater. Sci. Eng.*, 1132, 1, 012029, 2021.
30. Azeez, S., Mashinini, M., Akinlabi, E., Sustainability of friction stir welded AA6082 plates through post-weld solution heat treatment. *Proc. Manuf.*, 33, 27–34, 2019.
31. Mimouni, O., Badji, R., Hadji, M., Kouadri-Henni, A., Finite volume modeling of laser assisted friction stir welding of 2017A-T451 aluminum alloy for enhanced sustainability of welded joints. *AIP Conference Proc*, vol. 1769, 100014, 2016.
32. Majeed, T., Wahid, M.A., Alam, M.N., Mehta, Y., Siddiquee, A.N., Friction stir welding: A sustainable manufacturing process. *Mater. Today Proc.*, 46, 6558–6563, 2020.
33. Al-Wajidi, W., *Analysis of Friction Stir Welding Process: Sustainability and Optimization*, PhD Thesis University of Guelph, Guelph, ON, Canada, 2019.
34. Suresh, S., Elango, N., Venkatesan, K., Lim, W.H., Palanikumar, K., Rajesh, S., Sustainable friction stir spot welding of 6061-T6 aluminium alloy using improved non-dominated sorting teaching learning algorithm. *J. Mater. Res. Technol.*, 9, 5, 11650–11674, 2020.
35. Azeez, S. and Akinlabi, E., Sustainability of manufacturing technology: Friction stir welding in focus. *Prog. Ind. Ecol.*, 12, 4, 419–438, 2018.
36. Sharma, R.K., Sodhi, G.P., Bhakar, V., Kaur, R., Pallakonda, S., Sarkar, P., Singh, H., Sustainability in manufacturing processes: Finding the environmental impacts of friction stir processing of pure magnesium. *CIRP J. Manuf. Sci. Technol.*, 30, 25–35, 2020.
37. Hart, S.L. and Dowell, G., A natural-resource-based view of the firm: Fifteen years after. *J. Manage.*, 37, 5, 1464–1479, 2011.
38. di Lorenzo, R. and Fratini, L., A cost model for the friction stir welding process. *ESAFORM 2005*, pp. 1073–1076, 2005.
39. Boitsov, A.G., Kuritsyn, D.N., Siluyanova, M.V., Kuritsyna, V.V., Friction stir welding in the aerospace industry. *Russ. Eng. Res.*, 38, 12, 1029–1033, 2018.
40. Tipaji, P.K., *E-Design Tools for Friction Stir Welding: Cost Estimation Tool*, MSc Thesis University of Missouri-Rolla, Rolla, MO, United States, 2007.
41. Toumpis, A., Galloway, A., Cater, S., Molter, L., A techno-economic evaluation of friction stir welding of DH36 steel. *10th International Friction Stir Welding Symposium*, 2014.
42. Mononen, J. and Hänninen, M., H., Cost comparison of FSW and MIG welded aluminium panels. *Weld. World*, 47, 11–12, 2003.
43. Shrivastava, A., Krones, M., Pfefferkorn, F.E., Comparison of energy consumption and environmental impact of friction stir welding and gas metal arc welding for aluminum. *CIRP J. Manuf. Sci. Technol.*, 9, 159–168, 2015.

44. Dawood, H., II, Mohammed, K.S., Rajab, M.Y., Advantages of the green solid state FSW over the conventional GMAW process. *Adv. Mater. Sci. Eng.*, 2014, 105713, 2014.
45. Hossfeld, M. and Roos, E., A new approach to modelling friction stir welding using the CEL method. *International Conference on Advanced Manufacturing Engineering and Technologies NEWTECH (Stockholm)*, 2013.
46. Mofeed Alam, M., Jha, A.K., Mukherjee, S., Panda, S., Chakraborty, S.S., A review on friction stir welding—A green manufacturing technology, in: *Lecture Notes in Mechanical Engineering*, vol. 46, 2021.
47. Sejani, D., Li, W., Patel, V., Stationary shoulder friction stir welding–low heat input joining technique: A review in comparison with conventional FSW and bobbin tool FSW. *Crit. Rev. Solid State Mater. Sci.*, 47, 6, 865–914, 2021.
48. Sharma, C., Upadhyay, V., Verma, V., Tripathi, A., Sharma, S., Tensile behaviour of friction stir welded joints of different aluminium alloys. *J. Eng. Res.*, 9, 1–15, 2021.
49. Morishige, T., Kawaguchi, A., Tsujikawa, M., Hino, M., Hirata, T., Higashi, K., Dissimilar welding of Al and Mg alloys by FSW. *Mater. Trans.*, 49, 5, 1129–1131, 2008.
50. Masoudian, A., Tahaei, A., Shakiba, A., Sharifianjazi, F., Mohandesi, J.A., Microstructure and mechanical properties of friction stir weld of dissimilar AZ31-O magnesium alloy to 6061-T6 aluminum alloy. *Trans. Nonferrous Met. Soc. China (English Edition)*, 24, 5, 1317–1322, 2014.
51. Malarvizhi, S. and Balasubramanian, V., Influences of tool shoulder diameter to plate thickness ratio (D/T) on stir zone formation and tensile properties of friction stir welded dissimilar joints of AA6061 aluminum-AZ31B magnesium alloys. *Mater. Des.*, 40, 453–460, 2012.
52. Mahamud, M., II, Ishak, I., Halil, A.M., Study of dissimilar welding AA6061 aluminium alloy and AZ31b magnesium alloy with ER5356 filler using friction stir welding. *IOP Conf. Ser. Mater. Sci. Eng.*, 238, 1, 012002, 2017.
53. Baghdadi, A.H., Sajuri, Z., Keshtgar, A., Sharif, N.M., Rajabi, A., Mechanical property improvement in dissimilar friction stir welded Al5083/Al6061 joints: Effects of post-weld heat treatment and abnormal grain growth. *Materials*, 15, 1, 288, 2022.
54. Prabhu, S., Sri, M.N.S., Anusha, P., Saravanan, G., Kannan, K., Manickam, S., Improvement of mechanical behavior of FSW dissimilar aluminum alloys by postweld heat treatments. *Adv. Mater. Sci. Eng.*, 2022, 1–8, 2022.
55. Dickerson, T., Shi, Q., Shercliff, H.R., Heat flow into friction stir welding tools. *4th International Symposium on Friction Stir Welding*, USA, 2003.
56. Li, H., Gao, J., Li, Q., Galloway, A., Toumpis, A., Effect of friction stir welding tool design on welding thermal efficiency. *Sci. Technol. Weld. Join.*, 24, 2, 156–162, 2019.
57. Hattel, J.H., Schmidt, H.N.B., Tutum, C., Thermomechanical modelling of friction stir welding. *ASM Proceedings of the International Conference: Trends in Welding Research*, 2009.

58. Sharma, C., Upadhyay, V., Narwariya, B.S., Tensile properties of dissimilar friction stir weld joints of Al-2024 and Al-7039 alloys. *Mater. Res. Express*, 6, 2, 026524, 2019.
59. Feng, T. and Zhang, X., Three dimensional model for heat transfer and plastic flow of friction stir welding. *Hanjie Xuebao/Trans. China Weld. Inst.*, 34, 7, 105, 2013.
60. Khandkar, M.Z.H., Khan, J.A., Reynolds, A.P., Prediction of temperature distribution and thermal history during friction stir welding: Input torque based model. *Sci. Technol. Weld. Join.*, 8, 3, 165–174, 2003.
61. Chao, Y.J., Qi, X., Tang, W., Heat transfer in friction stir welding-experimental and numerical studies. *J. Manuf. Sci. Eng.*, 125, 1, 2003.
62. Zhu, X.K. and Chao, Y.J., Numerical simulation of transient temperature and residual stresses in friction stir welding of 304L stainless steel. *J. Mater. Process. Technol.*, 146, 2, 263–272, 2004.
63. Lienert, T.J., Stellwag, W.L., Grimmett, B.B., Warke, R.W., Friction stir welding studies on mild steel. *Weld. J.*, 82, 1, 1s–9s, 2003.
64. Abdallah, T., *Sustainable Mass Transit*, Imprint: Elsevier, Oxford, UK, 2017.
65. Bevilacqua, M., Ciarapica, F.E., Forcellese, A., Simoncini, M., Comparison among the environmental impact of solid state and fusion welding processes in joining an aluminium alloy. *Proc. Inst. Mech. Eng. B J. Eng. Manuf.*, 234, 1–2, 140–156, 2020.
66. Seuring, S., A review of modeling approaches for sustainable supply chain management. *Decis. Support Syst.*, 54, 4, 1513–1520, 2013.
67. Zhao, F., Bernstein, W.Z., Naik, G., Cheng, G.J., Environmental assessment of laser assisted manufacturing: Case studies on laser shock peening and laser assisted turning. *J. Clean. Prod.*, 18, 13, 1311–1319, 2010.
68. W.M.W. Thomas, E.D. Nicholas, J.C. Needham, M.G. Murch, P. Temple-Smith, C.J. Dawes, Friction stir welding process developments and variant techniques. Google Patents, 1991.
69. Garg, A. and Bhattacharya, A., Effect of tool size on AA6061-T6 double-sided friction stir welds. *Mater. Des. Process. Commun.*, 3, 5, 1–8, 2021.
70. Chen, J., Ueji, R., Fujii, H., Double-sided friction-stir welding of magnesium alloy with concave-convex tools for texture control. *Mater. Des.*, 76, 181–189 2015.
71. Rahmatian, B., Dehghani, K., Mirsalehi, S.E., Effect of adding SiC nanoparticles to nugget zone of thick AA5083 aluminium alloy joined by using double-sided friction stir welding. *J. Manuf. Process.*, 52, 152–164, 2020.
72. Lyu, X., Li, M., Li, X., Chen, J., Double-sided friction stir spot welding of steel and aluminum alloy sheets. *Int. J. Adv. Manuf. Technol.*, 96, 5–8, 2875–2884, 2018.
73. Thomas, W.M., Threadgill, P.L., Nicholas, E.D., Feasibility of friction stir welding steel. *Sci. Technol. Weld. Join.*, 4, 6, 365–372, 1999.
74. Thomas, W.M., Staines, D.J., Watts, E.R., Norris, I.M., The simultaneous use of two or more friction stir welding tools. *Int. J. Join. Mater.*, 17, 1, 52, 2005.

75. Kumari, K., Pal, S.K., Singh, S.B., Friction stir welding by using counter-rotating twin tool. *J. Mater. Process. Technol.*, 215, 1, 132–141, 2015.
76. Liu, X.M., Zhang, X.M., Yao, J.S., Yin, Y.H., Xia, P.Y., Optimization of rotation speed ratio in dual-rotation friction stir welding by finite element method. *Int. J. Adv. Manuf. Technol.*, 91, 5–8, 2869–2874, 2017.
77. Wang, F.F., Li, W.Y., Shen, J., Wen, Q., dos Santos, J.F., Improving weld formability by a novel dual-rotation bobbin tool friction stir welding. *J. Mater. Sci. Technol.*, 34, 1, 135–139, 2018.
78. Wahid, M.A., Sharma, N., Shandley, R., Friction stir welding process effects on human health and mechanical properties. *IJFEM*, 1, 1, 42, 2020.
79. Choi, D.H., Ahn, B.W., Lee, C.Y., Yeon, Y.M., Song, K., Jung, S.B., Formation of intermetallic compounds in Al and Mg alloy interface during friction stir spot welding. *Intermetallics (Barking)*, 19, 2, 125–130, 2011.
80. Bakavos, D. and Prangnell, P.B., Effect of reduced or zero pin length and anvil insulation on friction stir spot welding thin gauge 6111 automotive sheet. *Sci. Technol. Weld. Join.*, 14, 5, 443–456, 2009.
81. Ikuta, A., North, T.H., Uematsu, Y., Characteristics of keyhole refill process using friction stir spot welding. *Weld. Int.*, 32, 6, 417–426, 2018.
82. Cox, C.D., Gibson, B.T., Strauss, A.M., Cook, G.E., Effect of pin length and rotation rate on the tensile strength of a friction stir spot-welded al alloy: A contribution to automated production. *Mater. Manuf. Process.*, 27, 4, 472–478, 2012.
83. Wu, S., Sun, T., Shen, Y., Yan, Y., Ni, R., Liu, W., Conventional and swing friction stir spot welding of aluminum alloy to magnesium alloy. *Int. J. Adv. Manuf. Technol.*, 116, 7–8, 2401–2412, 2021.
84. TWI, Friction stir welding of airframe structures, 2022. https://www.twi-global.com/who-we-are/who-we-work-with/industry-sectors/aerospace/joining-of-airframe-structures/friction-stir-welding-of-airframe-structures.
85. TWI, Creating a stir in the rail industry, 2001. https://www.twi-global.com/technical-knowledge/published-papers/creating-a-stir-in-the-rail-industry-november-2001.

8

Plant-Based Biosorbents for Heavy Metal Removal From Wastewater

Narmadha V. and Siddhi Sreemahadevan*

Department of Biotechnology, PSG College of Technology, Coimbatore, Tamil Nadu, India

Abstract

The general composition of wastewater includes dissolved, suspended, and particulate matter of organic and inorganic nature. However, the recent surge in metal and dye production, combined with substandard farming practices, has led to dangerously high levels of heavy metals such as cadmium, arsenic, and lead in our water sources. Heavy metals are very well known to induce detrimental health reactions on living beings owing to their toxicity. While remediation of heavy metals using chemical methods is the go-to solution generally adopted, biological methods making use of plants (phytoremediation) and microorganisms (microbial bioremediation) have been gaining popularity due to their lesser environmental impact and their attributed sustainability. A few aquatic plants, namely, water hyacinth and duckweed have demonstrated the ability to absorb heavy metals and accumulate the same in their roots, stems, and leaves. Therefore, this article intends to review the predominantly used physical and chemical techniques and the most promising phytoremediation techniques applied for heavy metal pollution abatement in wastewater, along with their advantages and deficiencies. Since live growth of plants is an uncontrollable mode of heavy metal remediation, a few controlled methods involving the application of plant-based materials such as biosorbents and the mechanisms of biosorption of heavy metals are also discussed here.

Keywords: Phytoremediation, sustainability, biosorption, heavy metals, aquatic plants

**Corresponding author*: sid.bio@psgtech.ac.in

S. Thirumalai Kumaran and Tae Jo Ko (eds.) Sustainable Machining and Green Manufacturing, (155–176) © 2024 Scrivener Publishing LLC

8.1 Introduction

Industrialization and urbanization cause severe pollutions in the environment. A metal that has a specific gravity above 5 is called heavy metal. Heavy metal, a toxic pollutant present in the aquatic environment in industrial regions, causes serious effects on human health and the environment because of its toxic nature and recalcitrance. Heavy metal contamination in the environment is attributed not only to anthropogenic causes but to natural causes as well [1]. Natural sources consist of weathering of rocks and minerals, volcanic eruption, etc., and anthropogenic sources include the consumption of fertilizers, mining processes, electroplating, photographic units, etc., causing heavy metal contamination in the biosphere [2]. In addition, a huge number of these metals enter the environment through wastewater from industries. Some heavy metals are lethal even in the lowest concentration that is attributed to their nature of toxicity and non-biodegradability [3]. Exposure to heavy metals results in altering key functions in humans and the environment. However, not all heavy metals are toxic, as some of them are essential to our body in appropriate amounts. Depending on the heavy metals, ingestion of heavy metals causes various effects on human health such as respiratory illness and nerve damage.

Many approaches have been studied to recycle industrial wastewater before it gets into rivers or lakes. Such methods are chemical precipitation, coagulation, flocculation, membrane separation technology, and ion exchange technology [4–7]. However, these chemical methods are inefficient and expensive and produce secondary wastes that may cost high for its disposal [8]. Biosorption can be a better alternative to chemical methods. The advantages of using biosorption are high efficiency in removing metals, reuse of biosorbent after the metal removal process, low amount of secondary wastes generated, and most importantly the process and the biosorbent are eco-friendly [9, 10]. Therefore, biochar offers an economic yet potential adsorbent for heavy metal mitigation. Biochar can also be used as fertilizer after it is utilized for the removal of heavy metals. Plant-based biochar has demonstrated a higher efficiency in the adsorption of heavy metals from water bodies. However, the usage of biochar for heavy metal adsorption results in the loss of the same, the regeneration of which is very difficult. Hence, biochar-immobilized alginate beads have been used for better stabilization and regeneration for continuous wastewater treatment [11]. This review paper examines the different traditional physical and chemical methods available for heavy metal mitigation and focuses on the efficiency of biosorbent-based alternative methods of phytoremediation.

8.2 Physical and Chemical Techniques for Heavy Metal Removal

The most commonly applied physical and chemical techniques for the abatement of heavy metals are listed.

8.2.1 Chemical Precipitation

Heavy metal concentration can be reduced in wastewater by precipitating with chemicals. In this method, chemicals are allowed to react with heavy metals in a solution at an appropriate pH for some period of time, resulting in the formation of insoluble precipitates. After the precipitates are formed, they can be removed from the water using filtration [12]. On adding appropriate precipitants and adjusting to favorable pH, hydroxides, carbonates, and sulfides of metals are removed as precipitates [13]. Hydroxide precipitation is commonly practiced, since it is cheaper and less dangerous than sulfide precipitation.

8.2.2 Coagulation

Coagulation is the process of coagulating metal ions by neutralizing their charges by adding suitable chemicals [14]. Reducing surface charges makes the ions approach each other to form aggregates by means of Van der Waals force. The electrocoagulation process is the same as the conventional coagulation process, but the way of generating the aggregates is different. Electrocoagulation generates aggregates by producing coagulants from the anode. Electrocoagulation is cheaper and effective than chemical coagulation treatment [14].

8.2.3 Membrane Separation

Membrane separation technologies include microfiltration, nanofiltration, ultrafiltration, and reverse osmosis. Based on the size of the particles needed to be separated, an appropriate membrane separation technique is practiced. pH, metal ion concentration, pore size of the membrane, and applied pressure are the factors that affect the membrane separation process [15]. The main disadvantages of membrane separation are membrane fouling and a higher operational cost.

8.2.4 Ion Exchange Method

An ion exchange process involves exchanging ions of similar charges between solutions and insoluble materials. Resins, three-dimensional matrix, or hybrid materials can be used as ion exchange materials [16]. The materials can be organic, inorganic, natural, synthetic, anionic, cationic, or amphoteric. This process is reversible, hence the ion exchange solids can be reused.

8.2.5 Adsorption

Adsorption is the surface phenomenon where the solute molecules get adsorbed onto the adsorbent. Adsorption is a cost-effective, efficient, and flexible method, and hence it is more commonly preferred. Many materials such as carbon nanotubes [17], fly ash [18], and zeolites [19] are used as adsorbents. The important factors that direct the adsorption process are pH, temperature, and contact time. However, the cost of adsorbents and the huge investment required are deterrents to the large-scale application of the adsorption process.

8.3 Biological Methods for Heavy Metal Removal

Bioaccumulation and biosorption are the alternative methods to the traditional chemical methods for managing heavy metals in wastewater. These methods overcome the disadvantages of chemical methods. Bioaccumulation is a process whereby heavy metals are accumulated into the parts of living organisms. Bioaccumulation of heavy metals using bacterial cells [20] and plants [21] has been employed for the mitigation of heavy metal-polluted wastewater. However, bioaccumulation using live systems has some limitations such as (i) the living organisms may fall prey to the toxic concentrations of heavy metals and (ii) maintaining the optimum conditions of the living system is expensive [22]. Another biological method available is biosorption, the process of passive metal ion binding by dead organisms. This in turn is potentially a sustainable and economical method to deal with heavy metals in polluted water systems [23]. The important beneficial factors for using biosorbents widely are low cost and better availability [24].

8.3.1 Phytoremediation

Bioaccumulation using live plants has been carried out to find the rate of uptake of heavy metals by simply growing the plants in water that contains heavy metals. There have been reports on *Eichhornia*, *Pistia*, and *Salvinia* that were allowed to grow in water containing Cr, which resulted in a greater accumulation of Cr in the roots. The plants in the increasing order of Cr accumulation were *Eichhornia* > *Salvinia* > *Pistia* [21]. In a study conducted in constructed wetlands, *Salvinia* removed heavy metals, and the efficiencies were Cr (2.021 to 1.052 ppm), Pb (2.974 to 1.924 ppm), Cu (1.092 to 2.035 ppm), and Cd (0.251 to 0.018 ppm) [25]. Furthermore, *Typha latifolia*-treated Cr containing tannery effluent showed that the root uptake of Cr was higher and found to be 7 mg ml^{-1} [26]. Water hyacinth when grown in wetlands containing Cd, Pb, Zn, Ni, and Cu resulted in a maximum root uptake of these metals in the following order: Cu > Zn > Ni > Pb > Cr. The amounts of heavy metals adsorbed by the plant were 0.24 kg ha^{-1}, 5.42 kg ha^{-1}, 21.62 kg ha^{-1}, 26.17 kg ha^{-1}, and 13.46 kg ha^{-1} for cadmium, lead, copper, zinc, and nickel, respectively [27].

8.4 Biochar

Biochar is one of the low-cost and effective adsorbents (Figure 8.1). The char is prepared by pyrolysis of the biological material between the temperatures 350°C and 800°C under oxygen-limited conditions [28]. The resulting char is rich in carbon, which contributes to its enhanced adsorption capacity. The characteristics of the biochar are heavily dependent on the operating temperature of pyrolysis and the quantity of biomass. Furthermore, the attributes of the biochar that depend on the biomass are

Figure 8.1 Biochar [30].

high carbon content, porosity, surface area, and the presence of functional groups [29]. Tang *et al.* [29] also concluded the benefits of biochar that include alleviation of climate change, improvement of crop production, remediation of polluted environment, and recycling of agricultural wastes into value-added products.

8.4.1 Mechanism of Biochar Adsorption

The possible mechanisms involved in biochar adsorption are depicted in Figure 8.2.

8.4.1.1 Precipitation

Precipitation of heavy metals is thought to be one of the mechanisms participating in biochar adsorption. The presence of insoluble Cd aggregates such as $CdCO_3$, $Cd_3(PO_4)_2$ was discovered on the surface of the biochar prepared from water hyacinth after an adsorption experiment with Cd. The adsorption was possible as a result of the alkalinity of the biochar being stimulated by the inorganic minerals that in turn promoted Cd precipitation [31]. The crystals of $PbCO_3$ and $Pb_3(CO_3)_2(OH)_2$ generated during the process showed the contribution of precipitation to the same [32].

8.4.1.2 Surface Complexation

Due to the low pH_{pzc} (the pH of point zero charge), the hydroxylate group of SiO_2/Al_2O_3 will be generated, which would provide the sites to complex with minerals such as K^+, Ca^{2+}, Na^+, and Mg^+ in biochar. At low pH_{pzc}, the surface of the adsorbent becomes negative, which induces it to complex

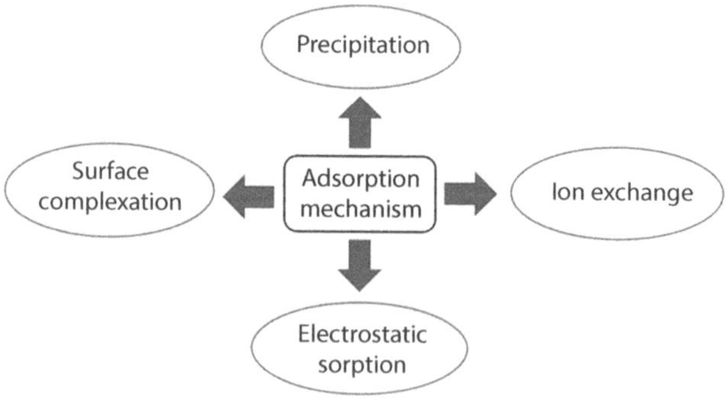

Figure 8.2 Mechanisms involved in the adsorption on biochar.

with cations. During sorption, upon application of Cd, the exchange of ions between Cd^{2+} or $Cd(OH)^+$ and cations leads to the complexation of Cd with SiO_2/Al_2O_3 [31]. Higher carboxylic functional groups of the *Alternanthera philoxeroides* biochar showed a higher adsorption capacity, indicating that the mechanism of surface complexation influenced the adsorption of Pb(II) [32].

8.4.1.3 Ion Exchange

Heavy metals can form insoluble complexes by means of ion exchange with biochar surface materials. For example, there are reports of formation of insoluble Cd complex such as $CdCO_3$ due to the exchange of ions between Cd and surface complex such as $CaCO_3$ [31]. Therefore, the adsorption capacity is attributed to the ion exchange between alkaline earth cations and Pb(II) in the carboxylic group (-COOMe), where Me is either of K(I), Na(I), Ca(II), or Mg(II) [32].

8.4.1.4 Electrostatic Sorption

It has been found that the interaction between cadmium and the delocalized lone pair of π electrons of the biochar brought about the formation of a Cd-Cπ bonding on the biochar surface [31], making electrostatic sorption, a proposed mechanism for biochar adsorption.

8.4.2 Immobilized Biochar

The disadvantage of using biochar as adsorbent is the necessity to separate solid and liquid phase by centrifugation and filtration techniques, which is not recommended at an industrial level [33]. Also, biochar application is restricted to column experiment due to the difficulties in the regeneration of biosorbents such as loss of biomass, need for separation of biosorbent after adsorption, low mechanical strength, and less control over biosorbent particles [34]. To overcome the disadvantages of biochar, immobilized biochar is used. The different methods of immobilization of these biosorbents are entrapment (in a porous but a strong matrix), cross-linking (immobilization effected by chemicals), encapsulation (entrapment in a membrane-like structure), and adsorption (attachment on inert carriers involving weak forces) [35]. The immobilization of biochar showed greater metal removal potential when used in fluidized or packed bed reactor, and the advantages of the same include size control, regeneration and reuse of biosorbents, minimal clogging of biosorbents into the column, and easy separation of biosorbents from the effluents [36].

8.5 Plant-Based Biochar

Various plants such as duckweed [37], *Typha latifolia* [38], seaweeds, seagrasses [39], and some other plants in different forms have been identified to absorb heavy metal. Biochar [31] and alginate bead forms of plant biochar [40] have been examined largely to determine their heavy metal removal efficiency in contaminated wastewater.

Desalted and untreated *Spartina alterniflora*, which was harvested in different seasons, was examined for its metal removal efficiency, and the biochar produced from the winter season showed high Cd^{2+} removal efficiency. Desalination and harvest time of *Spartina alterniflora* affected its biochar properties, viz., surface area, porosity, and Cd^{2+} adsorption [41]. Invasive species such as ragweed (*Ambrosia artemisiifolia L.*) and horseweed (*Conyza canadensis L.*) when subjected to pyrolysis at different thermal conditions were analyzed for their metal removal efficiency. However, ragweed biochar prepared at 450°C was found to have a greater adsorption capacity of 139 $mg{\cdot}g^{-1}$ for Cd and 358.7 $mg{\cdot}g^{-1}$ for Pb as compared to horseweed [42]. Here, ion exchange, precipitation, and complexation were deduced to be the major mechanisms participating in the adsorption of Cd and Pb on biochar.

Chemical modification of hydrochar of *Lepironia articulata* into biochar using KOH in the ratio of 1:3 (hydrochar-to-KOH) created a porous structure that in turn enhanced the surface area from 69.23 $m^2\ g^{-1}$ (hydrochar) to 820.54 $m^2\ g^{-1}$ (biochar) [43]. It was concluded that high adsorption efficiency could be due to an increase in functional groups, enhanced extent of cation exchange, and better availability of surface area on the chemically treated biochar.

Biochar of tobacco stem produced from different temperature and residence time was analyzed for the adsorption capacity for lead, cadmium, and copper in competitive mode wherein the three metals competed for sorption sites on biochar [44]. Lead showed the most competitive nature than Cd and Cu, but Cu showed constant adsorption capability on biochar with or without minerals.

Alkali-activated biochar derived from camphor (*Cinnamomum camphora*) leaves showed increased oxygenic functional groups, large surface area, and improved pore structure [45]. Removal of Pb(II) was reduced owing to the presence of copper ions in competitive mode. The adsorption rate of Pb(II) onto alkali-activated biochar appeared faster than the biochar without alkali activation. In addition, biochar produced at a high pyrolysis temperature attained equilibrium very quickly.

Biochar of Chinese herb residues including *Gastrodia elata* and *Angelica sinensis* generated at 300°C and 600°C was compared for its extent of adsorption, and it was found that Pb^{2+} adsorbed better when the herbs were pyrolyzed at 600°C [46]. The study further reported that the presence of phosphorus and sulfur in biochar might have released PO_4^{3-} and SO_4^{2-} to form precipitates with heavy metals that facilitate the elimination of the same.

Biochar derived from *Phytolacca acinosa* was studied to evaluate the mitigation of heavy metals, viz., Pb, Cu, Ag, and Cd [47]. The whewellite present in the biomass of *P. acinosa* was decomposed into calcite after pyrolysis process and the calcite in biochar precipitated with heavy metals aiding in heavy metal removal.

Activated and non-activated biochar of *Ipomoea fistulosa* was compared for its removal efficiency for cadmium metal [48]. KOH-activated biochar showed almost absolute removal of cadmium ions at the end of 6 h, suggesting that activation of biochar by KOH would improve the surface area by removing the partially clogged pores of biochar that might have increased the adsorption capacity and the adsorption rate.

Canna indica-derived biochar generated at different thermal conditions from 300°C to 600°C was analyzed for heavy metal removal efficiency [49]. The surface area availability and the content of carbon in the biochar were found to be increased with temperature hike, whereas other elements such as N, H, and O were found to be lowered with increasing pyrolysis temperature. In addition to this, the pH and the ash content of the biochar were seen to be increased with increasing process temperature. The biochar of *C. indica* made at 500°C showed greater removal efficiency for cadmium than those prepared at other temperatures. A moderate surface charge of this biochar indicated that the primary mechanism responsible for the adsorption of Cd^{2+} to the biochar was not electrostatic interaction.

Alternanthera philoxeroides-derived biochar at 600°C demonstrated maximum Pb(II) adsorption efficiency, 5.3 times more than that of activated carbon [32]. *Alternanthera philoxeroides* biochar was identified as pH-independent adsorbent during adsorption of lead, while activated carbon (AC) was a pH-dependent adsorbent.

Biochar derived from a variety of *Miscanthus sacchariflorus* called Geodae-Uksae 1 was employed for Cd removal, and the toxicity of Cd after the adsorption process was analyzed using *Daphnia magna* [50]. Biochar prepared at 500°C demonstrated a better adsorption capacity and removed the acute toxicity to *D. magna*. The toxicity test can be conducted with *D. magna* due to simple operation, less time, space, equipment, acclimatization time, and facility requirements and also *D. magna* is more reliable and functional and gives reproducible results [51].

8.5.1 Biochar From *Eichhornia crassipes*

Eichhornia crassipes, generally known as water hyacinth, has been a largely explored water plant for its application as a biosorbent for heavy metal pollution abatement. Oxidized biochar of the long root of water hyacinth produced at 800°C for 3 h was investigated for its ability to adsorb heavy metals, and they were found to be 0.57 mmol g^{-1} for Pb^{2+}, 0.41 mmol g^{-1} for Cu^{2+}, 0.44 mmol g^{-1} for Cd^{2+}, and 0.48 mmol g^{-1} for Zn^{2+} [52]. The biochar could be regenerated, and the removal efficiency remained the same for 10 cycles of adsorption and desorption processes.

Root biochar of *E. crassipes* produced at 500°C for 4 h was checked for its ability to adsorb heavy metals, namely, Pb(II), Zn(II), Cu(II), and Cd(II), for which the removal efficiency at 298K was found to be 39.09 mg g^{-1}, 45.05 mg g^{-1}, 48.20 mg g^{-1}, and 44.04 mg g^{-1}, respectively. Therefore, the results reported a superior heavy metal removal compared to other biochar generated at 200°C, 300°C, and 400°C [53].

The application of biochar pellets prepared from water hyacinth leaves and immobilized with microalga *Chlorella* sp. showed a high surface hydrophilicity and therefore a higher immobilization efficiency [54]. Immobilization efficiency and bioaccumulation capacity of algae were found to be directly proportional to the removal efficiency of Cd (II). *E. crassipes*-derived biochar at 700°C for 2 h at a heating rate of 3°C min^{-1} loaded with 30% (w/w) of ZnO nanoparticles removed about 80% of chromium(VI) from a synthetic solution [55]. Chromium(VI) was reduced to chromium(III) during the process, and Cr(III) was majorly adsorbed onto the ZnO/biochar.

Water hyacinth-derived biochar at 450°C exhibited high cation exchange capacity than those produced at 250°C, 350°C, and 550°C [31]. Moreover, the ability of biochar pyrolyzed at 450°C toward adsorbing cadmium ions was found to be 70.31 mg g^{-1}, which was quite higher than other sorbents such as bark char, activated carbon [56], and the biochar derived from dairy manure [57].

Spartina alterniflora and water hyacinth-derived biochar showed greater Cu(II) removal capacity at 600°C–700°C pyrolysis temperature and pH 6 during adsorption [58]. Table 8.1 summarizes other plant material biochar used for heavy metal removal.

Table 8.1 Biochar derived from other plant materials.

S. no.	Feedstock	Heavy metals	Pyrolysis conditions	Removal efficiency or adsorption capacity	Reference
1	Rice plant stalk (*Oryza sativa* L.), reed stalk (*Phragmites* Trin.), cassava stalk (*Manihot esculenta* Crantz)	Cd, Cr, Cu, Ni, Pb, Zn	550°C for 4 h	Cassava stalk biochar showed greater removal efficiency of Cr, Cu, and Pb. Zn was removed greater by rice stalk biochar while Cd and Ni by cassava and rice stalk biochar than reed biochar	[59]
2	*Chaenomeles sinensis* seeds	Cr(VI) and Cu(II)	300°C, 450°C, and 600°C for 2 h	Seed biochar produced at 450°C showed greater adsorption capacity for Cu(II) of 105.12 mg g^{-1} and Cr(VI) of 93.19 mg g^{-1}	[60]
3	Oil seed rape straw, *Miscanthus* straw pellet, and wheat straw pellet	Cd(II)	550°C and 700°C	Biochar of oil seed rape straw and *Miscanthus* straw pellet pyrolyzed at 550°C showed utmost removal of Cd(II)	[61]
4	Corn stalk biochar	Pb^{2+}	400°C, 450°C, 500°C, 550°C, and 600°C	Biochar generated at 450°C showed greater adsorption capacity of 49.70 mg g^{-1}	[62]

(Continued)

Table 8.1 Biochar derived from other plant materials. (*Continued*)

S. no.	Feedstock	Heavy metals	Pyrolysis conditions	Removal efficiency or adsorption capacity	Reference
5	Rice husk	Cu(II)	Produced at 500°C for 1 h and impregnated with KOH: biochar ratio of 3; activation at 800°C for 30 min	Highest adsorption capacity of Cu(II) was 265 mg g^{-1}	[63]
6	*Tectona* and *Lagerstroemia speciosa*	As(III) and As(V)	800°C for 1 h (heating rate of 10°C min^{-1})	1. As(III) was 666.7 µg g^{-1} and 454.54 µg g^{-1} for *Tectona, Lagerstroemia speciosa* biochar 2. As(V) was 1,250 µg g^{-1} and 714.28 µg g^{-1} for *Tectona, Lagerstroemia speciosa* biochar	[64]
7	Pine tree residues	Cd	300°C, 400°C, 500°C, and 600°C for 4 h	Biochar at 600°C demonstrated greater adsorption of Cd	[65]

(*Continued*)

Table 8.1 Biochar derived from other plant materials. (*Continued*)

S. no.	Feedstock	Heavy metals	Pyrolysis conditions	Removal efficiency or adsorption capacity	Reference
8	Reed (*Phragmites australis*)	Cd(II)	500°C for 2 h (heating rate of 10°C min^{-1})	Maximum adsorption capacity was 206.18 mg g^{-1} in the presence of tannic acid. Addition of tannic acid, fulvic acid, and humic acid enhanced the adsorption by 1.3%–7.4%	[66]
9	*Gingko* leaf, peanut shell, *Metasequoia* leaf	Cu and Pb	800°C for 3 h	*Gingko* leaf-derived biochar showed maximum removal capacities of both Cu and Pb	[67]
10	Rice straw	Cd^{2+} and Ni^{2+}	400°C and 700°C for 4 h (heating rate of 15°C min^{-1})	Biochar made at 700°C showed better removal of Cd^{2+} and Ni^{2+} in both single and binary metal test	[68]
11	Peanut hull	Cr(VI)	220°C for 12 h by Teflon-lined autoclave	Maximum adsorption capacity of 142.86 mg g^{-1} when amino-functionalized and magnetized	[69]

(*Continued*)

Table 8.1 Biochar derived from other plant materials. (*Continued*)

S. no.	Feedstock	Heavy metals	Pyrolysis conditions	Removal efficiency or adsorption capacity	Reference
12	Straws of sunflower, maize and wheat, sunflower straw husk, orange shell, chestnut shell	Pb^{2+}	300°C for 45 min (heating rate of 10°C min^{-1})	Maize straw showed adsorption capacity of 78.6 mg g^{-1}	[70]
13	Fallen phoenix tree leaves	Cr(VI)	500°C for 2 h	Magnetic biochar showed adsorption capacity of 27.2 mg g^{-1}	[71]
14	*Eucalyptus* sawdust	Cr(VI)	160°C, 190°C, 220°C, and 250°C for 1 h	0.05 N KOH-modified biochar (220°C) demonstrated adsorption capacity of 45.88 mg g^{-1}	[72]
15	Cones of larch and spruce	Cd^{2+}, Pb^{2+}, Co^{2+}	500°C and 600°C for 5, 10, and 15 min	The optimum conditions for pyrolysis of conifers cones were found to be 500°C for 5 min	[73]
16	Rambutan peel	Cu(II)	300°C, 400°C, 500°C, 600°C, and 700°C for 3 h (heating rate of 5°C min^{-1})	Biochar derived at 600°C was found effective	[74]

(*Continued*)

Table 8.1 Biochar derived from other plant materials. (*Continued*)

S. no.	Feedstock	Heavy metals	Pyrolysis conditions	Removal efficiency or adsorption capacity	Reference
17	Rice straw	Cu and Zn	600°C for 4 h	1. In single metal, 56.5 and 38.6 mg g^{-1} for Cu and Zn 2. In binary metal adsorption, 40.2 and 7.9 mg g^{-1} for Cu and Zn	[75]
18	Rape straw	Cd(II)	600°C for 1 h	$KMnO_4$-modified biochar showed highest adsorption rate, which is 14.46 g $(mg{\cdot}h)^{-1}$	[76]

8.5.2 Heavy Metal Removal Using Biosorbent-Immobilized Alginate Beads

Immobilizing biomass or biochar into alginate beads provides control in the size, regeneration, and mechanical strength of the adsorbent. Biomass powder of water hyacinth was immobilized to study the Cu(II) removal. Metal removal capacity was found to be 61 mg g^{-1} with the biomass loading of 0.5 mg L^{-1} [23]. Water hyacinth (WH) biomass powder of 2 g was immobilized into alginate beads and found to have a constant removal efficiency of Cu(II) for the five cycles of adsorption and desorption in a column experiment [10]. Bark powder of *Hemidesmus indicus*, also known as Indian sarsaparilla, was immobilized to study the biosorption of lead and other metal ions such as zinc and chromium by both batch experiment and column experiment [22]. Batch experiments revealed various parameters such as metal-bead contact period, bead biomass loading, and initial concentration of metal and co-metal ions. Column experiments determined the retention capacity of beads for Pb ions, regeneration and reuse of beads, and elution of metal ions. This study showed that immobilized biomass of *Hemidesmus indicus* was a potential biosorbent and could be regenerated. Biochar beads and magnetically modified biochar beads derived from water hyacinth were studied to compare the metal removal efficiency of Cd [40]. Magnetically modified biochar beads showed greater efficiency in removing Cd(II) than biochar alginate beads, and the adsorption efficiency was found to be 49 mg g^{-1}. Immobilized *Moringa oleifera* seed powder was used to check for its ability to remove Cu^{2+}, Zn^{2+}, Co^{2+}, Mn^{2+}, and Ni^{2+} from electroplating wastewater in both batch and continuous column mode [77].

8.6 A Comparison of Techniques for Removing Heavy Metals

The disadvantages of the conventional methods are inefficiencies in removing heavy metals, difficulties in sludge disposal, and being expensive [78]. The methods involving ion exchange, adsorption, and membrane filtration are commonly used for treating heavy metal-containing water. However, the ion exchange method is expensive because the regeneration of resins would cost higher and the exhausted resins may cause secondary pollution, and the membrane filtration has some disadvantages such as membrane fouling, process complexity, and being expensive, whereas adsorption shows better efficiency in treating wastewater containing even low concentrations of heavy metals [79]. Adsorption using biological feedstock called biosorption has been demonstrated to remove heavy metals as it overcomes the abovementioned disadvantages of the other treatment methods [16]. However, regeneration is difficult when biochar is used as adsorbents. Therefore, biochar may be immobilized in alginate beads and then the treatment process may be optimized [23].

8.7 Conclusion

The impact of industrialization has resulted in significant heavy metal pollution, posing a threat to both the environment and human well-being. To meet the standard regulation, many techniques have been developed. Adsorption is the efficient method among the conventional methods, and biosorption using natural sorbent such as plant-derived biochar has been proven to show efficient removal of heavy metals. These plant-derived products are eco-friendly, and therefore, disposal of biosorbents will not be a problem to the environment.

References

1. He, Z.L., Yang, X.E., Stoffella, P.J., Trace elements in agro ecosystems and impacts on the environment. *J. Trace Elem. Med. Biol.*, 19, 125, 2005.
2. Srivastava, S., Agrawal, S.B., Mondal, M.K., A review on progress of heavy metal removal using adsorbents of microbial and plant origin. *Environ. Sci. Pollut. Res.*, 22, 15386, 2015.
3. Southichak, B., Nakano, K., Nomura, M., Chiba, N., Nishimura, O., *Phragmites australis*: A novel biosorbent for the removal of heavy metals from aqueous solution. *Water Res.*, 40, 2295, 2006.
4. Fu, F., Xie, L., Tang, B., Wang, Q., Jiang, S., Application of a novel strategy-Advanced Fenton-chemical precipitation to the treatment of strong stability chelated heavy metal containing wastewater. *Chem. Eng. J.*, 189–190, 283, 2012.
5. Johnson, P.D., Girinathannair, P., Ohlinger, K.N., Ritchie, S., Teuber, L., Kirby, J., Enhanced removal of heavy metals in primary treatment using coagulation and flocculation. *Water Environ. Res.*, 80, 472, 2008.
6. Torma, C.Z. and Cséfalvay, E., Nanofiltration and electrodialysis: Alternatives in heavy metal containing high salinity process water treatment. *Chem. Pap.*, 72, 1115, 2017.
7. Trochimczuk, W.A., New ion-exchange/coordination resins with carboxylate and phosphate functional groups. *Eur. Polym. J.*, 35, 1457, 1999.

8. Ahluwalia, S.S. and Goyal, D., Microbial and plant derived biomass for removal of heavy metals from wastewater. *Bioresour. Technol.*, 98, 2243, 2007.
9. Crini, G., Recent developments in polysaccharide-based materials used as adsorbents in wastewater treatment. *Prog. Polym. Sci.*, 30, 38, 2005.
10. Mahamadi, C. and Zambara, P., High Cu removal from water using water hyacinth fixed on alginate. *Environ. Chem. Lett.*, 11, 377, 2013.
11. Hu, M.Z.-C. and Reeves, M., Biosorption of uranium by *Pseudomonas aeruginosa* strain CSU immobilized in a novel matrix. *Biotechnol. Prog.*, 13, 60, 1997.
12. Ku, Y. and Jung, I.L., Photocatalytic reduction of Cr (VI) in aqueous solutions by UV irradiation with the presence of titanium dioxide. *Water Res.*, 35, 135, 2001.
13. Zazouli, M.A. and Yousefi, Z., Removal of heavy metals from solid wastes leachates coagulation-flocculation process. *J. Appl. Sci.*, 8, 2142, 2008.
14. Akbal, F. and Camcı, S., Comparison of electrocoagulation and chemical coagulation for heavy metal removal. *Chem. Eng. Technol.*, 33, 1655, 2010.
15. Maximous, N.N., Nakhla, G.F., Wan, W.K., Removal of heavy metals from wastewater by adsorption and membrane processes: A comparative study. *World Acad. Sci. Eng. Technol.*, 40, 599, 2010.
16. Bashir, A., Malik, L.A., Ahad, S., Manzoor, T., Bhat, M.A., Dar, G.N., Pandith, A.H., Removal of heavy metal ions from aqueous system by ion-exchange and biosorption methods. *Environ. Chem. Lett.*, 17, 729, 2019.
17. Tofighy, M.A. and Mohammadi, T., Adsorption of divalent heavy metal ions from water using carbon nanotube sheets. *J. Hazard. Mater.*, 185, 140, 2011.
18. Visa, M., Bogatu, C., Duta, A., Simultaneous adsorption of dyes and heavy metals from multicomponent solutions using fly ash. *Appl. Surf. Sci.*, 256, 5486, 2010.
19. Hong, M., Yu, L., Wang, Y., Zhang, J., Chen, Z., Dong, L., Zan, Q., Li, R., Heavy metal adsorption with zeolites: The role of hierarchical pore architecture. *Chem. Eng. J.*, 359, 363, 2019.
20. Banerjee, S., Gothalwal, R., Sahu, P.K., Sao, S., Microbial observation in bioaccumulation of heavy metals from the ash dyke of thermal power plants of Chhattisgarh, India. *Adv. Biosci. Biotechnol.*, 6, 131, 2015.
21. Unnikannan, P., Sundaramoorthy, P., Baskaran, L., Ganesh, K.S., Chidambaram, A.L.A., Assessment of chromium phytotoxicity in some aquatic weeds. *Bot. Res. Int.*, 4, 13, 2011.
22. Sekhar, K.C., Kamala, C.T., Chary, N.S., Sastry, A.R.K., Rao, T.N., Vairamani, M., Removal of lead from aqueous solutions using an immobilized biomaterial derived from a plant biomass. *J. Hazard. Mater.*, 108, 111, 2004.
23. Mahamadi, C. and Zambara, P., Adsorption of Cu (II) from aquatic systems using alginate-immobilized water hyacinth beads. *Eur. J. Sci. Res.*, 71, 581, 2012.
24. Park, D., Yun, Y.-S., Park, J.M., The past, present, and future trends of biosorption. Biotechnol. *Bioprocess Eng.*, 15, 86, 2010.

25. Ranjitha, J., Raj, A., Kashyap, R., Vijayalakshmi, S., Donatus, M., Removal of heavy metals from industrial effluent using *Salvinia molesta*. *Int. J. Chem. Tech Res.*, 9, 608, 2016.
26. Sudarsan, J., Annadurai, R., Kumar, K.S., Nithiyanantham, S., Heavy metal removal using different parts of *Typha latifolia*. *J. Bionanoscience*, 12, 87, 2018.
27. Liao, S.-W. and Chang, W.-L., Heavy metal phytoremediation by water hyacinth at constructed wetlands in Taiwan. *J. Aquat. Plant Manage.*, 42, 60, 2004.
28. Patra, J.M., Panda, S.S., Dhal, N.K., Biochar as a low-cost adsorbent for heavy metal removal: A review. *Int. J. Res. Biosci.*, 6, 1, 2017.
29. Tang, J., Zhu, W., Kookana, R., Katayama, A., Characteristics of biochar and its application in remediation of contaminated soil. *J. Biosci. Bioeng.*, 116, 653, 2013.
30. Avello Bioenergy, Biochar, 2021. http://www.avellobioenergy.com/en/products/biochar/ (accessed 6.14.21).
31. Zhang, F., Wang, X., Yin, D., Peng, B., Tan, C., Liu, Y., Tan, X., Wu, S., Efficiency and mechanisms of Cd removal from aqueous solution by biochar derived from water hyacinth (*Eichornia crassipes*). *J. Environ. Manage.*, 153, 68, 2015.
32. Yang, Y., Wei, Z., Zhang, X., Chen, X., Yue, D., Yin, Q., Xiao, L., Yang, L., Biochar from *Alternanthera philoxeroides* could remove Pb (II) efficiently. *Bioresour. Technol.*, 171, 227, 2014.
33. Kanamarlapudi, S., Chintalpudi, V.K., Muddada, S., Application of biosorption for removal of heavy metals from wastewater, in: *Biosorption*, vol. 18, p. 69, 2018.
34. Yan, G. and Viraraghavan, T., Heavy metal removal in a biosorption column by immobilized M. rouxii biomass. *Bioresour. Technol.*, 78, 243, 2001.
35. Górecka, E. and Jastrzębska, M., Immobilization techniques and biopolymer carriers. *Biotechnol. Food Sci.*, 75, 65, 2011.
36. Hu, M.Z.-C. and Reeves, M., Biosorption of uranium by *Pseudomonas aeruginosa* strain CSU immobilized in a novel matrix. *Biotechnol. Prog.*, 13, 60, 1997.
37. Sekomo, C.B., Rousseau, D.P., Saleh, S.A., Lens, P.N., Heavy metal removal in duckweed and algae ponds as a polishing step for textile wastewater treatment. *Ecol. Eng.*, 44, 102, 2012.
38. Sudarsan, J., Annadurai, R., Kumar, K.S., Nithiyanantham, S., Heavy metal removal using different parts of *Typha latifolia*. *J. Bionanosci.*, 12, 87, 2018.
39. Sudharsan, S., Seedevi, P., Ramasamy, P., Subhapradha, N., Vairamani, S., Shanmugam, A., Heavy metal accumulation in seaweeds and sea grasses along southeast coast of India. *J. Chem. Pharm. Res.*, 4, 4240, 2012.
40. Liu, C., Ye, J., Lin, Y., Wu, J., Price, G.W., Burton, D., Wang, Y., Removal of Cadmium (II) using water hyacinth (*Eichhornia crassipes*) biochar alginate beads in aqueous solutions. *Environ. Pollut.*, 264, 114785, 2020.

41. Xia, H., Kong, W., Liu, L., Lin, K., Li, H., Effects of harvest time and desalination of feedstock on *Spartina alterniflora* biochar and its efficiency for Cd2+ removal from aqueous solution. *Ecotoxicol. Environ. Saf.*, 207, 111309, 2021.
42. Lian, W., Yang, L., Joseph, S., Shi, W., Bian, R., Zheng, J., Li, L., Shan, S., Pan, G., Utilization of biochar produced from invasive plant species to efficiently adsorb Cd (II) and Pb (II). *Bioresour. Technol.*, 317, 124011, 2020.
43. Kaewsichan, L. and Tohdee, K., Adsorption of hexavalent chromium onto alkali-modified biochar derived from *Lepironia articulata*: A kinetic, equilibrium, and thermodynamic study. *Water Environ. Res.*, 91, 1433, 2019.
44. Zhou, N., Zu, J., Feng, Q., Chen, H., Li, J., Zhong, M., Zhou, Z., Zhuang, S., Effect of pyrolysis condition on the adsorption mechanism of heavy metals on tobacco stem biochar in competitive mode. *Environ. Sci. Pollut. Res.*, 26, 26947, 2019.
45. Wang, C., Wang, H., Cao, Y., Pb (II) sorption by biochar derived from *Cinnamomum camphora* and its improvement with ultrasound-assisted alkali activation. *Colloids Surf. Physicochem. Eng. Asp.*, 556, 177, 2018.
46. Liu, G., Huang, Y., Xu, L., Biochar from Chinese herb residues as adsorbent for toxic metals removal. *IOP Conf. Ser. Earth Environ. Sci.*, 61, 2017, DOI 10.1088/1755-1315/61/1/012147 012147.
47. Wang, S., Gao, B., Li, Y., Ok, Y.S., Shen, C., Xue, S., Biochar provides a safe and value-added solution for hyperaccumulating plant disposal: A case study of *Phytolacca acinosa Roxb.* (phytolaccaceae). *Chemosphere*, 178, 59, 2017.
48. Goswami, R., Shim, J., Deka, S., Kumari, D., Kataki, R., Kumar, M., Characterization of cadmium removal from aqueous solution by biochar produced from *Ipomoea fistulosa* at different pyrolytic temperatures. *Ecol. Eng.*, 97, 444, 2016.
49. Cui, X., Fang, S., Yao, Y., Li, T., Ni, Q., Yang, X., He, Z., Potential mechanisms of cadmium removal from aqueous solution by *Canna indica* derived biochar. *Sci. Total Environ.*, 562, 517, 2016.
50. Kim, W.-K., Shim, T., Kim, Y.-S., Hyun, S., Ryu, C., Park, Y.-K., Jung, J., Characterization of cadmium removal from aqueous solution by biochar produced from a giant miscanthus at different pyrolytic temperatures. *Bioresour. Technol.*, 138, 266, 2013.
51. Khangarot, B.S., Ray, P.K., Chandra, H., *Daphnia magna* as a model to assess heavy metal toxicity: Comparative assessment with mouse system. *Acta Hydrochim. Hydrobiol.*, 15, 427, 1987.
52. Lin, S., Huang, W., Yang, H., Sun, S., Yu, J., Recycling application of waste long-root *Eichhornia crassipes* in the heavy metal removal using oxidized biochar derived as adsorbents. *Bioresour. Technol.*, 314, 123749, 2020.
53. Li, Q., Tang, L., Hu, J., Jiang, M., Shi, X., Zhang, T., Li, Y., Pan, X., Removal of toxic metals from aqueous solution by biochars derived from long-root *Eichhornia crassipes*. *R. Soc. Open Sci.*, 5, 180966, 2018.
54. Shen, Y., Zhu, W., Li, H., Ho, S.-H., Chen, J., Xie, Y., Shi, X., Enhancing cadmium bioremediation by a complex of water-hyacinth derived pellets immobilized with *Chlorella* sp. *Bioresour. Technol.*, 257, 157, 2018.

55. Yu, J., Jiang, C., Guan, Q., Ning, P., Gu, J., Chen, Q., Zhang, J., Miao, R., Enhanced removal of Cr (VI) from aqueous solution by supported ZnO nanoparticles on biochar derived from waste water hyacinth. *Chemosphere*, 195, 632, 2018.
56. Mohan, D., Pittman Jr., C.U., Bricka, M., Smith, F., Yancey, B., Mohammad, J., Steele, P.H., Alexandre-Franco, M.F., Gómez-Serrano, V., Gong, H., Sorption of arsenic, cadmium, and lead by chars produced from fast pyrolysis of wood and bark during bio-oil production. *J. Colloid Interface Sci.*, 310, 57, 2007.
57. Xu, X., Cao, X., Zhao, L., Wang, H., Yu, H., Gao, B., Removal of Cu, Zn, and Cd from aqueous solutions by the dairy manure-derived biochar. *Environ. Sci. Pollut. Res.*, 20, 358, 2013.
58. Li, M., Lou, Z., Wang, Y., Liu, Q., Zhang, Y., Zhou, J., Qian, G., Alkali and alkaline earth metallic (AAEM) species leaching and Cu (II) sorption by biochar. *Chemosphere*, 119, 778, 2015.
59. Lu, N., Han, J.C., Wei, Y., Li, G., Sun, Y.Y., The heavy metal adsorption capacity of stalk biochar in an aqueous phase. *Appl. Ecol. Environ. Res.*, 18, 2569, 2020.
60. Hu, X., Song, J., Wang, H., Zhang, W., Wang, B., Lyu, W., Wang, Q., Liu, P., Chen, L., Xing, J., Adsorption of Cr(VI) and Cu(II) from aqueous solutions by biochar derived from *Chaenomeles sinensis* seed. *Water Sci. Technol.*, 80, 2260, 2019.
61. Lam, Y.Y., Lau, S.S.S., Wong, J.W.C., Removal of Cd(II) from aqueous solutions using plant-derived biochar: Kinetics, isotherm and characterization. *Bioresour. Technol. Rep.*, 100323, 8, 2019, https://doi.org/10.1016/j.biteb.2019.100323.
62. Liu, L., Huang, Y., Zhang, S., Gong, Y., Su, Y., Cao, J., Hu, H., Adsorption characteristics and mechanism of Pb(II) by agricultural waste-derived biochars produced from a pilot-scale pyrolysis system. *Waste Manage.*, 100, 287, 2019.
63. Cuong, D.V., Liu, N.-L., Nguyen, V.A., Hou, C.-H., Meso/micropore-controlled hierarchical porous carbon derived from activated biochar as a high-performance adsorbent for copper removal. *Sci. Total Environ.*, 692, 844, 2019.
64. Verma, L. and Singh, J., Synthesis of novel biochar from waste plant litter biomass for the removal of arsenic (III and V) from aqueous solution: A mechanism characterization, kinetics and thermodynamics. *J. Environ. Manage.*, 248, 109235, 2019.
65. Park, J.-H., Wang, J.J., Kim, S.-H., Kang, S.-W., Jeong, C.Y., Jeon, J.-R., Park, K.H., Cho, J.-S., Delaune, R.D., Seo, D.-C., Cadmium adsorption characteristics of biochars derived using various pine tree residues and pyrolysis temperatures. *J. Colloid Interface Sci.*, 553, 298, 2019.
66. Wang, Y., Li, Y., Zhang, Y., Wei, W., Effects of macromolecular humic/fulvic acid on Cd(II)adsorption onto reed-derived biochar as compared with tannic acid. *Int. J. Biol. Macromol.*, 134, 43, 2019.

67. Lee, M.-E., Park, J.H., Chung, J.W., Comparison of the lead and copper adsorption capacities of plant source materials and their biochars. *J. Environ. Manage.*, 236, 118, 2019.
68. Deng, Y., Huang, S., Laird, D.A., Wang, X., Meng, Z., Adsorption behaviour and mechanisms of cadmium and nickel on rice straw biochars in single- and binary-metal systems. *Chemosphere*, 218, 308, 2019.
69. Cai, W., Wei, J., Li, Z., Liu, Y., Zhou, J., Han, B., Preparation of amino-functionalized magnetic biochar with excellent adsorption performance for Cr(VI) by a mild one-step hydrothermal method from peanut hull. *Colloids Surf. Physicochem. Eng. Asp.*, 563, 102, 2019.
70. Qiao, Y., He, C., Zhang, C., Jiang, C., Yi, K., Li, F., Comparison of adsorption of biochar from agricultural wastes on methylene blue and Pb2+. *BioResources*, 14, 9766, 2019.
71. Shi, S., Yang, J., Liang, S., Li, M., Gan, Q., Xiao, K., Hu, J., Enhanced Cr (VI) removal from acidic solutions using biochar modified by Fe_3O_4@ SiO_2-NH_2 particles. *Sci. Total Environ.*, 628, 499, 2018.
72. Zhang, X., Zhang, L., Li, A., Eucalyptus sawdust derived biochar generated by combining the hydrothermal carbonization and low concentration KOH modification for hexavalent chromium removal. *J. Environ. Manage.*, 206, 989, 2018.
73. Saletnik, B., Zaguła, G., Grabek-Lejko, D., Kasprzyk, I., Bajcar, M., Czernicka, M., Puchalski, C., Biosorption of cadmium(II), lead(II) and cobalt(II) from aqueous solution by biochar from cones of larch (*Larix decidua* Mill. subsp. decidua) and spruce (*Picea abies* L. H. Karst). *Environ. Earth Sci.*, 574, 76, 2017, https://doi.org/10.1007/s12665-017-6916-y.
74. Selvanathan, M., Yann, K.T., Chung, C.H., Selvarajoo, A., Arumugasamy, S.K., Sethu, V., Adsorption of copper(II) ion from aqueous solution using biochar derived from rambutan (*Nephelium lappaceum*) peel: Feedforward neural network modelling study. *Water Air Soil Pollut.*, 299, 228, 2017, https://doi.org/10.1007/s11270-017-3472-8.
75. Park, J.-H., Wang, J.J., Kim, S.-H., Cho, J.-S., Kang, S.-W., Delaune, R.D., Han, K.-J., Seo, D.-C., Recycling of rice straw through pyrolysis and its adsorption behaviors for Cu and Zn ions in aqueous solution. *Colloids Surf. Physicochem. Eng. Asp.*, 533, 330, 2017.
76. Li, B., Yang, L., Wang, C.-Q., Zhang, Q.-P., Liu, Q.-C., Li, Y.-D., Xiao, R., Adsorption of Cd(II) from aqueous solutions by rape straw biochar derived from different modification processes. *Chemosphere*, 175, 332, 2017.
77. Radhakrishnan, K., Sethuraman, L., Panjanathan, R., Natarajan, A., Solaiappan, V., Thilagaraj, W.R., Biosorption of heavy metals from actual electroplating wastewater using encapsulated *Moringa oleifera* beads in fixed bed column. *Desalination Water Treat.*, 57, 3572, 2016.
78. Hegazi, H.A., Removal of heavy metals from wastewater using agricultural and industrial wastes as adsorbents. *HBRC J.*, 9, 276, 2013.
79. Fu, F. and Wang, Q., Removal of heavy metal ions from wastewaters: A review. *J. Environ. Manage.*, 92, 407, 2011.

9

Sustainability in Manufacturing: Welding's Role as a Frontier

P. Arunkumar, N. Muthukumaran, K. S. Ramaneedharan, N. S. Mithun, B. Sanjay, K. Solaiyappan, S. Gokul and B. Arulmurugan*

KPR Institute of Engineering and Technology, Coimbatore, India

Abstract

Sustainable manufacturing considers environmental, economic, and energy implications. It is a philosophical and environmental movement that aims to protect and enhance natural systems and resources over the long-term and to promote social and economic practices that uphold human well-being within natural systems. To achieve sustainability, it is often important to reduce our dependency on resource consumption and to alter the way we use resources, particularly lessening our impact on the environment. Welding is a technique used in the process of combining two or more pieces of metal, and it can be recycled to create a longer life span, making it a sustainable process. In light of the fact that various welding procedures produce a variety of contaminants in varying amounts and forms, the choice of technique can have an effect on the environmental impact of the welding. Sustainability can be achieved through material selection, welding method, and heat input values. These parameters were measured in terms of physical performance, economic impact, social impact, and environmental impact. Welding techniques reviewed in this survey are shielded metal gas welding (SMAW), TIG or gas tungsten arc welding (GTAW), MIG or gas metal arc welding (GMAW), submerged arc welding (SAW), friction stir welding (FSW), and laser beam welding (LBW). These welding techniques are discussed in terms of energy savings, material waste, resources and parameters, environmental benefits, and cost savings. It is possible to accomplish sustainability in the manufacturing sector through the use of automation techniques, virtual training, environmentally friendly equipment, alternative welding methods, usage of renewable resources, and the recycling of scrap. Since global warming is increasing at a distressing rate, the entire world is getting ready for improvements by means of using sustainable methods in

**Corresponding author*: enggarul@gmail.com

S. Thirumalai Kumaran and Tae Jo Ko (eds.) Sustainable Machining and Green Manufacturing, (177–192) © 2024 Scrivener Publishing LLC

all areas of engineering. The world is developing toward a green economy, which aspires the exact purpose of sustainable development without degrading the environment. Sustainability is essential for achieving this goal. Welding is critical to economic progress, infrastructure development, and other manufacturing sectors, so it is important to adapt to sustainable welding to move toward a green economy.

***Keywords*:** Sustainability, welding techniques, green welding, economy

9.1 Introduction

Manufacturing is creating finished goods from raw materials with the usage of a variety of tools and machinery. Manufacturing is an important part of every business, so companies must have the right manufacturing system and the best tools for the job. Due to the increase in the population of the world, manufacturing processes are becoming very difficult to manage; hence, the processes must be sustainable in the long run. Sustainability is a long-term systematic approach to managing natural resources and social and economic systems so that they are capable of meeting the needs of present and future generations without compromising the ability of natural systems to regenerate or sustain positive populations of indigenous species. Sustainability is important in any field, including welding, because of its necessary role in preserving resources and protecting the environment. The increasing prevalence of welding technologies has necessitated an ever-increasing need for sustainable practices, particularly with regard to reducing carbon dioxide emissions and improving the efficiency of materials used. There are a few factors used to assess the sustainability of a product. They are known as sustainability factors, and these factors ensure the sustainability of the manufactured goods. The sustainability factor is a measure of how the product influences the economy of a nation, environmental performance, physical strength, and social influence. The economical effect refers to the value that the product gives to the economy in the long-run. The value of the product is required to be held for as long as possible by itself. Environmental performance is very vital when talking about the sustainability value of a product, since the environment is an unavoidable factor in our lives and the product must not harm valuable natural resources including the health of fellow human beings; hence, it is important that it satisfies the standards that are set to improve sustainability. Physical strength means that the product must withstand its purpose while fulfilling other sustainable goals. Physical sustainability also means the ability to sustain the functionality, production, performance, and/or life cycle of machines or equipment over an extended period of time. Social influence means that it should be capable of creating awareness among the society to create a market around the component. All of these sustainability factors must be considered to determine whether the manufactured product can be called a sustainable product. Sustainability can be used in welding operations in a way such that welders can use renewable materials in their welding operations, such as aluminum, steel, and other metals that can be reused or recycled. This reduces the number of natural resources used for welding and can help reduce emissions from welding operations. In a general note, the welders can use more efficient welding techniques, such as purchasing welding equipment that uses less electricity or opting for higher-efficiency welding processes that use less energy. They can also develop and use more efficient techniques, such as developing techniques that include a specific area of the weld that is preheated before welding, which can help to reduce spatter and consumption of power. We must also approach the operations with a much more sustainable mindset, opting to use welding processes that are less resource-intensive and have less environmental impact, such as cold welding or submerged arc welding. By adopting a more sustainable approach to welding operations, welders can help reduce their operational impact and ensure that their welding processes are more responsible and sustainable. To maintain sustainability in manufacturing, it is important to ensure efficient use of energy and resources by utilizing effective practices such as eliminating waste, optimizing power settings, and using renewable energy sources.

Saad *et al.* [1] aimed to develop an overall agenda for the sustainability assessment of important manufacturing processes and propose efficient ways to ensure the accuracy of the outcomes. The ways that were deemed to be more efficient were based on three dimensions: economic, social, and environmental aspects. Mani *et al.* [2] presented a review of various manufacturing practices based on life cycle assessment (LCA), and reliable software was used to make data models. They stated that sustainability characterization was very crucial in problem-solving processes. Sustainable manufacturing procedures are used in the industry to reduce harmful emissions from manufacturing and welding. Sustainability factors to be considered are shown Figure 9.1.

These implications affect the human ecosystem, human health, and work efficiency, thus affecting the environment and economy. Sustainable welding is a subset of green manufacturing that aims to create a high-quality

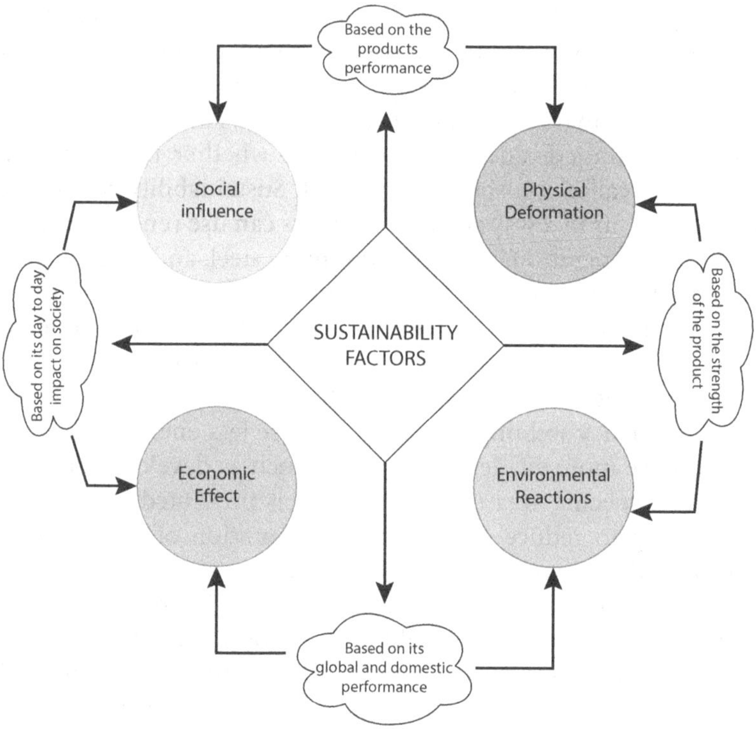

Figure 9.1 Sustainability factors.

weld with a pleasing appearance using as little resources as possible, including filler metals and shielding gas, while reducing energy usage. In addition, it boosted productivity and efficiency while decreasing weld joint costs. Thus, an industrial manufacturing process that involves sustainability analysis is essential to measure and reduce their negative effects. Sustainability impact categories are shown in Figure 9.2. Welding plays an important part in modern manufacturing, since it is utilized in numerous industries and applications, including the automobile industry, the construction industry, the shipbuilding industry, and turbine manufacture, among others. However, welding operations have a negative influence on the environment and the particular reason for the circumstance is that they require substantial amounts of resources [3]. In addition, the greenhouse gases (GHG) and fumes emitted by welding procedures are substantial and can create major health problems for welders. Sustainability issues in different welding techniques are discussed in detail.

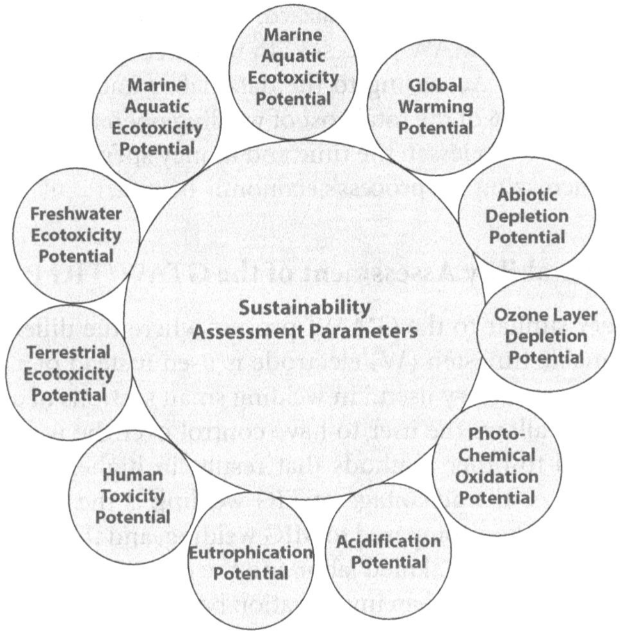

Figure 9.2 Sustainability impact categories.

9.2 Sustainability Assessment in Welding

9.2.1 Sustainability Assessment of the SMAW Process

Shielded metal arc welding is a labor-intensive welding process. It has an electrode roofed with flux that creates an arc using an external power supply. SMAW is also known as the manual metal arc welding process, and it is one of the most common welding processes used very widely around the world. SMAW has greater environmental implications in terms of acidification, global warming potential, and photochemical ozone generation for a 1-meter weld seam than the other welding procedures. The economic value of this type of welding has taken a downward turn due to improvements and development of new welding techniques that have way less effect on the environment and cost-efficient [4]. Electrical energy requirement, material consumption, fumes emission, molten slag formation, and dangerous working circumstances related to human health and occupational safety all contribute to the SMAW process's negative impact on the environment. The inhalation of welding fumes is a big risk for the welder when using SMAW. Proper planning and designing of the SMAW process can greatly reduce the dangers of this hazard. Alkahla and Pervaiz [5] investigated the SMAW or MMAW process with the three-level bottom-line sustainability approach. According to the data, labor and another overhead account for 80%–85% of the total cost of welding operations. Diverse optimization strategies can lessen the time and money spent on SMAW operations, hence increasing the process's economic benefit.

9.2.2 Sustainability Assessment of the GTAW/TIG Process

GTAW is very similar to the GMAW process where the difference is that a nonconsumable tungsten (W) electrode is used instead of a metal electrode. This process is very useful in welding small sections of a workpiece. In addition, this allows the user to have control over the welding process when compared to other methods that results in higher strength of the welded workpiece. The advantage of TIG welding is the weld is stronger and more precise when compared to MIG welding, and the disadvantage of TIG welding is it requires skilled labor and the cost of the welding is high.

Saad *et al.* [6] performed an investigation between various welding processes, and they revealed that GTAW was the lowest ranked on the basis of TOPSIS and GRA processes. Favi *et al.* [7] conducted a research on the environmental performances of various metal arc welding processes, and the results obtained showed that when compared to GMAW, GTAW had a higher environmental pollution rate that reduced its rankings. It is possible to weld using different types of GTAW, such as pulsed current gas tungsten arc welding (PCGTAW) and K-TIG welding [8–10]. These types of welding require less heat input and do not require multiple passes, both of which directly improve the sustainability of the process. In activated tungsten inert gas welding, the amount of time and resources that would have been spent on edge preparation is eliminated. Additionally, a single pass gives full penetration instead of the multiple passes that are necessary in regular TIG and MIG welding. This eliminates the need to employ filler materials and results in a substantial decline in the amount of shielding gas and energy consumed. Automated GTAW like TOPTIG and TIPTIG can also be used to improve the sustainability [11].

9.2.3 Sustainability Assessment of the MIG/GMAW Process

GMAW or MIG welding is a process in which an arc is formed between the workpiece and electrode in the presence of shielding gas. In gas metal arc welding, the electrode used here is consumable electrode and both the electrode and shielding gas are nourished through the welding torch. The main advantage of MIG welding is that welding is very less time-consuming. The disadvantage of MIG welding is that it is not very precise. Alkahla *et al.* [5] examined the microstructure and strength of weld joints made by cold wire - GMAW and compared it with the conventional GMAW process and stated that weld distortion was lower and the volume of filler metal was increased. Authors used the Weibull distribution to estimate fatigue life and found that CW-GMAW joints had a longer fatigue life with 99% reliability, especially as stress amplitude increased. Saad *et al.* [6] evaluated the welding processes by adopting a comprehensive model of sustainability, and they declared that the GMAW process had the lowermost score based on the COPRAS technique. In order to examine the environmental and social effects of various welding techniques, Chang *et al.* [12] investigated the sustainability of the GMAW process by applying the life cycle assessment and determined that GMAW had higher global impacts and led to health risks of workers. Favi *et al.* [7] revealed that GMAW presented the least environmental impact and performed very well among other processes. The GMAW process has the least impact because it uses the least amount of energy. Also, the most common shape for MIG welding is the U bevel configuration, which reduces the amount of filler material needed.

9.2.4 Sustainability Assessment of the SAW Process

The submerged arc welding process uses continuous electrode to form an arc in the workpiece. One of the major differences in the SAW process is that it uses a blanket of flux that is in powder form to create a protective shield. It is mainly suited for long and incessant weld in the workpiece. The mechanical properties majorly depend on the size of circulation of the flux. SAW can be automated by using the length of the weld, since it provides continuous electrode feed. Vimal *et al.* [13] proposed employing a complete factorial design for process parameter optimization for SAW, with the goal of reducing the environmental problems associated with SAW. They suggested adjusting the current, voltage, and welding characterization when using recycled slag. Nimkeer *et al.* [14] suggested to utilize SAW as a process to provide fresh flux which in turn improves sustainability. In the case of welding LPG valve bosses, Chucheep *et al.* [15] evaluated the GMAW and SAW welding techniques for quality and sustainability. Environmental and social impacts from GMAW welding told that the emissions and noise are greater than those from SAW welding. Slag, or solid waste, is produced during the SAW process and must be handled properly. Drakopoulos *et al.* [16] modeled and compared several kinds of welding procedures using SimaPro software. Results show that in terms of the three welding methods, flux core arc welding (FCAW) had the greatest impact on the environment, while SAW shows the lowest impact.

9.2.5 Sustainability Assessment of the FSW Process

Friction stir welding (FSW) is a solid-state welding technique that has a tool-pin arrangement used for joining workpieces without the need for heat. Frictional heat is produced by the rotation of the tool-pin over the workpiece. It has a nonconsumable electrode that works without melting the job. One of the major drawbacks of FSW is that solidification and liquid cracking are very much higher when compared with other welding techniques. A lot of parameters need to be considered in FSW like the rotational speed of the tool, temperature of the job, and melting points of both the tool and workpiece. The melting points of the tool and workpiece are a very crucial factor because they might melt the workpiece and thus spoil the efforts of the welder. But as a solid-state welding process, it has its own advantages, which include less pollution and the mechanical composition is guarded. Saad *et al.* [6] experimented with the sustainability examination of various welding processes by using a sustainable framework. Authors found that FSW was the most sustainable process due to not having the use for an electric arc, it had the highest ranking among other welding processes. The environmental effects of GMAW and FSW were calculated using the LCA method by Shrivastava *et al.* [17]. The authors found that when compared to GMAW, joining with FSW uses 42% less energy and roughly 10% less material when the same tensile force is required in the design. This actively demonstrates that FSW produces about 31% fewer greenhouse gas emissions than GMAW. The overall decrease in energy use can be attributed to lower energy usage and the involved pre- and post-processes during FSW. The mechanical characteristics and gaseous emissions of FSW and GMAW of 3-mm-thick Al-1030 were measured by Dawood *et al.* [18]. When compared to FSW, GMAW's CO and CO_2 emissions were around 3.7% and 1.6% higher, respectively. It was determined that when welding aluminum of the same thickness, FSW has better welding qualities than GMAW and is thus more environmentally friendly.

Donatus *et al.* [19] conducted a research on the corrosion taking place in aluminum when friction welding is employed. He observed the microstructure modification and the corrosion mitigation that directly increased the service life. Shen *et al.* [20] examined the microstructures of 5083/6082 friction stir welding (FSW) weld and found improvement in corrosion resistance. Liu and Wu *et al.* [21] examined how ultrasonic vibration affects the weld microstructure and the mechanical properties of 2024Al-T4 joints and attempted to explain UVFSW influence. At the same welding conditions, mechanical properties are enhanced. Therefore, the addition of ultrasonic vibration to traditional FSW helps contribute to its sustainability. Liu and Wu [22] found that UVFSW can eliminate the tunnel defect that is a major constraint in FSW. Kumaran *et al.* [23] did a comparative study between TIG and novel FWTPET (friction welding of tube-to-tube plate using an external tool) process. The authors found that the FWTPET process is capable of generating robust weld joints with lesser energy consumption. Wang *et al.* [24] examined the softening issue in 5083Al-H19 plates made with friction stir welding with and without additional water cooling. The authors found that, at a low rotation rate, additional water cooling effectively controls the softening issue. Majeed *et al.* [25] discovered that FSW is unsuitable for welding high-melting point materials. This requires tool materials with a high hot-hardness and puts the sustainability to the test. Life cycle assessment has been applied by Titmarsh *et al.* [26] to an aluminum and steel friction stir welding process. Results showed that FSW is the most environmentally friendly welding method, since it requires less energy and does not need inert gas, fillers, or consumables.

9.2.6 Sustainability Assessment of the Laser Beam Welding Process

Laser beam welding techniques are one of the major advancements in the welding field. The biggest pros to using LBW technique is that we can weld thermoplastics using the laser beam. Laser-assisted manufacturing processes have recently gained the interest of manufacturers due to their increased efficiency, higher sustainability, and shorter machining times. These advantages have been made possible as a result of recent advancements in the development of materials and manufacturing technologies. Sustainability proves to be a major factor during manufacturing processes when environmental impacts are evaluated. Laser beam welding is less time-consuming when observed against a conventional riveting process. The time difference plays a major role when it comes to energy savings. The time saved, alias energy saved, varies between 51% and 67% for investigated scenarios in aircraft parts manufacturing. Also, a slight reduction in costs was also observed in two structures of an aircraft due to lower energy consumption in the fusion of airframes [27]. Also, LBW produces about 53% less CO_2 when compared to riveting process in aircraft parts. A sustainable life cycle is ensured in the final product as well [28]. In general terms, laser beam welding does not need a large groove angle as in a conventional welding process. It contributes to more environmentally sound processing in joining of two components as well [29, 30].

9.3 Welding Processes Study on Sustainability Assessment

Welding processes are classified into two major types, namely, fusion and friction welding. But when considered as a whole, welding processes are critical in manufacturing and development industries. Welding has a very vast range of applications that range from small rods to huge bridges. It has become an essential part of our lives because it has uses in simple applications such as the chassis of an automobile body, aircraft frames, railway wagons, and ship buildings. So, it is obligatory to determine the sustainability of these processes to help the development of welding. Saad *et al.* [6] evaluated the comprehensive model for the evaluation of manufacturing processes. This framework is employed to identify the most sustainable welding technique among multiple possibilities for a particular application in order to maximize sustainability. The authors found that FSW was the most sustainable process due to not having the use for an electric arc, it had the highest ranking among three multicriteria decision-making methods (MCDM) along the lines of TOPSIS, GRA, and COPRAS method. Jamal

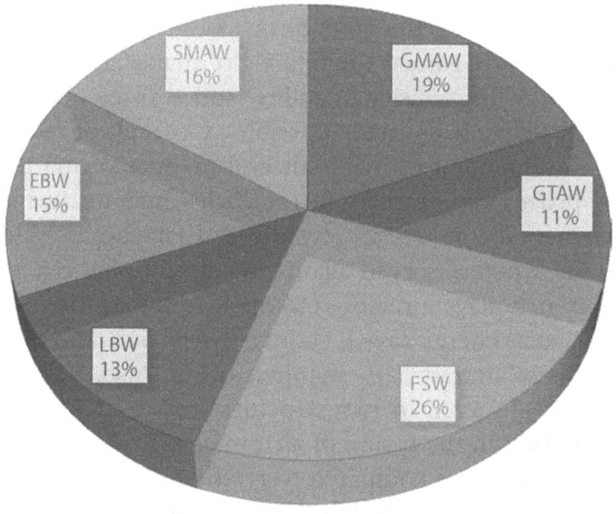

Figure 9.3 Sustainability analysis of welding processes.

et al. [31] also found FSW as the most sustainable process, since it came out with the highest social impact, physical performance, and environmental impact scores. Favi *et al.* [32] conducted research on the framework for a life cycle inventory of arc welding processes in order to evaluate the environmental performances of various processes for the purpose of comparison and decision-making. LCA is an important tool for measuring the effects that a given product or process has on the surrounding environment. LCA contains steps to (a) define objectives and scope, (b) conduct an inventory, (c) conduct a life cycle impact assessment, and (d) interpret the results. Singh *et al.* [33] investigated the sustainability assessment of GMAW technique and found a 38.81% improvement. Figure 9.3 represents a sustainability assessment of different welding processes.

9.4 5S Lean Strategy for Sustainability Manufacturing

The 5S strategy focuses on the tuning of the workplace and providing a calm and peaceful environment where most of the common issues are resolved primarily. The 5S stands for sorting, setting, shining, standardizing, and sustaining. Each of these terms has a very specific meaning toward the path of its goals. Together, when these 5S are implemented, it would result in improvement of the welding process. Here, sorting aims to analyze the entire process and be alert for all of the upcoming stages in that process. The entire process must be well sorted out for any dangers or unwanted events that may disturb the process, and all of the external and internal factors must be well sorted to ensure the wellness of the manufacturing process. The setting is the adjustment that is required to fulfill the manufacturing process. A well set manufacturing technique does not need any additional inspections during the duration of time required to complete the process. When the process is properly set, it must be carried out in an effective and creative way. Shining is a representation of the cleanliness of the worker that contributes toward the process/product. It ensures the health and hygiene of the workers that is one of the most important concerns in any given process; specifically in welding, the dust, smoke, and flux should not injure the worker and the worker must be well-equipped using necessary safety precautions. Standardizing is an aspect of every manufacturing technique that must be taken great notice of because standards are very vital in assessing the quality of the manufactured goods. All necessary information of the product must be thoroughly gone through to ensure the quality of the product. Every process that creates a product must be manufactured based upon the standards set for the process to make the product to extend its lifetime empirically and globally in both economic and social values. Lastly, the sustainability of the process and product determines the development of the process that in turn determines the development of the economy. Sustainability factors must be evaluated for all of the manufacturing processes for the reason that sustainability will calculate the development of the process. The sustainability of the supply chain is described as the incorporation of environmental, social, or corporate governance factors during the sourcing, processing, and delivery of raw materials to the market [34]. Supply chain assessment now includes sustainability criteria due to environmental concerns. Sustainability considers environmental, economic, and social factors. Due to natural resource depletion, firms are focusing more on supply chain sustainability. Izadikhah and Reza [35] developed a two-stage data envelopment analysis (DEA) method to assess supply chain sustainability in the welding sector. Sustainable manufacturing strategies (SMSs) can increase the welding process sustainability. Vimal *et al.* [36] examined how SMSs affect sustainable performance in shielded metal arc welding (SMAW). Their report recommended employee skill training and involvement and waste minimization and disposal SMSs. Waste minimization and disposal increased environmental performance and material conservation. The findings will inspire manufacturers to adopt sustainable production practices. The study expands welding research to sustainable process optimization and redesign. Titmarsh *et al.* [37] proposed a framework using Six Sigma to achieve the sustainable manufacturing requirements from the perspective of Industry 4.0 and its enablers. The authors came to the conclusion that utilizing a combination of Six Sigma tools, key performance indicators, and parts of 6R provides the user with an additional dimension to monitor the production status and make decisions that are more informed. Lean Six Sigma and 6R, when combined, provide a better potential for sustainable production in the future. This is in agreement with the modern manufacturing technologies of Industry 4.0 and the overarching goal of the manufacturing industry to decrease waste.

9.5 Conclusion

The manufacturing industries have shifted their attention toward designs that are sustainable and efficient in terms of energy and materials and have a minimal impact on the environment. It is important to ensure the efficient use of energy and resources by utilizing effective practices such as eliminating waste, optimizing power settings, and using renewable energy sources. Usage of an automated welding process to minimize waste and ensure consistent welds can be very pivotal in making it sustainable. Also, prioritizing using renewable energy sources, such as solar and wind, whenever possible can help with reducing valuable nonrenewable resources. To help in improving the world as a whole, sustainability is being adopted in all sorts of technologies by means of sustainability factors. Sustainability evaluation is a key decision-making tool in manufacturing and other industries. The increasing prevalence of welding technologies has necessitated an ever-increasing need for sustainable practices, particularly with regard to reducing carbon dioxide emissions and improving the efficiency of materials used. The use of modern energy-efficient welding equipment that highly limits fuel consumption and hazardous emissions is one way many welding professions are striving to become more sustainable. Additionally, introducing welding practices that minimize, reuse, as well as recycle waste materials is proving to be another way to reduce the environmental impact of welding. With these sustainable measures, welders are not only able to reduce their individual carbon footprints, but their efforts also contribute to the larger goal of achieving a more ecologically friendly industry. A comprehensive sustainability evaluation approach was utilized to analyze different welding processes' sustainability performance. Sustainable welding procedures minimize energy use, consumable waste, and weld quality to benefit the environment and conserve natural resources. Based on this study, manual metal arc welding poses a greater threat to the health of welders and the environment. Due to its high power consumption, reduced welding speed, and high shielding gas requirement, the GTAW process has high negative environmental indications next to SMAW. GTAW had the worst environmental impact due to its high power consumption and high cost, and its social impact remained unfavorable. GMAW welding process outperforms other metal arc welding processes. Due to its minimal energy requirements, GMAW has the lowest environmental impact. However, its economic and physical performance was much superior, while its social impact was identical to that of the GTAW process. SMAW has a smaller impact on the environment and society as a whole because of the lower levels of pollution and noise it produces when compared to GMAW. Due to increased efficiency, sustainability, and machining time, laser-based welding techniques are attracting manufacturers. Due to its lack of pollutants and superior physical performance compared to other varieties, friction stir welding had the least negative influence on the environment. Lean Six Sigma and 6R improve future sustainable production in line with Industry 4.0 manufacturing technology.

References

1. Mohammed Saad, H., Mohammad Nazzal, A., Basil Darras, M., A general framework for sustainability assessment of manufacturing processes. *Ecol. Indic.*, 97, 211, 2019.
2. Mani, M. and Madan, J., Sustainability characterization for manufacturing processes. *Int. J. Prod. Res.*, 52, 5895, 2014.
3. Kumaraguru, S., Rachuri, S., Lechevalier, D., Faceted classification of manufacturing processes for sustainability performance evaluation. *Int. J. Adv. Manuf. Technol.*, 75, 1309, 2014.
4. Chang, Y.J., Environmental and social life cycle assessment of welding technologies. *Proc. CIRP*, 26, 293, 2015.
5. Alkahla, I. and Pervaiz, S., Sustainability assessment of shielded metal arc welding (SMAW) process. *IOP Conf. Ser. Mater. Sci. Eng.*, 244, 1, 12001, 2017.
6. Mohammed Saad, H., Basil Darras, M., Mohammad Nazzal, A., Evaluation of welding processes based on multi-dimensional sustainability assessment model. *Int. J. Precis. Eng. Manuf.-Green Technol.*, 8, 57, 2021.
7. Favi, C., Campi, F., Germani, M., Comparative life cycle assessment of metal arc welding technologies by using engineering design documentation. *Int. J. Life Cycle Assess.*, 24, 2140, 2019.
8. Arulmurugan, B., Balaji, D., Rajkumar, S., Kamaraj, M., Mageshwaran, V., Sathishkumar, M., Manikandan, M., Arivazhagan, N., Influence of filler wire and welding process to mitigate the microsegregation of alloy C-2000 using continuous and pulsed current gas tungsten arc welding techniques. *J. Mater. Eng. Perform.*, 30, 6050, 2021.
9. Arulmurugan, B., Sathishkumar, M., Balaji, D., Muralikrishnan, K., Pranesh, S., Praveen, V., Praveen Kumar, K., Arivazhagan, N., Manikandan, M., Development of arc welding technique to preclude microsegregation in the dissimilar joint of Alloy C-2000 and C-276. *Proc. Inst. Mech. Eng. E J. Process Mech. Eng.*, 235, 1408, 2021.
10. Manikandan, M., Sasikumar, P., Arul Murugan, B., Sathishkumar, M., Arivazhagan, N., Microsegregation studies on pulsed current gas tungsten arc welding of alloy C-276. *Int. J. Sci. Eng. Res.*, 6, 33, 2015.
11. Patel, D. and Jani, S., ATIG welding: A small step towards sustainable manufacturing. *Adv. Mater. Process. Technol.*, 7, 514, 2021.
12. Chang, Y.J., Sproesser, G., Neugebauer, S., Wolf, K., Scheumann, R., Pittner, A., Finkbeiner, M., Environmental and social life cycle assessment of welding technologies. *Proc. CIRP*, 26, 293, 2015.
13. Vimal, K.E., Vinodhand, S., Raja, A., Optimization of process parameters of SMAW process using NN-FGRA from the sustainability viewpoint. *J. Intell. Manuf.*, 28, 1459, 2015.
14. Nimker, D. and Wattal, R., Recycling of submerged arc welding slag for sustainability. *Prod. Manuf. Res.*, 8, 1, 182, 2020.

15. Chucheep, T., Thangwichien, N., Mahathaninwong, N., Janudom, S., Yirong, C., Welding quality and sustainability of alternative LPG valve boss welding processes. *J. Sci. Technol.*, 41, 1146, 2019.
16. Drakopoulos, S., Salonitis, K., Tsoukantas, G., Chryssolouris, G., Environmental impact of ship hull repair. *Int. J. Sustain. Manuf.*, 1, 361, 2009.
17. Shrivastava, A., Krones, M., Pfefferkorn, F., Comparison of energy consumption and environmental impact of friction stir welding and gas metal arc welding for aluminum. *CIRP J. Manuf. Sci. Technol.*, 9, 159, 2015.
18. Dawood, H.I., Mohammed, K.S., Rajab, Y.M., Advantages of green solid state FSW over the conventional GMAW process. *Adv. Mater. Sci. Eng.*, 2014, 1, 2014.
19. Donatus, U., Corrosion behaviour of friction stir welded aluminium alloys: Unpredictability and protection. *J. Mater. Sci.*, 2, 1, 2017.
20. Shen, C., Zhang, J., Ge, J., Microstructures and electrochemical behaviors of the friction stir welding dissimilar weld. *China Environ. Sci.*, 23, 32, 2011.
21. Liu, X.C. and Wu, C.S., The influence of acoustic antifriction on heat generation and material flow in ultrasonic-assisted friction stir welding. *A MOE Key Lab for Liquid-Solid Structure Evolution and Materials Processing. Technol. Weld. Join.*, 3–4, 11, 567, 2633–2654, 2019.
22. Liu, X. and Wu, C.S., Elimination of tunnel defect in ultrasonic vibration enhanced friction stir welding. *Mater. Des.*, 90, 350, 2016.
23. Senthil Kumaran, S., Muthukumaran, S., Venkateswarlu, D., Balaji, G.K., Vinodh, S., Eco-friendly aspects associated with friction welding of tube-to-tube plate using an external tool process. *Int. J. Sustain. Eng.*, 5, 120, 2012.
24. Wang, B.B., Chen, F.F., Liu, F., Wang, W.G., Ma, Z.Y., Enhanced mechanical properties of friction stir welded 5083Al-H19 joints with additional water cooling. *J. Mater. Sci. Technol.*, 33, 1009, 2017.
25. Majeed, T., Wahid, M.A., Alam, M.N., Mehta, Y., Siddiquee, A.N., Friction stir welding: A sustainable manufacturing process. *Mater. Today Proc.*, 46, 6558, 2021.
26. Titmarsh, R., Assad, F., Harrison, R., Contributions of lean six sigma to sustainable manufacturing requirements: An industry 4.0 perspective. *Proc. CIRP*, 90, 589, 2020.
27. Lee, D.S., Fahey, D.W., Forster, P.M., Newton, P.J., Lim, L.L., Owen, B., Sausen, R., Aviation and global climate change in the 21st century. *Atmos. Environ.*, 43, 3520, 2009.
28. Kinnon, A.C., Cullinane, S., Browne, M., Whiteing, A., *Green Logistics: Improving the Environmental Sustainability of Logistics*, pp. 107–121, Kogan Page Publishers, 2010.
29. Mazumder, J., Laser welding: State of the art review. *JOM*, 34, 16, 1982.
30. Arulmurugan, B., Sathish Kumar, M., Kannan, T., Karuppiah, S., Kumaraguru, N., Ponsundar, E., Ragu, G., Manikandan, M., Investigation on mechanical and microstructure characteristics of nickel based C-2000 super alloy using laser beam welding. *Mater. Tod. Proc.*, 43, 3044, 2021.

31. Jamal, J., Darras, B., Kishawy, H., A study on sustainability assessment of welding processes. *Proc. Inst. Mech. Eng. B J. Eng. Manuf.*, 234, 501, 2020.
32. Favi, C., Campi, F., Germani, M., Mandolini, M., A data framework for environmental assessment of metal arc welding processes and welded structures during the design phase. *Int. J. Adv. Manuf. Technol.*, 105, 967, 2019.
33. Singh, V., Chandrasekaran, M., Devarasiddappa, D., Sustainability assessment of gas metal arc welding process of AISI 201LN using AHP-TLBO integrated optimization methodology. *J. Braz. Soc. Mech. Sci. Eng.*, 43, 1–20, 2021.
34. Vieira, P., Tenreiro, A., Oliveira, A.T., The increase of sustainability in cylinder manufacturing. *Clean Technol. Environ. Policy*, 12, 83, 2010.
35. Izadikhah, M. and Reza, F.S., Developing a linear stochastic two-stage data envelopment analysis model for evaluating sustainability of supply chains: A case study in welding industry. *Ann. Oper. Res.*, 1, 1, 2021.
36. Vimal, K.E.K., Vinodh, S., Raja, A., Modelling, assessment and deployment of strategies for ensuring sustainable shielded metal arc welding process–A case study. *J. Clean. Prod.*, 93, 364, 2015.
37. Titmarsh, R., Fadi, A., Robert, H., Contributions of lean six sigma to sustainable manufacturing requirements: An industry 4.0 perspective. *Proc. CIRP*, 90, 589, 2020.

10

Sustainable Development of Redundant Articulated Robot Components Using Simscape Multibody

M. Saravana Mohan[1*], P. S. Samuel Ratna Kumar[2,3] and P. M. Mashinini[3]

[1]*Department of Mechatronics Engineering, Kumaraguru College of Technology, Coimbatore, India*
[2]*Department of Mechanical Engineering, Kumaraguru College of Technology, Coimbatore, India*
[3]*Department of Industrial and Mechanical Engineering, University of Johannesburg, Johannesburg, South Africa*

Abstract

This chapter proposes the effects of aluminum alloy A380 on the joint dynamic sustainability of the redundant articulated robot (RAR) by changing the degrees of freedom (DOF). RAR components were designed using the Fusion 360 CAD model. The A380 alloy material is employed as a design material and then incorporated the model into MATLAB/Simscape Multibody. The change in degrees of freedom does not easily achieve a better dynamic property. To avoid such difficulties, Simulink simulation of the RAR system blocks scheme is used to identify the dynamic property, or torque, applied at robot joints for 7 degrees of freedom. This change in DOF helps to perform better compared to the normal DOF. Based on the dynamic property observations, the Simscape Multibody simulation results demonstrated the relevance of using A380 alloy while designing RAR components. Similarly, the mechanical and fatigue analysis using the Ansys research software was performed to determine the sustainability of the proposed material for redundant articulated robot components in real-time function. The mechanical strength analysis will show the maximum amount of stress induced and total amount of deformation. Fatigue analysis will propose the stress leads to failure, equivalent alternative stress, total life cycle, and damage safety factor for the RAR. Also, it shows that the change of material and adding an additional DOF will not affect

**Corresponding author*: saravana.moha@gmail.com

S. Thirumalai Kumaran and Tae Jo Ko (eds.) Sustainable Machining and Green Manufacturing, (193–220) © 2024 Scrivener Publishing LLC

the performance of the RAR. Rather, it improves the dynamic performance of the robot joint dynamics to perform more precise and accurate. This work proposes the use of Simscape simulation, and the Ansys analysis software can be used to perform the analysis before proposing a manufacturing method for developing a component. Thus, it shows that the methodology used in this work is more reliable and sustainable for the development of robotic components. Hence, the proposed work will be more advantageous for the robotic industries in achieving a better joint dynamic performance.

Keywords: Redundant articulated robot, sustainable material, A380 alloy, Simscape Multibody simulation, dynamic performance, mechanical strength analysis

10.1 Introduction

The increasing demands to manufacture low cost, high quality and selection of material in industrial robot plays a major role. These factors influence various advantages in the selection of cost-efficient electrical drives and other fixed/variable costs. Many researchers are now interested in choosing the right material for industrial robots, and aluminum alloys, particularly aluminum A380 die casting, continue to play an important role in this. Aluminum alloy A380 has some disadvantages; they are subject to fatigue and therefore its use in overloading conditions is dangerous [1, 2]. Robotic gripping and interaction abilities can be enhanced by making use of redundancy. These robots perform substantially better than with the traditional ways because to the application of grip redundancy [3]. High-degrees of freedom (DOF) manipulators may offer advantages such as improved obstacle avoidance skills, higher resilience with regard to mechanical failure, and the capacity for novel robot locomotion and gripping techniques [4]. ZK60 alloy performs well in terms of impact toughness and other mechanical qualities when aged at temperatures between 140°C and 200°C.

Die-cast A380 samples with hardness varying from 76 to 85 HRN consolidated at varied applied pressures. Tensile strength, yield strength, and elongation of the samples increased with pressure, and both alloys' hardness changed similarly with pressure, with hardness values rising as applied pressure rose. A linear link between the real data and the experimental data was also shown by the model fit to the experimental data. The SEM revealed various surface characteristics that were dispersed throughout the base materials when diverse pressures were applied.

The tiny granules were evenly scattered on micrographs of both metals and weighed 1,400 kg/cm^2. Effective dislocation blocking increases both metal strength and flexibility, considering the tiny mistake they made [3]. Al and Al alloys usually have higher room temperature-specific strengths than other materials, which makes them suitable substitutes for lightweight constructions in the automobile industry. One of the most important selection factors when using lightweight constructions is specific strength. The main factor affecting the tensile characteristics of the BTFSW joint is texture dispersion inside the SZ-Stir Zone. Because of the uneven texture distribution inside the SZ of BTFSW joints, the junction's stress characteristics are compromised by undesirable plastic deformation. A robust 0001 pattern with the surface angled away from the tensile direction by 48 degrees, which adds to the yield strength declination, is easily capable of inducing plastic deformation. The way the texture develops is influenced by the various tool rotation rates. A rise in the tool rotation speed from 400 to 600 rpm causes a large variation of the 0001 texture in the SZ resulting in the more complicated pattern of plasticized metal flow, which improves tensile characteristics [4]. Robotic gripping and interface skills can be improved by utilizing redundancy. The introduction of grip redundancy improves the performance of these robots significantly over traditional method. Only the book code, a library map, and some knowledge of the library's logical structure are given to the robot. The system not only integrates to complete its mission, but it also performs consistently and reliably as a whole. Online image segmentation, stereo vision, visual tracking, probabilistic matching, motion estimation, probabilistic selection, multisensory-based reaching and grasping, visual serving, and hybrid control are a few of the techniques that are available [5]. To save costs and simplify the design, a timing belt and a zero-backlash cable drive circuit were swapped out for the development of a cost-effective robotic arm manipulation research purpose. Instead of brushless or brushed motor with a low torque at low speeds, the stepper motors were opted for high torque at low speeds. The interaction with unstructured surroundings like a typical house or workplace, where intrinsic mechanical compliance safety is a critical design factor, was the intended application for these design choices. The cost-control choices covered in this paper were made to create reasonably priced compliant manipulators, a topic of study that has not gotten much attention yet but that, in our opinion, might have a big impact on how rapidly robots are incorporated into common homes and workplaces [6]. Superior obstacle avoidance, enhanced resilience against mechanical failure, and the capacity to conduct novel types of robot locomotion and grasping are among the potential benefits of high-DOF manipulators. Manipulators are redundancy-rich manipulators with an unlimited or nearly infinite degree of redundancy. In terms of shape and operation, these manipulators are similar to "snakes," "elephant trunks," or "tentacles." These robots are possibly superior for activities in severely confined and unique locations because to their highly articulated structures. Hyper-redundancy can also be used to create new types of robotic locomotion that are similar to that of worms, slugs, and snakes. Furthermore, hyper-redundant robots may be more resistant to mechanical failure [7].

Kinematic and structural dimensions, as well as the motors and gearboxes, are all parameterized as design variables in this paper. The design of the 5 DOF manipulator was optimized. This would also provide us with more information on how to optimize other parameters in the mechatronics system [8]. This study used MATLAB to do structural optimization or topology optimization to lower the mass of the system, which was then confirmed in Ansys.

This will also assist us in optimizing the topology of the proposed manipulator design. To consolidate the separate optimum topologies, the sufficient dynamic loading conditions, a weighted density parameter approach was proposed. The volume fraction of 0.5 was used to attain the optimal topology using MATLAB-based method.

Von Mises stress and deflection were calculated from simulations and determined to be 2.485 MPa and 0.00762 mm, respectively. The best topologies, on the other hand, had a Von Mises stress of 3.07366 MPa and a deflection of 0.007917 mm among the different angular position topologies. These results were also obtained for two distinct topologies. As a result, the proposed technique is found to be suitable for dynamic loading [9]. The structural parts of the IRB 120 robot are thoroughly detailed in this thermal behavior experiment document. The robot's materials included cast aluminum alloy, plastics, and high-alloyed steel. The structural anatomy of the IR 120 is described in detail in this publication [10]. The drive optimization approach is used to construct a robotic arm that is both efficient and effective. Rather than only improving the material, it will also assist us in optimizing the driving system. In contrast to weight optimization targets, the Pareto optimum method requires following the optimization process, and the developer and/or group will have to choose between a number of criteria, enabling early decisions induced by uncertainty to be postponed and delivering litheness to the process. The traditional method of relying on equation models to monitor the functioning of the developing system is judged ineffective due to the interdisciplinary character of the design endeavor. Interdisciplinary communication is thought to benefit from geometry-centric models, such as vibrant representations with more visible physical behavior and a genuine appearance. In the example study of a 6-DOF TAU haptic device, only parametric (CAD) geometries and kinematic models were handled, but the incorporation of specialized techniques and multi exchange of factors and model parameters was highly beneficial [11]. Role of torque in drive selection for various DOF manipulators clearly explains the importance of optimizing the torque. This would result in reduced system weight and cost without affecting the performance. An ideal method for classifying modules that may be used to build sequential manipulators with any number of degrees of freedom while staying within bounds. To identify the worst torque demands and drive-train choices for robotic arms with any range of degrees of freedom, an optimization issue is presented. The iterative process is examined and contrasted before being used to choose actuators for a large collection of manipulators. Three different phases are advised depending on the thorough practice. A trade-off must be made between the largest number of robotic applications and the minimum modular inventory in the library. The method is additionally used to determine potential modular combinations for every given number of degrees of freedom, in addition to the optimized number of modular divisions [12].

Three different 7-axis manipulator kinematic designs were discussed with DH parameter, which lays the foundation for our kinematic simulation in the upcoming days. Isotropy is defined as the difference between the workspace's lowest and highest singular values in a new global isotropy index (GII). It can be used with either the mass matrix or the Jacobian matrix, summarizing the overall performance of a mechanism by a scalar. By post-multiplying the matrix by a diagonal scaling matrix, the physical units of the matrix can be standardized and customized to meet a particular set of needs. By pre-multiplying the matrix by a diagonal scaling matrix of free variables, it is also possible to calculate optimal actuator scaling factors (transmission ratios) along with geometric parameters. Additionally, a unique optimization technique that is a variation on the global search approach is suggested. In order to remove other parameters from contention, it continually uses the worst configuration of one parameter until only the best configuration is left.

Progression, finite time ending, and a universal outcome are all guaranteed by the method. The optimization of either the GII or another local measure is demonstrated using two different iterations of the procedure. With ratios up to 1,850,000 observed, the method continuously exhibits dramatic improvements over global searches: 1. On a Sparc 5 workstation, an optimization that took about 3 days to complete using the culling technique would take 15,000 years to complete using global search. Consequently, the culling procedure enables the solution of high-dimensional problems that would otherwise be too technologically difficult to tackle with simple computer technology. Kinematic GII and maximum inertia of a 5-bar linkage-based planar haptic interface are optimized. It is shown that keeping the robot in an "M" or elbow-out position is ideal, and that the ideal workspace and robot architecture have left/right symmetry. To get an overall optimal design, kinematic and dynamic performance is traded off using a sensitivity analysis. Kinematic improvements of the Stewart, Spider, and Twin-Elbow hybrid serial/parallel devices demonstrate the Twin-Elbow A.

Typical result of an optimization for r B at any given value. The given ratio is an average since the parameter space was partitioned into 10 pieces to get around hardware (memory) restrictions. Because partitioning decreases the culling algorithm's effectiveness, the indicated improved ratio is conservative [13]. MATLAB and its Toolbox advantages in simulating manipulators in calculating various joint parameters like angle, torque, etc. were clearly examined for this work in which they have design 6-DOF manipulator [14]. The 2R manipulator was designed and verified using MATLAB/Simulink in which the respective torque values of each link were calculated and the mechanical modeling was completed in SolidWorks. When you take into account how different combinations of the two angles will affect the value of the potential energy of the second link and how the potential energy of a link will affect the torque value of that link, it is easy to see why. The dissipated energy during the robot's movement in the first posture (elbow down) are lower in our instance and for the same trajectory than they would be if the robot moved in the second posture (elbow up), according to the results of the comparative study of the two postures [15]. Industries employ industrial robots for pick-and-place and a variety of other tasks. To ascertain the maximum shear stress, dynamic characteristics, and mode shapes of a 6-axis industrial robot, this work aims to estimate the stresses and perform a modal analysis at various sites. By using the Ansys workstation, finite element analysis (FEA) is used to perform the optimal stress and modal analysis. The mesh size used for this investigation is 0.01 mm. The gripper is subjected to various loads in order to determine the highest stress intensity. Different cracks on the robot are taken into account for modal analysis. To identify the major weakness on the robotic construction, the modal shapes and natural frequencies for robots with and without cracks are examined [16]. A novel form of equipment called a continuum-type hyper-redundant robotic manipulator is inspired by the structure and functionality of an elephant trunk in its creation. For the ADAMS multibody dynamic simulation, a special type of multi-sectioned arm with a continuous backbone curve is taken into account. The design and operation of the bio-inspired arm are covered in this article, which is following by a simulation analysis using ADAMS and Ansys to determine the shape of each section's backbone curvature. Predicting the curvature of each component makes it easier for the arm to deal with the geometric nonlinearity that it experiences during actuation [17]. How to define, keep track of, and there has been discussion about hysteresis torsion in elastic robot joints. A detailed description of the nonlinear rotating spring with hysteresis and additional linear damping is provided in the proposed phenomenological model of elastic joint transmission. It is feasible to determine the direct and inverse joint transmission since the nonlinear hysteresis function has not been inverted. It has been predicted that the dynamic friction model will properly represent the input features of the joints' resistive behavior. The latter is used to monitor the responsive joint torque, an internal approach that is often invisible during joint operation. The elastic robot joint, which is basically a two-mass-type device, exhibits a complex dynamic behavior with memory [18, 19].

A bearing is a mechanical component that limits relative movement and, when properly lubricated, lowers resistance between moving components. For normal and light duty, endurance and racing applications, bi-metal aluminum-bearing manufacturers claim that their products provide long-term durability, dry-start protection, resistance to debris, corrosion resistance, and geometric correctness. The main objective of this research is to use a thrust-type seizure testing machine to calculate the safe load and seizure load of an aluminum alloy-bearing material, to use Ansys to calculate the stress for that material's safe load under the same operating conditions, and to compare those results with experimental result modifications and increases in all axes' coordinate points [20]. In metals with greater yield and ultimate tensile and smaller elastic modules, more vector node alterations occur because the trusses sustain more productive destruction. As a result, mechanical load is greater than it would be for S37 steel. Contrarily, because S37 steel has a larger elastic modulus and a lower tensile strength, trusses may stretch more easily with less stress placed on their nodes. When the rotational impact on roof lattice metals was compared, it became clear that S70 steel had moment changes that grew in exact correlation to the outcomes of force-induced mechanical vector modeling investigations. When the rotational influence on roof lattice metals was compared, it became clear that S70 steel had moment changes that grew in directly proportionate to the outcomes of force-induced mechanical vector modeling investigations. When referred to S37 roof lattice steel, S70 steel lattice trusses showed the greatest mechanical effect and moment vector changes, rotation effects between axis nodes, resultant vector changes, and bending [21].

10.2 CAD Modeling

Architects, video game designers, artists, manufacturers, and, of course, engineers employ a set of design tools together referred to as CAD, or computer-aided design. CAD systems assist their users by doing many intricate geometrical calculations in the background without the user even realizing it. Solid modeling is used in the computer program SolidWorks, which is used for computer-aided design and engineering. SolidWorks is used solely for the design of mechatronic systems. The application's initial uses include project management, planning, visual ideation, modeling, feasibility study, and prototyping.

The 5 DOF of industrial robot is modified into 7-DOF redundant robot as shown in Figure 10.1 and Figure 10.2 by modifying and adding manipulator links and joints and providing extra DOF to the robot so that it can be able to move extra path locomotion. So it can achieve more flexible work with higher efficiency. For our study, end effector is not shown for specific task application, the modified robot is common for all industrial

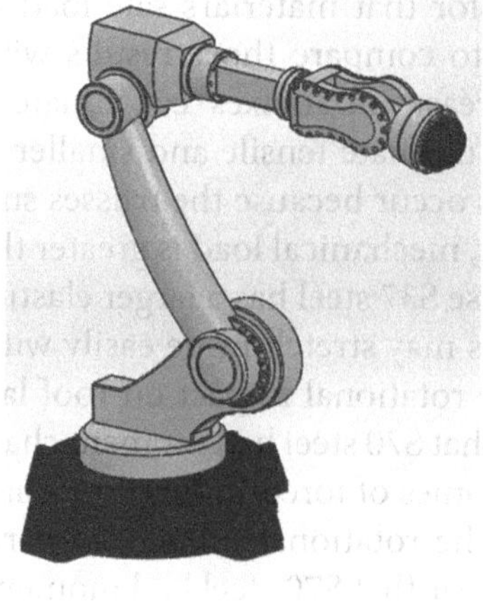

Figure 10.1 7-DOF industrial robot.

Figure 10.2 RAR robot.

robot applications and our main motto is to do a case study on various parameters.

After the detailed modification of a CAD model as given in Table 10.1. The material chosen is assigned for this redundant robot.

Table 10.1 Articulated robot structural dimension.

Sl. no.	CAD model part name	Dimension
1.	Base	Radius – 150 mm
2.	Link 1	Horizontal (x) distance – 310 mm Vertical (y) distance – 230 mm
3.	Link 2	Horizontal (x) distance – 80 mm Vertical (y) distance – 295 mm
4.	Link 3	Horizontal (x) distance – 75 mm Vertical (y) distance – 260 mm
5.	Link 4	Horizontal (x) distance – 210 mm Vertical (y) distance – 130 mm
6.	Link 5	Horizontal (x) distance – 175 mm Vertical (y) distance – 78 mm
7.	Link 6	Horizontal (x) distance – 120 mm Vertical (y) distance – 80 mm

10.3 Assigning Aluminum A308 Alloy for RAR

The most popular die-casting material is A380 aluminum alloy because it combines several desirable qualities, including ease of casting, machining, efficient heat transmission, and others. A380 has proven simple to process,

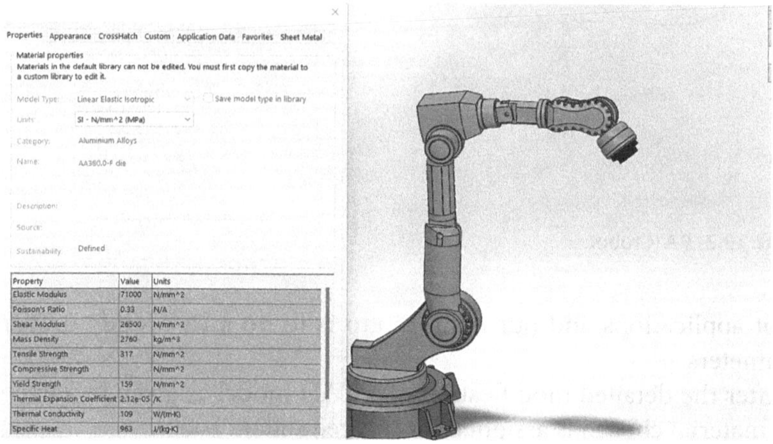

Figure 10.3 Articulated robot with assigned material property of A380.

Table 10.2 Aluminum A380 alloy material composition.

Material	Composition %
Al	89.5
Cu	4.0
Fe	1.3
Mg	0.10
Mn	0.50
Ni	0.50
Other	0.50
Si	9.5
Tin, Sn	0.35
Zinc, Zn	3.0

Table 10.3 Material property of aluminum A380 alloy.

Property	Value	Units
Elastic modulus	70900	N/mm^2
Poisson's ratio	0.332	N/A
Shear modulus	26499	N/mm^2
Mass density	2759	Kg/m
Tensile strength	316.5	N/mm^2
Yield strength	158.9	N/mm^2
Thermal expansion coefficient	$2.11e^{-05}$	/K
Thermal conductivity	108	W/(m.K)
Specific heat	962	J/(kg.K)

however, because of the high silicon concentration, it is a little rough. Figure 10.3 indicates the RAR of applied aluminum A380 material, and Table 10.2 and Table 10.3 indicate the material composition and properties of aluminum A380.

10.4 Kinematics and Dynamic Studies

Robot dynamics is the study of manipulator motion in relation to the speed at which the robot's configuration changes over time. The dynamic conditions are often calculated using the time-consuming equations, however, in this case, the dynamics of a system are studied using simulation methodology. In this part, the manipulator linkages are investigated. SimMechanics' second-generation platform came in handy. The between joint frames, for example, dynamic variables can be sensed. The robot's design parameters were altered using the ability of joint blocks to sense their surroundings has been demonstrated. The stiffness matrix was constructed with the aid of MATLAB. Rotational matrices are widely used in robotics. MATLAB is frequently used to tackle issues regarding rotation via matrix solutions. A graphic interface, flexible block modules, and models for modelling and simulating dynamic systems are all included in Simulink. The users may export simulation data for additional analysis and integrate mathematical methods in MATLAB. A graphic interface, flexible block modules, and models for modeling and simulating dynamic systems are all included in Simulink. Given its integration with MATLAB, users may export simulation data for additional analysis and include MATLAB methods into models [22].

10.5 Assigning DH Parameters

The uniform transition four fundamental transformations are used to create Matrix Ai in Equation (10.1). It defines how the tool is positioned and oriented in relation to the reference frame specified by Equations (10.2) and (10.3). Applying the DH parameter coordinates for the RAR is shown in Figure 10.4 the DH parameter values for RAR are obtained as shown in Table 10.4.

$$Ai = T\,(z, d)\; T\,(z, \Theta)\; T(x, a)\; T(x, \alpha) \tag{10.1}$$

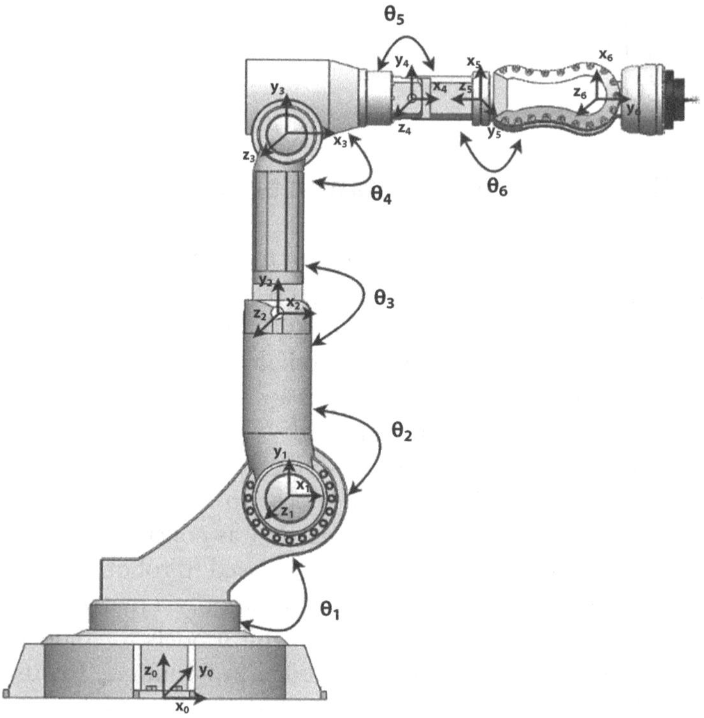

Figure 10.4 DH parameter coordinates of RAR.

Table 10.4 DH parameter table of the RAR.

Frame	a_i	α_i	d_i	θ_i
1	a_1	$\frac{\pi}{2}$	d_1	θ_1
2	0	0	0	θ_2
3	0	0	0	θ_3
4	a_4	0	0	θ_4
5	0	$\frac{\pi}{2}$	0	$\frac{\pi}{2}+\theta_5$
6	0	$\frac{\pi}{2}$	d_6	$\frac{\pi}{2}+\theta_6$

$$Ai = \begin{bmatrix} R_i^{i-1} & O_i^{i-1} \\ 0 & 1 \end{bmatrix} \tag{10.2}$$

$$Ai = \begin{bmatrix} Cos\theta i & -Cos\alpha iSin\theta i & \sin\alpha i\sin\theta i & aiCos\theta i \\ Sin\theta i & Cos\alpha iCos\theta i & -\sin\alpha iCos\theta i & ai\sin\theta i \\ 0 & \sin\alpha i & Cos\alpha i & di \\ 0 & 0 & 0 & 1 \end{bmatrix} \tag{10.3}$$

The homogeneous matrix of the RAR for each joint is shown in the Equations 10.4 to 10.9.

$$A1 = \begin{bmatrix} c_1 & 0 & s_1 & a_1c_1 \\ s_1 & 0 & -c_1 & a_1s_1 \\ 0 & 1 & 0 & d_1 \\ 0 & 0 & 0 & 1 \end{bmatrix} \tag{10.4}$$

$$A_4 = \begin{bmatrix} c_4 & -s_4 & 0 & a_4c_4 \\ s_4 & c_4 & 0 & a_4s_4 \\ 0 & 0 & 1 & 0 \\ 0 & 0 & 0 & 1 \end{bmatrix} \tag{10.5}$$

$$A_2 = \begin{bmatrix} c_2 & -s_2 & 0 & 0 \\ s_2 & c_2 & 0 & 0 \\ 0 & 0 & 1 & 0 \\ 0 & 0 & 0 & 1 \end{bmatrix} \tag{10.6}$$

$$A_5 = \begin{bmatrix} -s_5 & 0 & c_5 & 0 \\ c_5 & 0 & s_5 & 0 \\ 0 & 1 & 0 & 0 \\ 0 & 0 & 0 & 1 \end{bmatrix} \tag{10.7}$$

$$A_3 = \begin{bmatrix} c_3 & -s_3 & 0 & 0 \\ s_3 & c_3 & 0 & 0 \\ 0 & 0 & 1 & 0 \\ 0 & 0 & 0 & 1 \end{bmatrix} \tag{10.8}$$

$$A_6 = \begin{bmatrix} -s_6 & 0 & c_6 & 0 \\ c_6 & 0 & s_6 & 0 \\ 0 & 1 & 0 & d_6 \\ 0 & 0 & 0 & 1 \end{bmatrix} \tag{10.9}$$

Homogeneous transformation matrix of joint 7 position

$$T = \begin{bmatrix} c_1 ss_{2345} + s_1c_6 & c_1c_{2345} & -c_1c_6s_{2345} + s_1s_6 & c_1[c_{2345}(d_6 + a_4) + a_1] \\ s_1s_6s_{2345} - c_1c_6 & s_1c_{2345} & -s_6\left[s_1s_{2345} + c_1\right] & s_1[c_{2345}(d_6 + a_4) + a_1] \\ -s_6[c_{2345}] & s_{2345} & c_6c_{2345} & s_{2345}(d_6 + a_4) + d_1] \\ 0 & 0 & 0 & 1 \end{bmatrix} \tag{10.10}$$

Joint 7 position is indicated in Equations 10.10–10.13.

$$x = c_1 \, [c_{2345} \, (d_6 + a_4) + a_1] \tag{10.11}$$

$$y = s_1 \, [c_{2345} \, (d_6 + a_4) + a_1] \tag{10.12}$$

$$z = s_{2345} \, (d_6 + a_4) + d_1 \tag{10.13}$$

10.6 Simscape Multibody Simulation

SimMechanics is a 3D mechanical and electromechanical multibody simulation environment. Blocks representing bodies, joints, constraints, and force elements can be used to model the multibody system. SimMechanics then formulates and solves the problem. Entire mechanical system equations of motion based on a computer-aided design (CAD) model the equations that were created as a result of the preceding section's kinematic modeling will not be employed. SimMechanics was used to analyze the system's performance. It has the ability to modify its structure and optimize its performance. The parameters involved in the system and the examined outcomes inside the SimMechanics environment can be derived in significantly shorter amount of time. The SolidWorks 3D CAD model can be imported to SimMechanics platform. The dynamics of the system are displayed using a 3D animation that is generated automatically by Mechanics Explorer in MATLAB. The CAD model was created in SolidWorks, which was stated in the preceding part. It is used for the purpose of simulation SimMechanics is an environment for multibody simulations (Figure 10.5).

The proposed RAR 3D CAD model was exported as XML and STL files from the SolidWorks environment to the MATLAB/SimMechanics environment. Second-generation SimMechanics connection. The MATLAB command was used to run the model's XML file. The robot's CAD model was transformed into a diagram with connecting blocks that symbolize. The primitives for the input joints to get the output through are assigned in the joint blocks specified in the workspace block diagram [22–24].

10.7 Torque Results Using Simscape Multibody

Torque is used by the robot's motor to turn the wheels. The follower gear is turned by the driver gear's torque. Inches are used to measure distance, and ounces are used to measure force for tiny motors. The distance and force are measured in feet and pounds, respectively, for big motors. The rotational force is expressed in feet per pound (ft-lbs) or inches per ounce (in-oz).

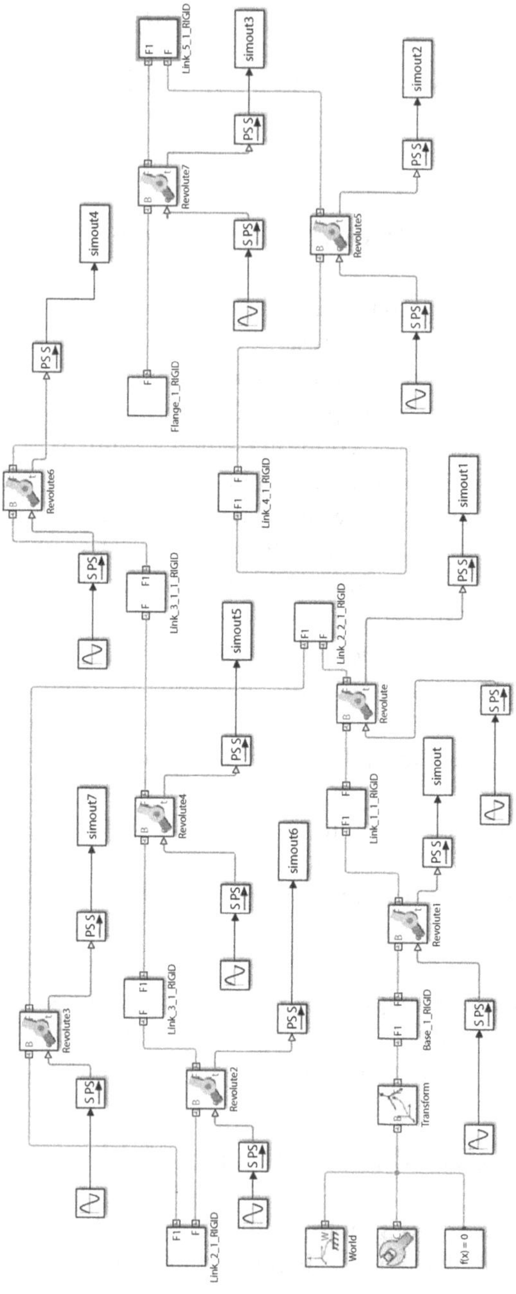

Figure 10.5 Simscape Multibody diagram for joint torque identification.

The selection of an electrical drives depends profoundly on the torque between the joints, which also plays a significant role in the robot development. The minimum torque between the joints helps to reduce the overall cost of a robot and makes it easier for the engineer to select the minimum and suitable drives for their robots. Finally, it leads to an increase in the profit share of a company and increases the efficiency of production. The torque acting at joints 1 and 2 are 457.10 Nm and 696.70 Nm (Figure 10.6). Torque of 50 Nm and 136.20 Nm are depicted in Figure 10.7 for joints 3 and 4. The torque observed at joints 5, 6, and 7 is 42.60 Nm, 10.92 Nm, and 3.725 Nm, respectively, as shown in Figure 10.8 and Figure 10.9.

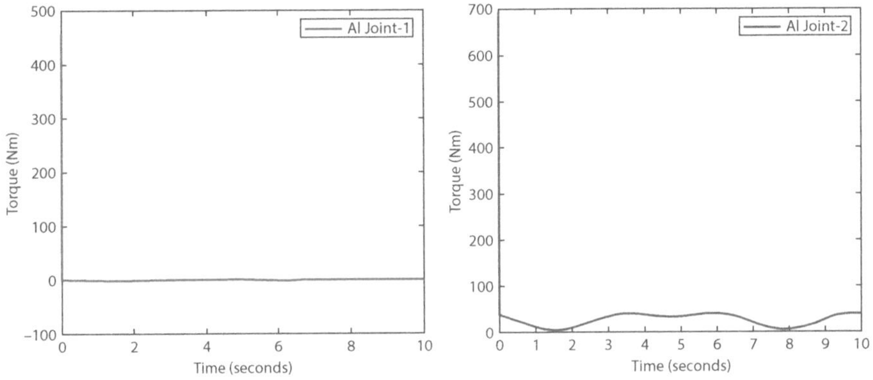

Figure 10.6 Torque at joints 1 and 2, aluminum A380.

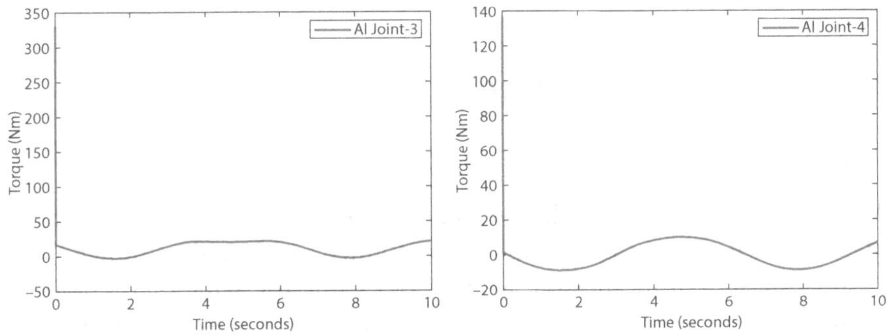

Figure 10.7 Torque at joints 3 and 4, aluminum A380.

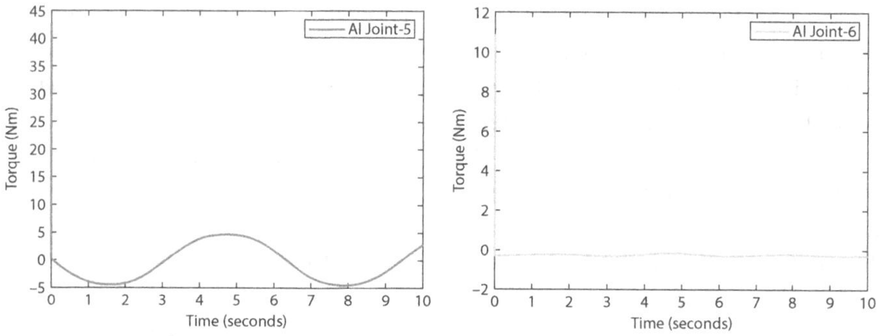

Figure 10.8 Torque at joints 5 and 6, aluminum A380.

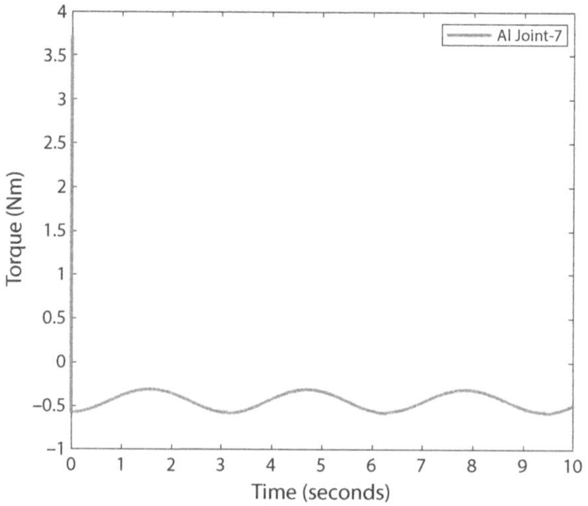

Figure 10.9 Torque at joint 7, aluminum A380.

10.8 Static Analysis Under Twisting Moment

10.8.1 Mesh Properties

A more precise mesh is made possible through mesh management. Ansys Mechanical gives you the option to customize local meshes rather than a global mesh that meshes the whole CAD using the same methodology. Local scaling, refinement, and sphere of influence defeaturing of the geometry are some examples of local meshing controls. Delaunay tetrahedral meshes are unstructured, meaning that the connectivity between adjacent

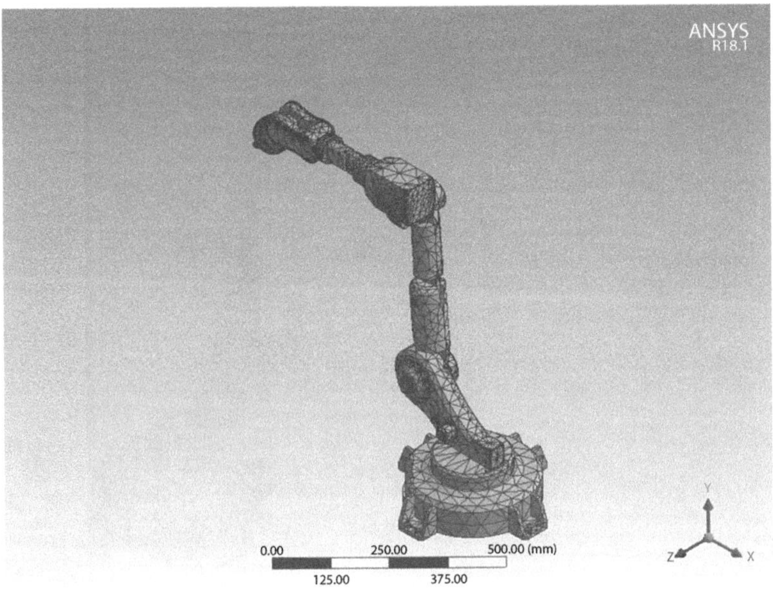

Figure 10.10 Complete tetrahedron and triangular prism mesh.

elements is uneven. Commercial software was used to create an unstructured tetrahedral mesh of the fluid domain (Figure 10.10).

A polyhedron with two triangular bases and three rectangular sides is referred to as a triangular prism. It is a three-dimensional form with two base faces, three side faces, and connections between them at the edges. It is referred to as a right triangle prism if the sides are rectangular; otherwise, it is referred to as an oblique triangular prism. A uniform or regular triangular prism is one that has square sides and equilateral bases. The two bases of this prism are parallel and congruent to one another, just like in other prisms. It contains 5 faces, 6 vertices, and 9 edges altogether.

10.8.2 Postprocess of Max Stress

The Von Mises theory asserts that when the greatest distortion energy per unit volume in the item meets the yielding distortion energy per unit volume, the component subjected to the real loading (biaxial or triaxial) fails. In other words, this represents the total strain energy held in the body's various differential volumes. The total strain energy held by the body may be calculated by adding this strain energy over all differential volumes, also known as integration across the entire volume. Volumetric energy is the

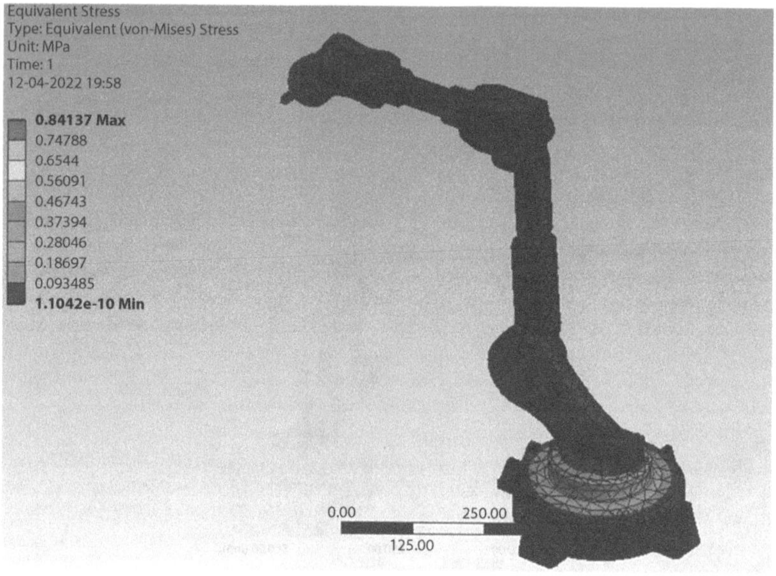

Figure 10.11 Stress variation using A380.

portion of this total energy that is used to alter the volume of the material (also known as volumetric strain). The remaining energy, often referred to as deviatoric energy, is utilized to change the form of the substance. The total stress component entering the distortion energy is related to the Von Mises stress (Figure 10.11).

10.8.3 Postprocess of Deformation

The model's deformation results in three coordinates by selecting the total deformation option (X, Y, and Z). To display the deformation result of your physical model in that direction, choose an X, Y, or Z coordinate in directional deformation (Table 10.5, Figure 10.12).

10.9 Work Envelope of RAR

A robot's work envelope is its range of motion. This form is produced when a manipulator stretches upward, below, ahead, and backward. These distances are determined by a robot's arm length and axis arrangement. Each axis contributes its own range of motion. A robot arm's workspace is the collection of all spots it can reach. This is determined by a variety of factors, including the arm's size (Figure 10.13).

Table 10.5 Different stress and deformation results based on moment.

Joints	Max stress (MPa)	Total deformation (mm)	Max principle stress (MPa)
1	1.2557	3.66 e^-3	0.7659
2	28.842	0.300	32.624
3	22.423	0.276	18.326
4	4.780	0.001	3.457
5	9.233	0.060	9.506
6	1.827	0.006	1.979
7	0.020	1.74 e^-6	0.020

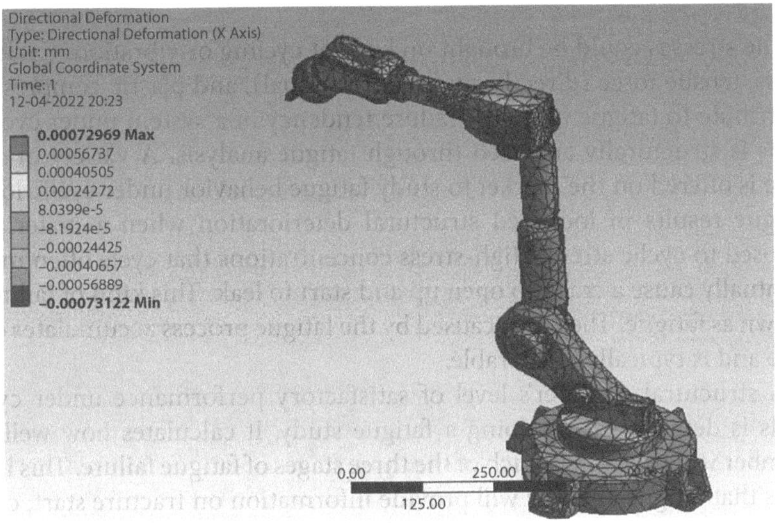

Figure 10.12 Directional deformation variation using Al A380.

10.10 Fatigue Report of RAR

Fatigue is the gradual, brittle, and localized structural damage caused by the material weakening brought on by cyclic stress. Even though the intensity of repeated alternating or cyclic loads is substantially lower than the usual strength, each loading cycle will slightly widen the fracture once it has begun.

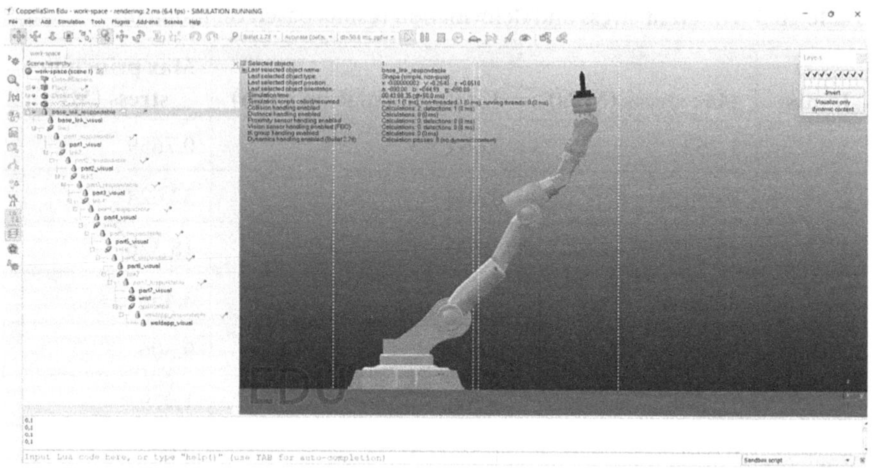

Figure 10.13 Work envelope of RAR using CoppeliaSim.

The stresses could be brought on by heat cycling or vibration. Cycles of stress, tensile force (directly applied or residual), and plastic compression contribute to fatigue [25]. The failure tendency of a system under cyclical loads is structurally analyzed through fatigue analysis. A variety of software is offered on the market to study fatigue behavior under cyclic loads. Fatigue results in localized structural deterioration when a material is exposed to cyclic stress. High-stress concentrations that cycle often might eventually cause a crack to open up and start to leak. This kind of failure is known as fatigue. The harm caused by the fatigue process accumulates over time and is typically irreparable.

A structural member's level of satisfactory performance under cyclic loads is determined by doing a fatigue study. It calculates how well the member will perform in each of the three stages of fatigue failure. This indicates that fatigue analysis will provide information on fracture start, crack propagation, and ultimately failure likelihood for a particular material.

The plumbing industry uses the S-N approach of fatigue analysis, which employs the local strain or strain-life (e-N) method. The S-N curve is used by the Caesar II program's fatigue analysis as an input, and it is compared to the pipe stresses to determine a safe time frame before failure. The S-N curves for each material are specified in specifications like ASME Sec VIII-Div 2.

The Goodman diagram (Figure 10.14) represents the fatigue life test in a more practical manner than the prior method, "stress-life," did. Based on the qualities of the material, it is used to predict the fatigue life at a

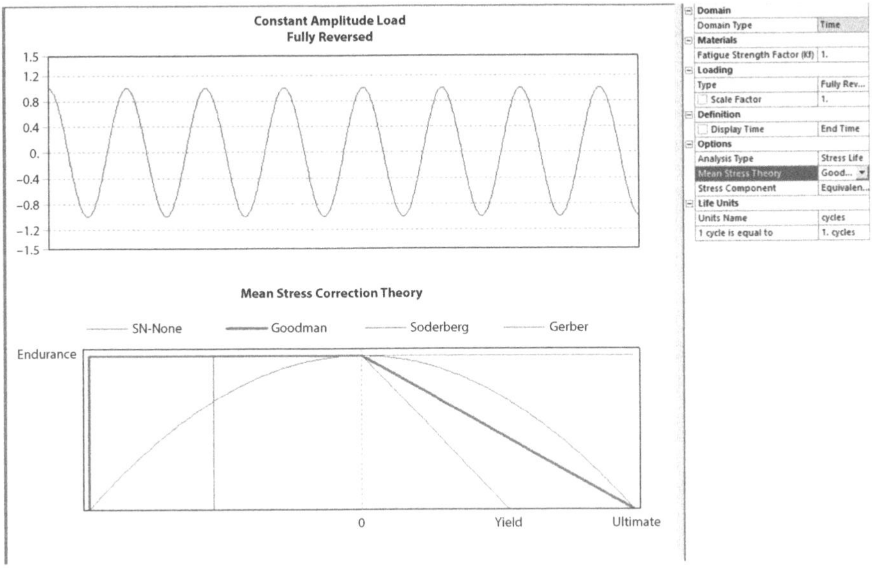

Figure 10.14 Graph of Goodman criteria.

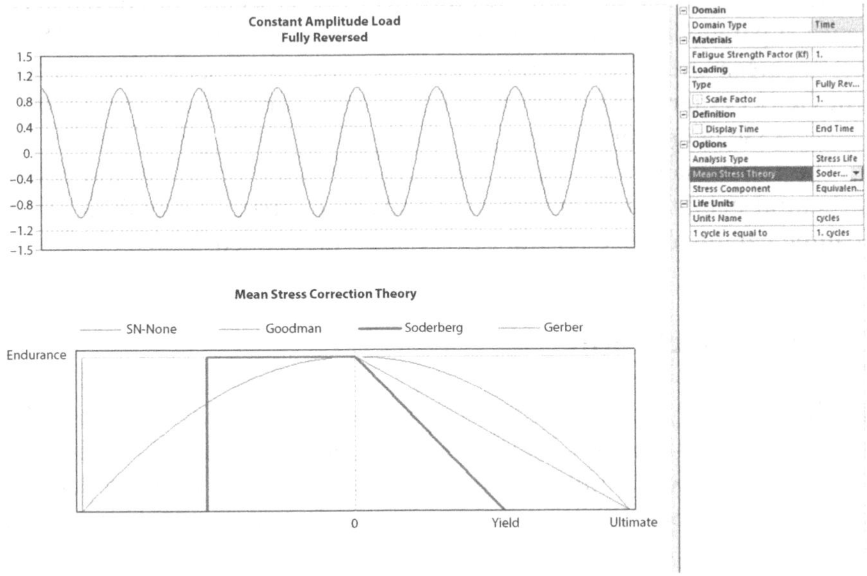

Figure 10.15 Graph of Soderberg criteria.

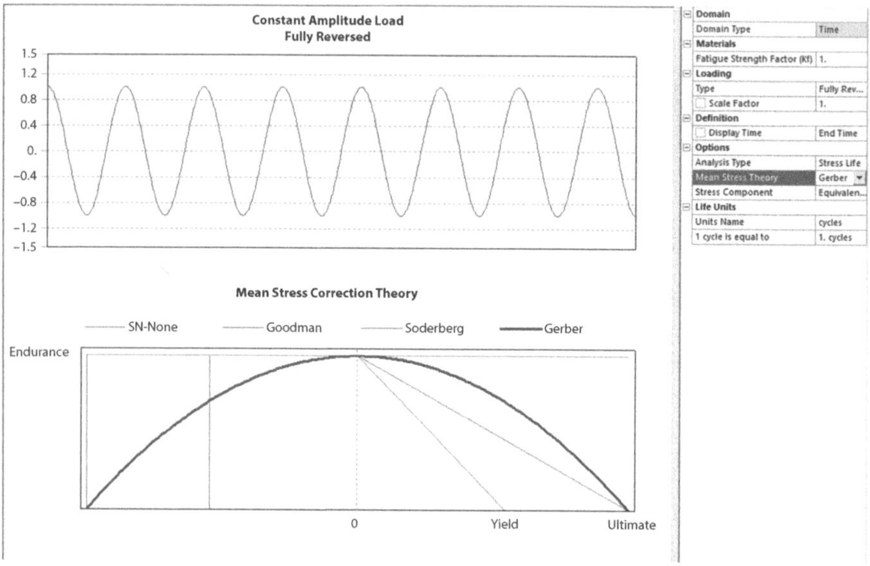

Figure 10.16 Graph of Gerber criteria.

particular stress ratio (alternating and mean loads). The Soderberg criterion (Figure 10.15) or line is the straight line connecting (Se) on the stress amplitude axis and (Syt) on the mean stress axis. It is believed that the triangle region located beneath the Soderberg line is secure for eternal existence. A parabolic curve with a join on the stress amplitude axis and a "Sut" on the mean stress axis is known as the Gerber parabola (Figure 10.16). The Gerber criteria classify the region below this curve as safe. Figure 10.17 shows A380 fatigue properties.

The robot manipulator links built out of aluminium A380 has better structural stability, according to the results of the structural analysis, if the loading parameter (800 N) and FOS (1.331) was applied.

10.11 Conclusion

In this chapter, RAR's model was designed using SolidWorks, and the materials' parameters were derived from earlier research. The parameters of the materials were employed as a design material to the CAD model. The degrees of freedom have been increased from the conventional articulated

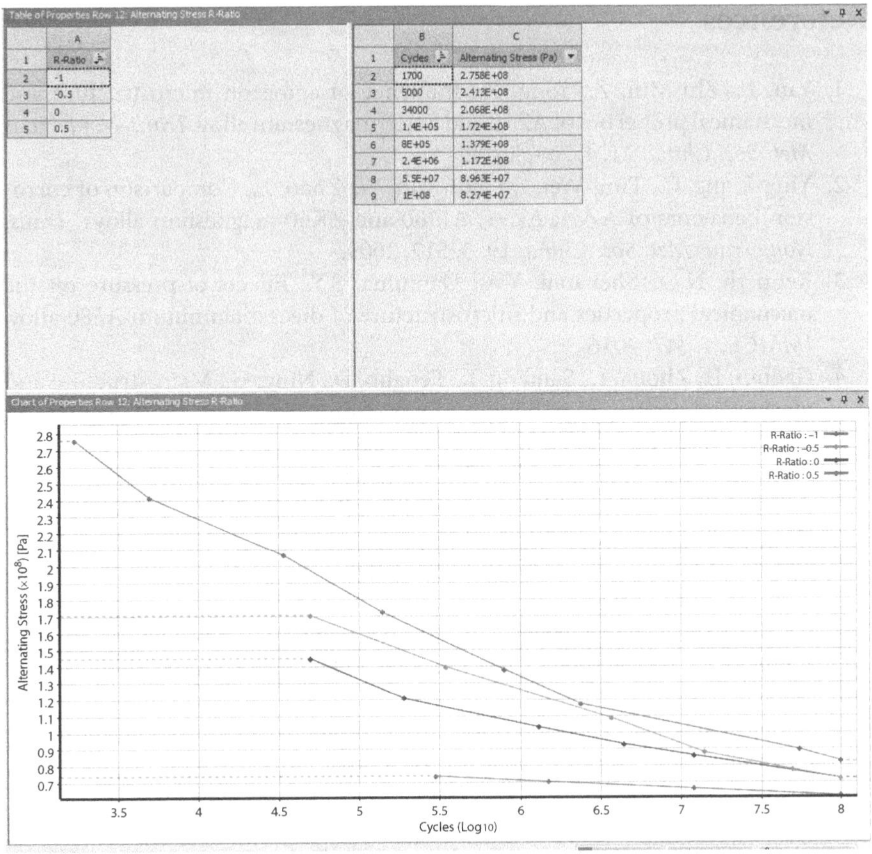

Figure 10.17 A380 fatigue properties.

robot to achieve sustainable dynamic properties. Thus, reliable material aluminum A380 was selected, and better torque value was obtained using Simscape Multibody simulation. Actuator selection was simple feasible due to lesser torque requirement. Due to less torque value, the power requirement results in low cost of manufacturing and utilization. The body built of aluminum A380 has better structural stability, according to the results of the structural analysis. Aluminum A380 for the loading scenario (800 N) and FOS (1.331). Based on the above findings, A380 holds a better life cycle even after the DOF was increased to 7, and A380 holds good for RAR sustainable manufacturing.

References

1. Yan, L., Zhi-Min, Z., Yong, X., Influence of aging on microstructure and mechanical properties of AZ80 and ZK60 magnesium alloy. *Trans. Nonferrous Met. Soc. China*, 21, 4, 739, 2011.
2. Ying-Liang, C., Ting-Wei, Q., Hui-Min, W., Zhao, Z., Comparison of corrosion behaviors of AZ31, AZ91, AM60 and ZK60 magnesium alloys. *Trans. Nonferrous Met. Soc. China*, 19, 3, 517, 2009.
3. Kenneth, N.O., Shekarau, Y.A., Danjuma, S.Y., Effects of pressure on the mechanical properties and microstructure of diecast aluminum A380 alloy. *JMMCE*, 4, 347, 2016.
4. Gaohui, L., Zhoua, L., Sanfeng, L., Fengbo, D., Ning, G., Microstructure and mechanical properties of bobbin tool friction stir welded ZK60 magnesium alloy. *J. Mater. Res. Technol.*, 9, 4, 9348, 2020.
5. Prats, M., Sanz, P.J., Del Pobil, A.P., The advantages of exploiting grasp redundancy in robotic manipulation. *Intell. Serv. Robot.*, 1, 4, 321, 2008.
6. Quigley, M., Asbeck, A., Ng, A., A low-cost compliant 7-DOF robotic manipulator. *IEEE International Conference on Robotics and Automation*, Shanghai, China, p. 6051, 2011.
7. Chirikjian, G.S. and Burdick, J.W., Design and Experiments with a 30 DOF robot. *IEEE International Conference on Robotics and Automation*, p. 113, 1993.
8. Zhou, L. and Bai, S., A new approach to design of a lightweight anthropomorphic arm for service applications. *ASME. J. Mech. Robot.*, 7, 3, 031001, 2015.
9. Srinivas, G.L. and Javed, A., Topology optimization of industrial manipulator-link considering dynamic loading. *Mater. Today Proc.*, 18, 3717, 2019.
10. Dumitraşcu, C., Nicolescu, A.F., Cristoiu, C., Experimental investigation of an irb 120 robot thermal behavior for heat sources and thermal field distribution identification. *Proc. Manuf. Syst.*, 14, 3, 97, 2019.
11. Zhou, L., Bai, S., Hansen, M.R., Design optimization on the drive train of a light-weight robotic arm. *Mechatronics*, 21, 3, 560, 2011.
12. Singla, E., Singh, S., Singla, A., Drive-train selection criteria for n-dof manipulators: Basis for modular serial robots library. *Int. J. Nonlinear Sci. Numer. Simul.*, 22, 2, 169, 2021.
13. Angeles, J., Ranjbaran, F., Patel, R.V., On the design of the kinematic structure of seven-axes redundant manipulators for maximum conditioning. *Proceedings 1992 IEEE International Conference on Robotics and Automation*, Nice, France, vol. 1, p. 494, 1992.
14. Iqbal, J., Islam, R.U., Khan, H., Modeling and analysis of a 6 DOF robotic arm manipulator. *Can. J. Electrical Electron. Eng.*, 3, 6, 300, 2012.
15. Gouasmi, M., Ouali, M., Fernini, B., Meghatria, M.H., Kinematic modelling and simulation of a 2-R robot using solidworks and verification by MATLAB/simulink. *Int. J. Adv. Robot. Syst.*, 9, 6, 245, 2012.

16. Sahu, S., Choudhury, B.B. *et al.*, Stress and modal analysis of six-axis articulated robot using ANSYS, in: *Information and Communication Technology for Competitive Strategies*, S. Fong (Ed.), pp. 303–311, Springer, Singapore, 2018.
17. Nagarajan, A., Rajesh Kanna, S.K., Manoj Kumar, V., Multibody dynamic simulation of a hyper redundant robotic manipulator using ADAMS ansys interaction. *2017 International Conference on Algorithms, Methodology, Models and Applications in Emerging Technologies (ICAMMAET)*, Chennai, India, p. 1, 2017.
18. Ruderman, M., Bertram, T., Iwasaki, M., Modeling, observation, and control of hysteresis torsion in elastic robot joints. *Mechatronics*, 24, 5, 407, 2014.
19. Meng, Y., Gao, H., Yan, Y., Gao, L., Effects of phase difference and stress ratio on biaxial tension crack propagation 2 behavior of rolled ZK60 magnesium alloy. *Mater. Today Commun.*, 240, 101159, 2020.
20. Balthilak, A., Sudharsan, S., Santhosh, B., Investigation of safe load of aluminium alloy bearing material experimentally and comparing it with that of ANSYS. *Second International Conference on Current Trends in Engineering and Technology-ICCTET 2014*, Coimbatore, India, p. 282, 2014.
21. Bilgin, Z., Semih, T., Kürşat, K., Investigation of force and moment effect of st 37 and st 70 roof lattice steels in ansys program. *Middle East J. Sci.*, 4, 1, 23, 2018.
22. Saravana Mohan, M. and Anbumalar, V., Modelling and simulation of multi spindle drilling redundant SCARA robot using SolidWorks and MATLAB/SimMechanics. *Rev. Fac. Ing.*, 81, 4, 2016.
23. Lingala Purandhara, S.S., Neeraj, M., Swadhin Sri Aurobindo, M., Vasanthakumar, D., Saripella Koushik, V., Sumit, S., Design and analysis of a robotic arm underdifferent loading conditions using FEA simulation. *Mater. Today Proc.*, 50, 5, 759, 2022.
24. Haitao, L., Zhengcang, C., Yuquan, L., Hongguang., W., Rigid-flexible coupling dynamics simulation of 3-RPS parallel robot based on ADAMS and ANSYS. *Appl. Mech. Mater.*, 290, 91, 2013.
25. Blatherwick, A.A. and Lazan, B.J., *Fatigue Properties of Extruded Magnesium Alloy Zk 60 Under Various Combinations of Alternating and Mean Axial Stresses*, Air Force Avionics Laboratory, WADC Technical Report, 53–181, 1979.

11

Implementation of Green Manufacturing Practices in Automobile Fields: A Review

Sampath Boopathi

Department of Mechanical Engineering, Muthayammal Engineering College, Namakkal, Tamil Nadu, India

Abstract

One of the workable methods to lessening the effects of global warming on the planet is green manufacturing. Waste disposal and recycling are big issues nowadays and have a negative impact on the environment. Manufacturing generates a lot of trash and pollution, which directly impact society by causing bad weather and severe health problems. One of the healthiest methods for producing goods for the automobile industry is the green manufacturing process. The emphasis was sharply on waste reduction in production as the paradigm switched to sustainable manufacturing. Green production limits the influence on the environment by taking a 360-degree view of the manufacturing system. Resources left over from traditional production techniques are hardly ever used by society. Mercedes, BMW, Audi, and Volkswagen, four German automakers, offer distinctive III–V fiber composites for interior and exterior uses. A safe and clean society is endangered by gases, garbage, and toxic materials. Every Indian resident has a basic right to access clean air, fresh water, and environmentally friendly transportation infrastructure. It has been discovered that the fundamental activities covered by sustainable manufacturing are lowering wastewater, emissions, energy use, and physical waste. Based on India's status as a center for global automobile manufacture, domestic market shares in percentage, and other factors, the Indian automotive sector was profiled. The research has considered green design, green purchasing, green transportation, and green packaging. To establish a template for the procedures of waste minimization, decreased emissions, and increased production efficiency in the green manufacturing regime, two models have been developed. Two models were built using multinomial logistic regression analysis and relate to management and employee attitudes about turning green.

Email: boopasangee@gmail.com; ORCID: https://orcid.org/0000-0002-2065-6539

S. Thirumalai Kumaran and Tae Jo Ko (eds.) Sustainable Machining and Green Manufacturing, (221–248) © 2024 Scrivener Publishing LLC

Keywords: Green manufacturing, waste disposal, sustainability

11.1 Introduction

The thesis primarily focuses on comprehending green manufacturing practices employed in the Indian automotive industry and their social impact. Today's corporate leaders are considering their three-manufacturing system because pollution has grown to be a serious issue. On the other side, a customer who has formed an eco-friendly perspective when making a purchasing decision avoids goods that are made in a way that is harmful to the environment. The amount of pollution has grown exponentially as a result of the number of automobiles on the road. Green manufacturing entails streamlining the production process, striving for production efficiency, and eliminating waste that might otherwise contaminate the environment. Before pollution reaches an excessive degree, the green manufacturing system needs to be implemented and monitored.

The consumer has grown environmentally conscious and is now picky about his or her mode of transportation. The government has even established requirements that automakers must follow in order to minimize their negative environmental impact. However, if required adjustments were made to the system of making cars, the environmental impact might be decreased. Internal combustion engines are contributing to a rise in the depletion of fossil fuels. Concern has been raised about the increase in traffic accidents, congestion, and traffic bottlenecks. Aside from end-of-life vehicle, factors affecting the car manufacturing industries' greenness include the raw materials used, the production method, etc. [1]. The best use of natural resources is the goal of green manufacturing (GM). The ultimate objective of GM is to ensure that society is not impeded by the rising waste and emissions produced by the aforementioned businesses. Due to the waste produced by the automotive manufacturing industry, General Motors is particularly concerned about protecting the environment [2]. The idea of GM demonstrates how to reuse waste materials to some level in order to reduce the cost of natural resources. It is essential for using the least amount of water possible while manufacturing, which is projected to be the major problem in the future [3].

There are two levels displayed, and they are as follows: processing of materials; manufacturing. Post-use, which includes recycling, remanufacturing, product recovery from waste, disposing of the product, and so forth [3]. The technique of sequencing has an impact on how efficiently cars are made. Customers want a wide selection of cars that are affordable and easily accessible for spare parts. For the efficient use of resources and the seamless running of processes, the sequencing operation is essential. Planning is essential for maximizing the utilization of natural resources [4]. Future competitions will not be based on resources like labor, capital, etc. The rivalry from green technology will nevertheless exist. The primary areas of attention will be system optimization for waste and energy as well as emission reduction. To protect the environment from pollution, government regulations are getting harsher [5].

Unsustainable energy practices are employed extensively in the vehicle manufacturing sector. In turn, this causes a significant amount of greenhouse gas (GHG) to be produced, which pollutes the environment. Environmentally conscious production is a brand-new idea in the manufacturing industry [6]. "Development that serves the requirements of the present without compromising the potential of future generations" is the definition of sustainability. The "triple bottom line" concept in business and sustainability considers social, environmental, and financial performance, encompassing a business's impact on people, the planet, and profits, beyond just financial success. All industries, regardless of product, consider the aforementioned three factors simultaneously [7]. Although implementing green practices would initially cost more, the current climate and environment will benefit future generations. The environment-friendliness of the product a consumer is consuming must be guaranteed [8].

Vehicle use is responsible for 80% of the environment's impact. Making the car lighter is crucial at the design stage. Automakers must prioritize after-sales in order to satisfy customers when there is less profit. It is planned to combine new energy sources and improve fuel economy. Reducing the negative consequences of human activity on the environment, especially in poor nations, is one of the goals of GM [9]. Between 1980 and 1990, GM became a concept in Germany. Process quality was first the main focus; later, goods were enhanced to produce the least amount of waste and emissions. After that, the emphasis in sustainable manufacturing switched from being on processes to being on products. The development of automobiles has had a profound impact on society from many angles and significantly altered how people live all over the world. It is generally accepted that using cars contributed to a wide range of air pollution that led to smog, GHG, environmental change, and human diseases. In the 21st century, environmental security and green arrangements are becoming more and more necessary [10]. Automobile OEMs and part manufacturers contributed significantly to the growth of the nation's rural and urban economies. However, the production of GHG, traffic accidents, congestion, lack of open space, poor air quality, parking lots, and urban sprawl have a negative impact on both health and the environment. Environmentally friendly products inspire consumer behavior. To help them green their products, industries have developed a wide range of programs. By utilizing green and clean energy sources, such as solar, wind, and other renewable energy sources, the depletion of fossil fuels can be halted or replaced in the future [11].

Emissions and trash are disrupting the climate and contributing to global warming, which is bad for society. To conserve the environment, there must be a trade-off between automotive development and sustainability. The relevant government must enact strict restrictions for the utilization of natural resources [12]. The manufacturing system is viewed from all angles according to the notion of GM. Recycling waste, properly disposing of industrial waste, and minimizing environmental damage are all emphasized. In GM, important locations where pollution originates are identified, and the appropriate steps are then implemented to eliminate pollution. Adopting a sustainable/ecological manufacturing system would be essential to achieving optimal resource usage with minimal negative environmental effects. Ensuring that resources are not exhausted purely for profit, leaving nothing for the community to rely on, is now practically necessary for the long-term growth of civilization. With minimal environmental impact, GM aims to strike a balance between increasing production capacity and conserving resources for future generations. Although it is possible to claim that a growing population is the main cause of these problems, it is also impossible to ignore the role that production-related businesses have played.

Traditional manufacturing techniques leave very little behind for society to use and discharge wastes into land and aquatic environments. The production process cannot easily be changed because of the rigidity of these systems. Therefore, trash disposal needs to receive more focus in order to limit environmental damage. Sustainable development is the process of meeting present-day needs without compromising the availability of natural resources for future generations. It is best to apply the principles of GM in such a situation. It takes a 360-degree approach to the manufacturing system to limit environmental effect—from concept design to waste disposal. The automotive sector is characterized by fierce competition, overloading, extreme market penetration, expensive labor, and fixed expenses. Corporate strategies allow a small number of international corporations to control the automotive sector. The rate of motorization is an additional crucial factor. It can appear like an automobile has a relatively limited life span. However, a longer product life cycle eases the depletion of natural resources from an environmental standpoint. Ecological sustainability is referred to as being green. It covers a range of issues, including waste management, energy usage, and contamination of the air, water, and land.

The emphasis on minimizing waste during production became the paradigm for sustainable manufacturing in the 1980s. The progress of the automobile over time reflected both societal and personal lifestyles. The transition from gasoline and diesel to electric automobiles on the road shows that a person is conscious of and attentive to natural resources. The auto industry is coming up with a plan to address the situation and guarantee that once its product is in use, the environment will not be affected. With transportation becoming more accessible, it has become advantageous for people to travel long distances. Yet, everything is bad for the environment. Manufacturing sectors can differentiate their products, raise the caliber of their manufactured items, and make minor adjustments to the manufacturing process to cut costs associated with implementing new products and methods that have a lower environmental impact. The 1990s saw the development of GM, a novel idea. There has not been much work done in this area.

GM's goals are to achieve zero environmental pollution, zero hazard to the operator's and end-user's health, maximum recycling, and minimal disposal. Studies on genetic modification have been split into two categories: works and the idea of optimal use of natural resources. Environmental consequences are lessened via sustainable manufacturing at the system, process, and product levels. It must integrate minimal energy use, waste reduction, potential employee safety, and health improvement while preserving product quantities. Premanufacturing, manufacturing, and consumption were the main focuses of traditional manufacturing, which produced excessive waste and increased landfill disposal. The four steps of the manufacturing strategies for the complete life cycle—premanufacturing, manufacturing, usage, and post-use—led to a decrease in landfills. To maximize benefits, apply the 6Rs process (remanufacture, redesign, reduce, recycle, recover, and reuse). Less advantage is gained by new industries that have embraced sustainable production. The management must strategically implement sustainable manufacturing. It necessitates system-level process and product integration and optimization. It is necessary to acknowledge effective methods for boosting manufacturing's economic benefit [13]. Without reducing use of time, energy, and resources, economy cannot be optimized. We can produce goods that are profitable while using fewer natural resources. Energy is one of the best ways to gain money in this area. We can significantly reduce the quantity of waste created if we carefully examine the manufacturing processes. Since the end of World War II, Toyota Motor has been experimenting with lean production methods. Utilizing its tools and processes, the lean manufacturing philosophy successfully produced cars and other items with fewer labor and raw materials. Lean production causes excessive wasteful material use, which raises fuel and energy consumption and directly raises environmental emissions. Additional wastes such as inventory waste, product repair waste, unnecessary product relocation, complexity, personnel hanging around due to delays, etc. will all ultimately result in increased emissions and waste. The phrase "development that fulfils the demands of the present without compromising the ability of future generations to meet their own needs" is used to characterize sustainable development. The environment's capacity to meet existing and future needs is constrained by the state of technology and social organization. A sustainable world is not even marginally more ecologically conscious than the one we live in now. The breakeven criterion must be reached for the company to survive. There must be a minimal impact on the environment, as well as resource and energy savings. Additionally, it must be efficient and safe for workers, clients, and communities. GM should be viewed as a chance to increase market shares both locally and globally. The manufacturers will be better able to comprehend GM technology and tactics if they have a deeper understanding of them [14].

By using additive manufacturing (AM), valuable products can be created directly from construction files. It is a technique for joining materials to create objects using 3D model data, sometimes layer by layer. Within the automotive industry, AM technologies have worked wonders to bring new shapes to life. AM reduces waste by reducing manufacturing steps, inventory, and the number of important parts required for assembly. Twenty-five percent of the remaining automotive waste is dumped underground as "auto shredder residue." Plastics are inexpensive, lightweight, and durable materials that may be molded into a variety of goods. Over the past several years, the use of plastics in the automobile industry has grown significantly in order to produce lighter vehicles. Four percent of the total gas and oil used in the globe is used to make plastic. Because of its strength, malleability, and durability, steel is one of the most important elements used in automobiles. Manufacturers have been reducing its use because it still requires a lot of work. Due to the aforementioned characteristics, it is challenging to manufacture things utilizing steel. Due to its endurance, malleability, and hardness, steel is a vital part of automobiles. Because it is still cumbersome, manufacturers have been restricting its use for these reasons. Steel presents a number of manufacturing challenges as a result of the aforementioned qualities [15]. In addition, 92.3% of automobile buyers have already seen engine coolant, 62% believe that water works just as well, but 62% have never purchased the product. When acquired, coolant was missing from 53.8% of the autos.

A real problem because the recommended refined or deionized water is not frequently used. Deionized water, glycols, corrosion inhibitors, biocides, and other ingredients like color, defoamers (also known as antifoam or foam suppressors), buffers, or caustics are all used in the production of engine coolants. The desirable engine coolants must have an alkaline pH, ideally one that is higher than 7.5. Despite this, a greater percentage of car owners still favor water for cooling their engines. Because of malfunctioning cooling systems, the majority of car owners have experienced overheating, burned gaskets, and part breakage [16]. Diverse levels of recycling are used. The direct material flow system, which disposes of all the materials used in the vehicle at the end of its useful life, is the least expensive. The most important aspect is closed-loop recycling, which involves remanufacturing a product's parts into similar manufactured goods. Programs for recycling materials aim to maximize their worth by ensuring that each resource has gone through as many different quality levels as feasible [17].

All Indian inhabitants have a basic right to clean air, fresh water, and environmentally friendly transportation infrastructure. If they are pollution-free, many environmental resources, including public parks, urban parks, rivers, mountains, woods, and beaches, support a healthy and joyful way of life for society. The growth of the automotive industry is essential for the nation's GDP, economic development, and fast-paced lifestyle that allow people to travel vast distances in comfort and little time. Carbon dioxide (CO_2) and other GHGs have an impact on the environment either directly or indirectly. Human health is unaffected by atmospheric CO_2 levels below a certain point, however, productivity and sleep may be affected. The release of CO_2 into the atmosphere is caused by the burning of fossil fuels, such as oil, natural gas, and coal, as well as solids, other biological materials, garbage, and trees.

Transportation-related GHG emissions result from the burning of fossil fuels by cars, trains, planes, trucks, and ships. Petroleum-based energy accounts for more than 90% of the energy utilized in transportation. Hydrofluorocarbons and sulfur hexafluoride are artificial potent GHGs released via various production processes. Top management, standard green adaptation, green purchasing, green packaging, green marketing, green disposal, recycling of used automotive components, and the 6Rs (reuse, recycle, repair, reduce, recover, and redesign) are among the green design initiatives. Manufacturing is the key to generating wealth and promoting social well-being. Sustainability is one of the top 3 issues facing the sector, according to the World Economic Forum's 11 Dimensions of Sustainability: The Triple Bottom Line research.

Due to governmental restrictions and the safety of modern life, sustainable goods design and development are urgently needed. Economic viability is essential for the industry's continued existence. But if permanent harm from GHG emissions and toxic waste is not taken properly, the business or organization will not last for a long time. A researcher has created a few models that describe how management and employee attitudes toward GM and other factors that affect sustainability are explained. After a thorough review of the literature and the creation of a questionnaire, a thorough survey of numerous automotive OEM and automotive component manufacturing businesses was conducted.

11.2 Green Manufacturing Production

Surface Mount Technology (Holdings) Ltd. has conserved energy, retained and recycled paper, and decreased pollutants. Battery uses minimal lead, cadmium, and mercury, and invests in waste, emissions, and water management. Green IT places a strong emphasis on equipment deployment and power efficiency. Energy-efficient slim clients (lightweight computers) are taking the place of individual PCs, and the IT sector is utilizing virtualization package systems to run many operating systems on a single workstation. Telecommuting and remote computer administration cut transportation emissions. The Constraint of Harmful Constituents standards are being followed in the production of electronics, which increases efficiency through environmentally responsible design and production. Measurement, analysis, and improvement of processes are done using quality and Six Sigma approaches. There are restrictions on the use of some metals, including lead, mercury, cadmium, hexavalent chromium, and polybrominated biphenyl ethers. Other countries, including China, Japan, and South Korea, have enacted similar laws [9]. Sustainable manufacturing has been improved by the use of AM, which benefits the environment and increases corporate profitability. In the manufacturing industry, metal additive forming and AM were both discussed [18].

After industrial production, industrial CO_2 emissions can be decreased in a variety of ways, such as by reprocessing raw materials or substituting biomass for fuel. Enhancing the capacity for encapsulating and storing such emissions minimizes industrialized secretions and increases energy efficiency of activities [19].

11.2.1 Green Marketing

GM must be viewed as a chance to expand regional and international markets and take in this exciting environment. An improved understanding of green production techniques and strategies will enable producers to realize that, unlike various competing goods and methods, keeping green favorably benefits all other possible ends.

11.2.2 Government Support and Regulations

More rules and sanctions are being developed by the EU, Asia, the United States, and South America for nations that do not reach their environmental goals. Eco-friendly production techniques gain greater and more support when manufacturers adopt a global perspective rather than just being an option. Potential GM is ensured when scientific enablers and this economic knowledge are combined. Based on the aforementioned factors, existing producers interested in improving environmentally friendly practices may be financially feasible [14].

11.2.3 Identifying Problems in Automobile Fields

In order to reduce emissions, production costs, waste, and resource usage, GM appears to be the industry's future trend. One of the financial points to be made in this research is the cost of quality (COQ). It estimates the expenditures incurred as a result of manufacturing subpar materials.

In the context of the expanding manufacturing sector, the author analyzed GM practices used by the Indian car industry. The cost–benefit analysis of introducing GM in the Indian automobile industry is another area he examines. The author's conclusions offer suggestions for increasing energy efficiency, lowering emissions, and increasing industrial efficiency.

11.2.4 Green Manufacturing: Automotive Research Focus

Manufacturing generates a lot of trash and emissions, which have a negative impact on society by causing bad weather and major health problems. Water is frequently utilized in large quantities in the vehicle manufacturing industry. My research topic will support the idea that GM with zero waste and low emissions can have a positive impact on the environment. My research will be focused on lowering waste, harmful materials, energy use, and carbon emissions because they use a lot of energy. This research will also explore more on recycling, reuse, and moderate emissions. Analysis of the financial advantages of green activities is considered. To cut energy use and conserve fossil fuel energy, renewable energy is also advised. GM aims to make operations more efficient, use less resources, and benefit the environment. Enterprise resource planning (ERP) and supply chain management are two examples of practices driven by technology. The size of a nanoparticle, but the savings will add up to a significant amount of value overall.

11.2.5 Green Manufacturing Efficiency

Power, time, and commodities are all forms of riches. A great way to build money is to produce comparable produced things with few resources and abilities. There is a lot of waste produced that may be eliminated from the processes and the manufacturing.

11.3 Green Manufacturing in the Automobile Field

11.3.1 Green Manufacturing for Reduction of Emission

NOx emission reduction due to amendments of the GM concept is illustrated in Figure 11.1. Lowering greenhouse gas emissions inside the delivery/shipping event will reduce carrier exhaust emissions. You can achieve this in the following ways: utilizing the mode of transportation that uses the least amount of fuel per delivered unit of weight, improving the fuel efficiency of the transportation means, and reducing the amount of time the engine runs overall by either carefully planning the path of the vehicle or shortening the time it runs [14]. Car manufacturers frequently used

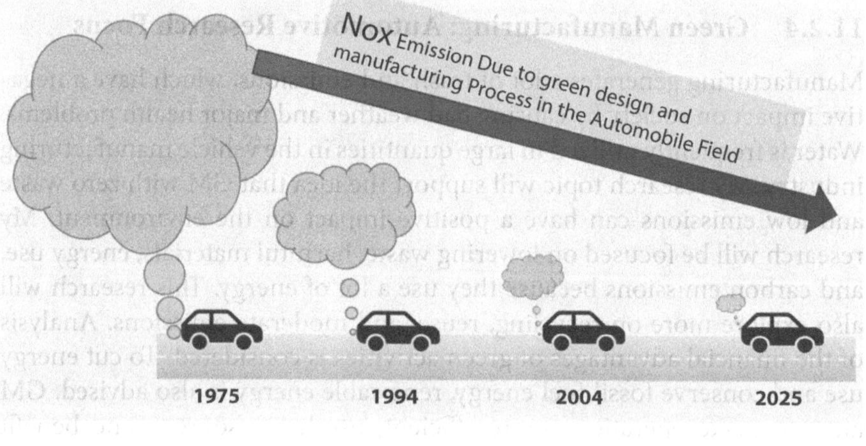

Figure 11.1 NOx emission reduction due to amendments of the GM concept.

catalytic evangelists as part of their conformance strategies in response to ecological standards put forward by several countries in the 1970s. Several auto manufacturers used three-way facilitators in the early 1980s as part of a clogged-loop secretions control strategy. On-Board Pinpointing (OBP) system creation and use have become a focal element of community strategy. Engine issues and control device events are discovered by the OBP-I system using electrical techniques. Measurement Tool is also used to evaluate the performance of the engine and the amount of motorized discharges [11].

The majority of manufacturing organizations are reorienting themselves toward the goal of expanding the usage of lightweight raw materials. The primary obstacle to employing lightweight raw materials is their high cost. The pursuit of cost savings through the development of novel manufacturing processes, know-how, and raw materials is given emphasis. The fuel economy improves by 7% for every 1% reduction in a vehicle's total load of 10%. Approximately 20 kg of CO_2 are released for every kilogram of weight that is dropped in a motor vehicle. Auto manufacturers have been experimenting with steel alternatives made of aluminum, magnesium, composites, and foams in order to achieve lightweight production without compensating for rigidity. Sustainability will continue to be a major issue for motorized production, encouraging further significant reductions in the overall ecological impact of cars globally. Implementing improved contaminant control techniques for autos, such as the after-therapy components and devices or the catalytic apostles, is a concern [17].

Manufacturing, technology, and digital strategy have created a crucial methodology for modernizing the equipment fabrication industry and increasing the fundamental affordability of organizations. The production and controlling methods have increased due to the digitization of creating, which includes raw material selection and product design. The volume and weight of components have decreased, and produce growth and manufacturing have both improved. Range Rover Sport Hybrid and Range Rover Hybrid from Jaguar Land Rover were introduced. The top-ranked outstanding diesel SUV hybrids in the world reduced CO_2 emissions by 26% compared to counterparts with conventional powertrains and offered 44.1 mpg on combined electric and diesel power. The Passenger Vehicle Business Unit and Commercial Vehicle Business Unit action plans for this approach were developed, according to Tata Motors sustainability report 2015-16. Its approach to reducing environmental change and promoting low carbon expansion is 3-fold [22]. BMW looks for ways to lower CO_2 emissions and boost fuel efficiency in its vehicles. Guidelines for automotive and device secretion are described by the Ecological Safety Bureau.

A conformance program to use a secretion control plan for portable sources is confirmed by the Fresh Air Act. For the model years 2017–25, new regulations on fossil fuel and carbon efficacy were made. Additionally, it has received recognition for its high-efficiency exterior lighting, solar-clear glass and glazing, dynamic seat aeration, and active cabin ventilation. Dynamic transmission is a method that quickly raises the communication fluid's temperature to a usable high range by using waste heat from the vehicle. By reducing parasitic losses associated with fluid diffusion, such losses—which are associated with friction and water stickiness—can increase the overall transmission efficiency [6].

The cost of fabrication for part fabrication in an automotive industry has decreased by approximately 10.8% as a result of the use of tailored lean and environmentally friendly strategies. Lean practices ensure environmental benefits, but on the other hand, environmentally friendly practices typically result in improved lean services. The relationship between lean dissipates and their ecological (green) influence is depicted. By reducing defective parts, Lean Manufacturing (LM) helps to reduce the need for energy, raw materials, and storage space for them. Additionally, reducing other litters such excess processing, waiting, transportation, mobility, and overproduction reduces excess energy, raw materials, emissions, and space.

11.3.2 Green Manufacturing to Minimize Automobile Waste

The proper drainage system must receive waste from auto dealerships and workplaces, not storm drainpipes that run to canals and ponds. Exhaust secretions are the main source of contamination resulting from delivery or conveyance (comprising mostly unpredictable carbon-based amalgams, CO, CO_2, and NOx). There is a chance that each procedure will result in environmental pollution from the seepage of hydraulic fluid, gasoline, or oil.

BMW has committed a portion of its innovation to efficient heat management as a way to reduce fuel consumption and CO_2 emissions. Waste heat is utilized in cars to speed up cold starts and for a variety of additional heating purposes. Even yet, a fuel-efficient ignition engine converts only approximately one-third of the energy from fossil fuels into the vehicle's propulsion. Utilizing fossil fuels more efficiently could lower CO_2 emissions and fuel consumption. GM practice for automobile waste management is depicted in Figure 11.2. Today, a very small amount of energy can be obtained by preheating the engine. Additional improvements will cut petroleum consumption and CO_2 by a number of fractions. Mahindra Rise (2012) placed a strong emphasis on lowering water consumption

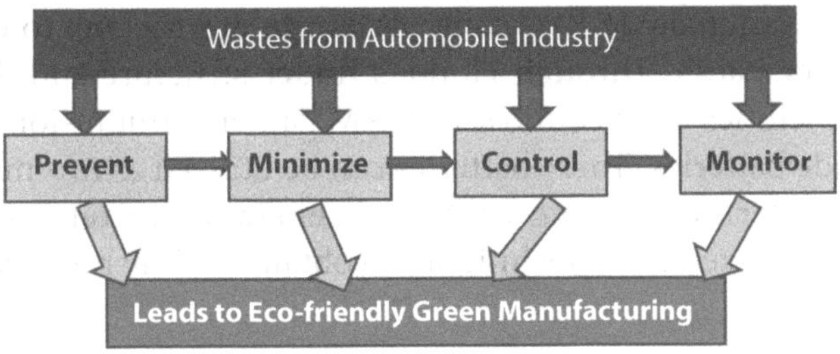

Figure 11.2 Green manufacturing practices for automobile waste management.

by roughly 13.45%. It also made the initiative to treat wastewater with enzymes so that it could be reused. Mahindra created a system to collect and purify Chennai's wastewater before returning it to its users [5, 7]. The liberalization of India has made it easier for foreign business, particularly in the automotive sector. In a nation like India, hazardous waste ends up in landfills or trash cans. Every year, approximately 4.4 million tonnes of hazardous waste need to be removed. Severe physical health issues may have developed if the garbage is not systematically removed. Most of the time, recycling, reuse, and reduction are used to manage inorganic waste. The primary contributors to the reduction and recycling of inorganic materials are cutting-edge methods and distinguishing structures. Contracts for the management of paint shop chemicals, plastic, and metal refuse are issued to authorized dealers.

General Motors offers more than 100 landfill-free amenities worldwide and recycles 90% of the waste from its global production. The company processes and recycles by-products to save approximately $1 billion in revenue each year. GM is also the second-largest consumer of landfill gas in manufacturing. Volkswagen received a prize from the Leadership in Energy and Environmental Design competition of the US Green Building Council. For its environmentally friendly paint shop, it conserves 50 million gallons of water annually. This water is used by the nearby hydroelectric dam to produce clean energy. Utilizing a rainwater collection system and low-flow toilets for urinals has helped to reduce water consumption. The highest-quality insulation and white reflecting roof are used to maintain 72,000 KW of power. LED lighting on the plant's facade uses 68% less energy than conventional lighting [17].

The total amount of water used by Mahindra and Mahindra has decreased by approximately 1,34,696 m^3. It also invests in renewable energy. It has a 269-KWP Solar Energy Plant installed at a tractor stockroom, which meets 27% of its energy needs. Since 2013, Honda has collaborated with Societe Nationale de Recycling d'Affinage des Metaux to collect and recycle battery packs. Through Honda's trader structure and Authorized Treatment Facilities in 22 countries, SNAM gathers lithium-ion and nickel metal hydride batteries. These batteries can be utilized as raw materials for brand-new batteries, reducing waste and generating income. Honda also employs modernized reuse techniques to obtain healthy raw materials.

11.3.3 Green Manufacturing: Resource Utilization

Two main groups of fibers can be distinguished: man-made and natural. In motorized manufacturing, organic fibers significantly increase the capacity to offer more remarkable better value, sustainability, renewability, and lower pricing. Plant fibers as a whole contain roughage, while mammal fibers contain proteins (fur, silk, and fabric). Plant fibers include foil, plant seeds, produce fruit, burrow, grass, grain chaff, and wood. They also include bast (or stem or soft sclerenchyma tissue) fibers. Bio-based amalgams will enhance machine-driven power and audio performance, reduce material mass and petroleum consumption, reduce manufacturing costs, enhance customer protection and unbreakable performance under extreme temperature variations, and improve the internal biodegradability of automobiles [20]. Renault's circular economy is used, as evidenced by the "drive the change" (2014) film. The corporation used its previous items to create new ones. The most effective utilization of natural resources is carried out in a circular economy. The circular economy's guiding concepts are repair and reuse. The French automaker annually refurbishes 16,000 fuel infusion structures, 20,000 gearboxes, and 30,000 engines. Ninety percent of the parts are recycled. According to Renault, the circular economy makes sound business and financial sense. Ford Motor Company invested $2 billion in a living roof that can hold 4 million gallons of storm water annually. By placing these wastes and pollutants on the roof, a corporation was able to prevent its trash, such as lubricant, grease, and chemicals, from contaminating the nearby stream. Geothermal energy is used by Ford Motor's Ohio plant to cool the facility's air. The method makes use of water from granite mines that are located on the factory property. The procedure costs $ 300,000 less to establish than a traditional freezing pillar and saves 4,300 metric tonnes of CO_2 annually [21]. Mongrelized kenaf fiber may have improved the dampening capabilities of the composite. Natural fibers are the best choice for protection because of their nonabrasive qualities. When distorted, organic fibers do not generate extreme ends like cut-glass grains do. They also suggest a way to calm the agitation brought on by machinery, wheels, exhaust structures, and appliances.

11.3.4 Green Manufacturing: Cost Minimization

Opportunity factors are mostly connected to COQ, which is stated as revenue lost. Opportunity costs can be added to uncertainty to produce more effective and lucrative customer satisfaction initiatives. Opportunity losses account for a lesser percentage of COQ, also known as profit not earned [4, 15]. Automation of business processes including virtualization (such as online electronic commerce) enables better utilization of business resources, lower costs, and better amenity provision for customers [22]. Advanced zero-carbon manufacturing techniques are available, thanks to small and zero-carbon know-how, but which expertise should be applied? A cohesive systems-thinking approach is essential. The facility's material, energy, and waste flows can be used to examine the impact of technology. The remarkable achievements of these pieces of knowledge have all been consistently utilized to advance industrial technology in recent centuries due to the rapid growth of knowledge in areas such as electromechanical assimilation, PC, data, command and control, etc. Numerical technologies, novel materials, close-to-net form production, clean fabrication, and garbage-free production are a few examples [23].

11.3.5 Implementation of Green Manufacturing in the Automobile Field

Opportunities to work in environmentally friendly ways are still available in India's branded and OEM automotive manufacturing industries. These approaches are not used and published in the public domain in GM, according to an analysis of the research publication materials. According to the survey, India's automotive manufacturing industries consume a lot less electricity generated by solar and wind energy. According to survey data, persons who work in or near manufacturing businesses are least concerned about serious health problems brought on by environmental pollution.

Method: The discussion that follows concentrates on the methodology to be used, going into detail into the many aspects of the inquiry design, research area, sample size, data collection kits, and data evaluation.
Design: To increase India's national production, the exploratory investigation focuses on creating models for making GM processes prevalent and ubiquitous.
Manufacturing: Administration policies and procedures will have an impact on the manufacturing sector to increase service flexibility and

knowledge sharing. This was built upon a number of past influences, and future inspirations can be investigated.

Data Collection: The study's primary focus is auxiliary data, which are supplemented with primary data. The following are the data collection tools: (a) Questionnaire: A questionnaire is developed primarily to assess the factors that contribute to model construction and value creation in the processes of GM. (b) Interviews: There are numerous stakeholder interviews undertaken. Getting knowledge about various services, laws, and regulations is the main goal of this. A consumer consultation is also conducted to assess their expectations and needs. (c) Secondary Data: government directives and rules, consulting specialists and examining academic journals.

11.4 Green Manufacturing Practices in the Automotive Field

11.4.1 Automotive Emission Control Practices

When a fuel is burned inside of an internal combustion engine, the bulk of its energy is released as a cloud of waste into the atmosphere. Figure 11.3 illustrates the relation between vehicle fuel and energy wastages. The fossil fuel economy could be significantly improved by using cutting-edge information. Modern internal combustion engines only convert one-third of the latent chemical energy in fossil fuels into motorized effort. Due to friction and garbage heat generated by the moving engine parts, the remaining

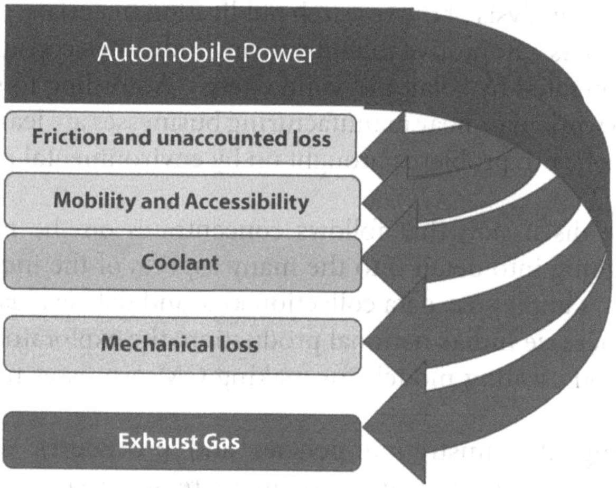

Figure 11.3 The relation between vehicle fuel and energy wastages.

energy is lost. All of these power loss measures present new opportunities for cutting-edge technologies to increase the productivity of the automobile [24].

In order to achieve these goals, elements such as well-being, excellence, eco-friendliness, security, user-friendliness, mass, human factors, expense efficacy, and manufacturability of car mechanisms must be considered in addition to technological, environmental, and financial restraints. Recent research on magnesium and magnesium composites has focused on its mass reduction, power-conserving, and ecologically constricting effects.

Global advancements drive motorized manufacturing to create lighter, more environmentally friendly, safer, and less priced autos. Losing weight reduces greenhouse gas emissions while also conserving energy. Car weight has evolved as the incredibly important benchmark for project effectiveness assessments since CO_2 radiation is directly proportional to fuel consumption. The size of the vehicle must be reduced by at least 10% to achieve a stipulated reduction in motorized mass.

A motor vehicle will save approximately 0.5 liters of gasoline per 100 km if its mass drops by 100 kg. In order to reduce the mass of the cars, high-strength steels, aluminum, and amalgams are currently being used. A better use of low-density magnesium and its additions could result in further decreases. Currently, 0.25% (3.8 kg) of magnesium and 8% (120 lb) of aluminum are used on average in vehicles in North America. Magnesium management, alloy development, and surface therapy still require more research. The most economical option for significantly lowering fossil fuel consumption and CO_2 emissions is the mass decrease [25–27].

11.4.2 Manufacturing Cost Reduction Practices

To achieve the desired business goals, both prototypes use managing techniques, garbage detection, and trash decrease techniques. Others devise and carry out "Green" Plans, which reduce electricity use and waste production. Lean and environmentally friendly strategies are superior to other company alternatives. Being more resource- and energy-efficient, lean businesses demonstrate significant ecological improvements. Few studies specifically show how lean and environmentally friendly practices help many of the best traditions reduce their corresponding garbage. Sustainable production techniques, such as those recommended by groups like the US Environmental Protection Agency, can be developed and adopted with the help of lean manufacturing.

11.4.3 Waste Reduction Practices

The powerful tool known as Value Stream Mapping (VSM) is used to find waste in a company. The VSM is a pen-and-paper tool that makes it easier to see and understand the flow of information when the product enters the value stream. These ideas are applied in a company that manufactures motorized elements with an emphasis on tube subassembly lines. To clean up the assembly and logistics department of the company, VSM and method time measurement are merged. The disc brake actuator manufacturing company currently uses this idea in their manufacturing processes. The current status of a fabrication line could be represented by VSM or another group of lean kits, like kaizen. It could see trash piled up in the middle [28].

Sort, straighten, sweep, standardize, and sustain are the acronyms for the 5Ss. The +1 denotes the importance of ecological security and well-being throughout the entire thing.

Sort: Typically, this starts with a tour of the office where the extra items are displayed. When the components are found and evaluated, they are either repositioned or, if required, removed.
Straighten: The purpose of this action is to designate a space for everything that is now in use and keep everything in its proper position or to organize the tools and products so that they are easy to locate and use when needed. This is done by identifying and positioning the equipment and items such that they may be recognized visually.
Sweep: This involves cleaning the equipment and work area. Maintenance work can be done to prevent any impending interruptions while the workspace and equipment are being cleaned.
Standardize: Sort, straighten, and sweep integration would result in an effective normalizing and improve task area pace, excellence, and security. This necessitates maintaining and managing the original three Ss, S, M, and R, as well as each of their individual subcommands.
Sustain: The goal of 5S +1 is to design a life pathway in which the abovementioned behaviors are not useful until they are sustained over an extended length of time. They must be carried out repeatedly in a consistent manner to reap rewards like money, wealth, and social capital [7, 18].

11.4.4 Resource Utilization Practices

The first step in end-of-life vehicle management is depollution, which is nothing more than draining fluids and removing hazardous materials (such batteries and airbags). Reusable spare and core parts are recovered.

The body's scraps are fed into a shredder. The appropriate utilization of resources is met in this manner [19].

11.4.5 ERP for Green Manufacturing Practices

11.4.5.1 Enterprise Resource Planning

According to the European Automotive Survey, which surveyed 300 companies with operations in Europe, European car and subassembly manufacturers are upbeat about the future. A decrease in the costs of manufacturing, warehousing, and transportation is of utmost importance from an industry's point of view because of the continuously expanding market needs and intensifying conflicts. An ERP system is designed to easily include operational strategy with a better road map, control many commercial activities, and access real-time records. The technological and operational inclusion of business purposes would be one of the major factors for an ERP implementation's potential cost savings. ERP systems in the automotive industry enable the planning, management, and control of logistic flows involved in the transportation of goods, cash, and information along the whole supply chain. The standard functions of an ERP system in logistics include provisioning, warehouse management, manufacturing planning, trades, delivery, and operations refurbishment and preservation. Software from Systems Applications and Products (SAP) is used by more than 3,000 automobile companies worldwide. Automotive companies who are SAP customers generate 77,000 autos every day. The motorized and component automotive trades are often represented by a different version, design, or sector of ERP systems. Cycle times, accounting, and operational costs can all be decreased with SAP ERP. The program makes it possible to standardize shipping procedures, which boosts productivity and lowers errors. Using Electronic Data Interchange (EDI) technology, all contracts may be immediately consigned for JIT and JIS workloads [26, 29, 30].

11.4.5.2 Supply Chain Management

One of the most popular green initiatives is called Green Supply Chain Management (GSCM), and it is especially popular in the car industry. Automobile manufacturers like Toyota, Honda, Ford, and others frequently use inverse logistics, one of the core components of the GSCM.

Consumer ignorance and a lack of administrative support for the lowest-level barrier constitute the highest level of difficulty. It has been observed that GSCM positively correlates with business success. GSCM is one of the key issues that businesses are focusing on at the moment. An ecological burden is produced by the acceleration of economic growth, particularly in the automotive industry. The benefits of green initiatives for supply chain management include cost savings and environmental protection. Automobile firms in several developing nations, like China or Malaysia, have used GSCM and picked up capabilities from international partners.

11.5 Case Study: Automobile Green Manufacturing Firm

Green manufacturing includes both the production of "sustainable" goods and the production of all goods in a sustainable manner. Producing goods that use non-polluting methods, save energy and common resources, and provide representatives, communities, and customers with financial stability and security embodies a commitment to environmental stewardship and social responsibility, fostering a harmonious balance between economic growth and ecological preservation. India's annual budget, or $80 billion, is lost to ecological deprivation. It includes products relating to social fairness, on-conventional energy, power efficiency, environmentally friendly building, and other "green" products. The ecology has gotten worse, causing air and water pollution for the locals. With a nominal price to GDP, the emissions can be reduced. The Environmental Management Systems standard ISO 14001 can be a great place to start for the green idea. GM program at the Michigan Manufacturing and Technology Centre complies with this criterion from every viewpoint.

11.5.1 Green Methodology

A conference was conducted with industrial ecologists who had the necessary knowledge and green concepts to address the green concept in the manufacturing industry. The conference was planned since the academic study will be kept private and the name of the sector must remain a secret. Research methodologies include eight eco-friendly elements that aid the manufacturing sector in achieving the green characteristics. The final score will determine how well each of the green methods performs. Businesses can evaluate their strengths and weaknesses and implement a green supply chain administration with the aid of these green concept aspects.

11.5.2 Implementation Procedures

A case study on the manufacturing sector, which creates fasteners for transmission lines for cars, was conducted. Several green-related elements have been observed in the industry, which are listed below, and are compatible with the standard practice considered in the sector. The implementation procedure of the GM process in the automotive industry is illustrated in Figure 11.4.

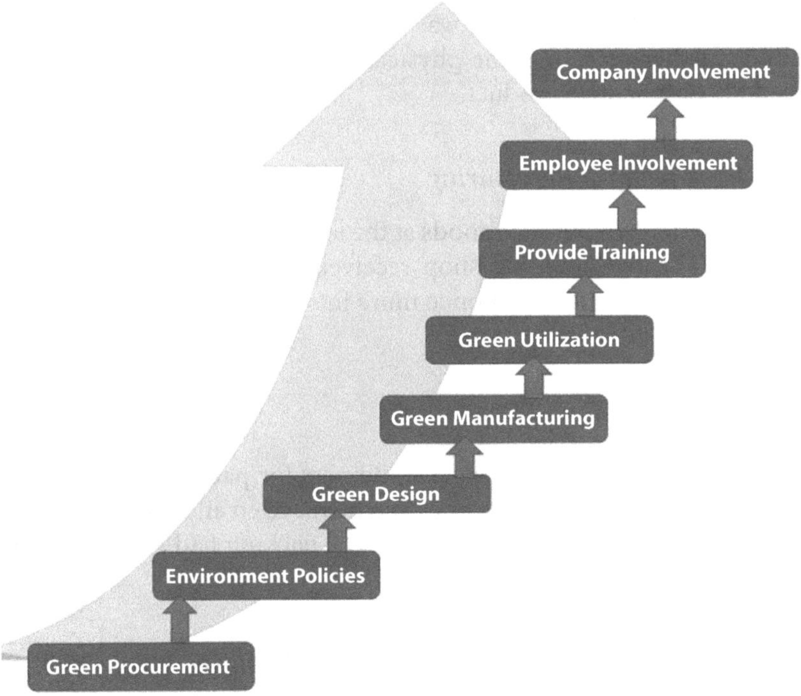

Figure 11.4 Implementation procedure of the GM process in the automotive industry.

11.5.2.1 Green Procurement

Suppliers of chemicals, oils, development materials, and packaging materials should be effectively directed by the green purchasing materials and spare part acquiring division. It can be flexible if we place more emphasis on the following issues: shipping, vendor familiarity evidence, paperless jobs, and eco-friendly vendors.

11.5.2.2 Environment Policies

Reduce environmental risk by using operational procedures and emergency preparedness programs, such those provided by the Department for Environment, Food, and Rural Economy in England and Wales and the UK's Ministry of Energy, Water, and Natural Resources.

11.5.2.3 Green Design

As a practical designer, eco-friendly design is also known as environmental design, environmentally viable design, or ecologically aware design. Here, 93.3% of the steel that was used was recycled scrap steel from the industry. Without further recycling, the physical qualities of the steel are lost and turned into a new steel product.

11.5.2.4 Green Manufacturing

The goal of GM is to produce goods at the lowest possible cost while reducing waste. The steel melting shop received the scrap material gathered during production and cast it once more into steel slabs for usage in other industries.

11.5.2.5 Green Utilization

Instead of utilizing plastic or steel containers for packaging, the fasteners gunny sacks were chosen, since they can be recycled and reused repeatedly. Volume, form, and material qualities of the packing had an impact on the product's transportation properties.

11.5.2.6 Provide Training for Improving Employee Involvement

Employers ask employees and laborers to create high-quality circles that highlight problem-solving motivated by an eco-friendly approach, increasing their awareness of the concept of being green. Senior supervisors from different businesses are asked to participate in discussion group sessions for the specialists, who will receive preparation on green business practices.

11.5.2.7 Customers Responsiveness Program

Meetings with buyers are held for environmentally friendly reasons. Green products are also subject to customer feedback and interaction to raise awareness of the GM and concept.

11.5.2.8 Automobile Industry Commitments

The Apex Administration is committed to long-term green policies. The long-planned budget is given preference over the green concept. Additionally, it keeps an eye on the procedures used in the sector. Analyzing the elements of a green concept, the most important element in a green design, has the greatest influence.

11.5.3 Outcomes

For the manufacturing sector to implement green supply chain management, there are eight green concept factors. This will lessen the adverse effect on the climate and contribute to environmental preservation. To preserve the environment for future generations, the product will be economical and environmentally responsible.

11.6 Case Study: Water Conservation Technologies

11.6.1 Technical Factors for Implementations

Bowl size: For proper water filling to flush, syphoned cleansing requires a larger stretched bowl than the washdown's compact design. While extended shapes are preferred in the United States, space-saving variants are more popular in Europe. As a result, European restrooms are used more frequently than US-type bathrooms.
Water saving: For solid waste or liquid trash, a double-flushing utensil with 3 or 6 liters of capacity can normally be used to flush down the toilet. In contrast, syphoned bathrooms require 6 gallons or more to flush properly. Few countries have 4.5-gallon flushing capacity.

Trap way: Sometimes, sprinkling occurs when dependable waste hits the water from the faucet for washing purposes. By covering the water with a little toilet paper, this will be lessened. After flushing, syphoned water leaves the bowl with less water than it did when it was half full.
Spare parts and maintenance: A flush valve, ABS plastic float valve, and top button flush are components of a washdown toilet that occasionally need to be replaced. A drift ball and chain are present in a few editions, and those need be changed more regularly than the ones that feature a side lever in place of the main lever.
Cleaning: Instead of using a plunger, which is more often used to plug a syphoned toilet, Europeans like to utilize a latrine brush as part of their shower decorations. Due to the gaps along the bowl rim, European washdown toilets frequently feature "watermarks."
Changing toilets: Siphoned toilets typically require a 12″ (305 mm) outlet from the wall to the center of the bowl. Considerably 8″–12″ (200-300mm) are needed for washdown adjustments, so see whether it is possible to make changes to current units.

11.6.2 Low-Flow Outlets

With a low-flow aerator, the stream's flow will slow to 1.5 gallons per minute or less. Approximately 2.2 gallons per minute are surged out of the typical faucet. A person would use approximately three quarters of a gallon of water to wash their hands for 20 seconds. When cleaning, water pressure is essential, and installing a low-flow spigot aerator does not compromise this requirement. The installation of an aerator will still allow the average household to save roughly 700 gallons of water annually, or the equivalent of 40 showers. Natural water resource utilization in the automobile sector to save the environment is illustrated in Figure 11.5.

11.6.3 Waterless Urinals

They can save up to 1,70,000 liters of water a year per urinal. Urine, which naturally contains primary water and is antibacterial, does not require additional water for it to gush. Water is not required for waterless urinals, and pricey sanitation fixtures are typically required for cleaning. The spread of contagious diseases is slowed down by touch-free procedures and the dry process of dehydrated urinals. Air entrap systems with sealing fluid, bacterial control, a layer, and a screen valve work to stop odors from developing inside urinal drainage lines. When using a conventional urinal, ammonia from the drainpipe is released into the bathroom, creating a dreadful odor

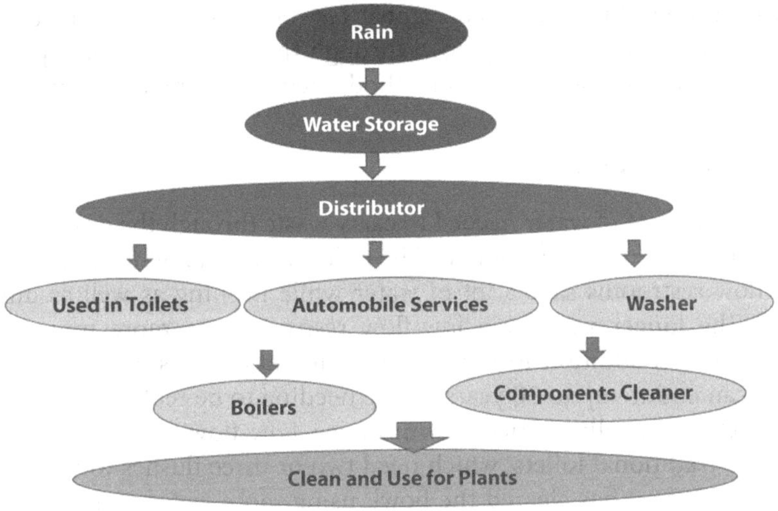

Figure 11.5 Water resource utilization in the automobile sector to save the environment.

and nasty stench. Zerodor has a barrier-like action that stops the bad odor from coming from the drainpipe. Additionally, it conserves water and keeps the restroom fresh. Using waterless urinals will cut emissions, expenses, and electricity use, all of which are related. Less water will need to be used to treat urine waste, which will affect the price and emissions. The collected concentrated urine will be utilized as fertiliser and to make hydrogen. More than 95% of the volume of human feces is made up of water. The usage of hard water and the ensuing scaling formation that results in pipe blocks and the requirement for a deodorizer will increase costs. A significant collection of diluted pee could harbor airborne pathogens as a result of excessive water use. Hand-operated flush valves on toilets may encourage the development of bacteria, which may lead to infectious diseases. Naphthalene and strong cleaning chemicals will contaminate the water, resulting in significant wastewater production. Rising nitrogen and phosphate concentrations in drainage may be decreased if urine is collected from sewage for use.

11.6.4 Washdown in Toilets

Low-flow toilet: The water after the container pours down into the box and washes up the person's excrements down the vent in a washdown combat restroom. It can be a "pushing" activity that does not empty the bowl. Streaming water basically pushes human feces off of the entrance way (usually 4 in diameter).

Siphoned toilet: In a syphoned combat restroom, water is pumped from the reservoir into the mixing bowl through the jet, and the rim "pushes" human waste out. The syphoned method often uses a smaller output.

11.6.5 Outcomes

The total amount of water needed to carry waste through the drainage system and away from your home will be reduced with a low-flow bathroom. Low-flow restrooms save a lot of water while flushing as well as during usage. The longer you use a low-flow restroom, the more money you should save in addition to recovering your initial investment. Low-flow toilets can endure up to 30 years before needing to be replaced. They may also disclose the selling price of your house. Low-flow toilets are an alternative to traditional toilets, which need two or three flushes to verify that heavy excrement has cleared the bowl, using more water and decreasing the effectiveness of the latrine. If you live in an older home, low-flow toilets may not function with the plumbing already in place. The likelihood of needing a sanitation upgrade to comply with low-flow lavatory increases with the age of the home. Additionally, low-flush toilets are expected in homes built in the last two decades. As a result, the establishment will take longer and cost more money than anticipated.

References

1. Paul, I.D., Bhole, G.P., Chaudhari, J.R., A review on green manufacturing: It's important, methodology and its application. *Proc. Mater. Sci.*, 6, 1644–1649, 2014.
2. Sanchez Rodrigues, V. and Kumar, M., Synergies and misalignments in lean and green practices: A logistics industry perspective. *Prod. Plan. Control.*, 30, 5–6, 369–384, 2019.
3. Ghazilla, R.A.R., Sakundarini, N., Abdul-Rashid, S.H., Ayub, N.S., Olugu, E.U., Musa, S.N., Drivers and barriers analysis for green manufacturing practices in Malaysian smes: A preliminary findings. *Proc. CIRP*, 26, 658–663, 2015.
4. Staeblein, T. and Aoki, K., Planning and scheduling in the automotive industry: A comparison of industrial practice at German and Japanese makers. *Int. J. Prod. Econ.*, 162, 258–272, 2015.
5. Gupta, S., Dangayach, G.S., Singh, A.K., Key determinants of sustainable product design and manufacturing. *Proc. CIRP*, 26, 99–102, 2015.

6. Mittal, V.K. and Sangwan, K.S., Development of a structural model of environmentally conscious manufacturing drivers. *J. Manuf. Technol. Manage.*, 25, 8, 1195–1208, 2014.
7. Pallaro, E., Subramanian, N., Abdulrahman, M.D., Liu, C., Sustainable production and consumption in the automotive sector: Integrated review framework and research directions. *Sustain. Prod. Consum.*, 4, 47–61, 2015.
8. Jawahir, I.S., Badurdeen, F., Rouch, K.E., Innovation in sustainable manufacturing education. *11th Glob. Conf. Sustain. Manuf*, pp. 9–16, 2013.
9. Rusinko, C.A., Green manufacturing: An evaluation of environmentally sustainable manufacturing practices and their impact on competitive outcomes. *IEEE Trans. Eng. Manage.*, 54, 3, 445–454, 2007.
10. Rehman, M.A.A. and Shrivastava, R.L., Green manufacturing (GM): past, present and future (a state of art review). *World Rev. Sci. Technol. Sustain. Dev.*, 10, 1-2-3, 17–55, 2013.
11. Hascic, I., De Vries, F., Johnstone, N., Medhi, N., Effects of environmental policy on the type of innovation: The case of automotive emission-control technologies. *OECD J. Econ. Stud.*, 2009, 1, 49–66, 2009.
12. Zhang, X., Badurdeen, F., Jawahir, I.S., On improving the product sustainability of metallic automotive components by using the total life-cycle approach and the 6R methodology. *Global Conference on Sustainable Manufacturing*, pp. 194–199, 2013.
13. Gao, M., Wang, Q., Wang, N., Ma, Z., Li, L., Application of green design and manufacturing in mechanical engineering: Education, scientific research, and practice. *Sustainability*, 14, 1, 237–246, 2022.
14. Deif, A.M., A system model for green manufacturing. *J. Clean. Prod.*, 19, 14, 1553–1559, 2011.
15. Hamer, G., Solid waste treatment and disposal: Effects on public health and environmental safety. *Biotechnol. Adv.*, 22, 1–2, 71–79, 2003.
16. Yusuf, T.A., A study on the factors influencing prefered use of water over the branded engine coolant among automobile users. *Int. J. Trend Res. Dev.*, 4, 6, 2394–9333, 2017.
17. Mayyas, A., Qattawi, A., Omar, M., Shan, D., Design for sustainability in automotive industry: A comprehensive review. *Renew. Sustain. Energy Rev.*, 16, 4, 1845–1862, 2012.
18. Garetti, M. and Taisch, M., Sustainable manufacturing: Trends and research challenges. *Prod. Plan. Control.*, 23, 2–3, 83–104, 2012.
19. Boopathi, S., An extensive review on sustainable developments of dry and near-dry electrical discharge machining processes. *J. Manuf. Sci. Eng.*, 144, 5, 1–37, 2022.
20. Ashori, A., Wood-plastic composites as promising green-composites for automotive industries! *Bioresour. Technol.*, 99, 11, 4661–4667, 2008.
21. Singh, R.K., Kumar Mangla, S., Bhatia, M.S., Luthra, S., Integration of green and lean practices for sustainable business management. *Bus. Strateg. Environ.*, 31, 1, 353–370, 2022.

22. Bose, R. and Luo, X., Integrative framework for assessing firms' potential to undertake Green IT initiatives via virtualization-a theoretical perspective. *J. Strateg. Inf. Syst.*, 20, 1, 38–54, 2011.
23. Mungai, E.M., Ndiritu, S.W., Rajwani, T., Do voluntary environmental management systems improve environmental performance? Evidence from waste management by Kenyan firms. *J. Clean. Prod.*, 265, 121636–121649, 2020.
24. Mahlia, T.M.I., Tohno, S., Tezuka, T., A review on fuel economy test procedure for automobiles: Implementation possibilities in Malaysia and lessons for other countries. *Renew. Sustain. Energy Rev.*, 16, 6, 4029–4046, 2012.
25. Sampath, B. and Myilsamy, S., Experimental investigation of a cryogenically cooled oxygen-mist near-dry wire-cut electrical discharge machining process. *Stroj. Vestnik/Journal Mech. Eng.*, 67, 6, 322–330, 2021.
26. Boopathi, S., Balasubramani, V., Kumar, R.S., Singh, G.R., The influence of human hair on kenaf and Grewia fiber-based hybrid natural composite material: an experimental study. *Funct. Compos. Struct.*, 3, 4, 045011–045021, 2021.
27. Boopathi, S., Cryogenically treated and untreated stainless steel grade 317 in sustainable wire electrical discharge machining process: A comparative study. *Environ. Sci. Pollut. Res.*, 1–10, 2022. (in Press) https://doi.org/10.1007/s11356-022-22843-x
28. Boopathi, S., An investigation on gas emission concentration and relative emission rate of the near-dry wire-cut electrical discharge machining process. *Environ. Sci. Pollut. Res.*, 29, 86237–86246, 2022.
29. Boopathi, S., Sureshkumar, M., Jeyakumar, M., Kumar, R.S., Subbiah, R., Influences of fabrication parameters on natural fiber reinforced polymer composite (NFRPC) material: A review. *Mater. Sci. Forum*, 1075, 115–124, 2022.
30. Boopathi, S., Balasubramani, V., Kumar, R.S., Influences of various natural fibers on the mechanical and drilling characteristics of coir-fiber-based hybrid epoxy composites. *Eng. Res. Express*, 5, 1, 15002–15020, 2023.

12

Minimization of Manufacturing Industry Wastes Through the Green Lean Sigma Principle

Sampath Boopathi

Department of Mechanical Engineering, Muthayammal Engineering College, Namakkal, Tamil Nadu, India

Abstract

In this chapter, three separate strategies that have developed over time, lean manufacturing, green technology, and Six Sigma, have been discussed. The primary goal is to boost organizational productivity by methodically eliminating numerous non-value-added tasks. Under the aegis of Green Lean Six Sigma (GLS), various tools, generally accepted processes, and connections between all three can be combined. The current study is describing whether GLS can improve the financial and societal aspects of sustainability in the manufacturing sector while reducing environmental emissions. Several GLS implementation enablers were discovered to be suitable for modeling and analysis. Managers can use the study to comprehend how different GLS enablers are interconnected and related, which will ultimately lead to the GLS program's successful implementation. The implementation barriers have been minimized by providing training, a knowledge base, and continuous improvements. Additionally, a thorough literature review revealed impediments that are specific to GLS. Additionally, it offers a thorough method for evaluating and interpreting environmental impacts. GLS tools have been added to the stepwise structure to help industry managers implement this sustainable strategy. Through the systematic integration of Six Sigma, Lean, and Green Technology tools in a manufacturing scenario, the study successfully illustrates the practical advantages of implementing the suggested GLS framework.

***Keywords*:** Green technology, lean manufacturing, Six Sigma, manufacturing

Email: boopasangee@gmail.com; ORCID: orcid.org/0000-0002-2065-6539

S. Thirumalai Kumaran and Tae Jo Ko (eds.) Sustainable Machining and Green Manufacturing, (249–270) © 2024 Scrivener Publishing LLC

12.1 Introduction

Any country's ability to flourish is largely dependent on how quickly its industrial sectors can develop operationally and technologically. The innovation of agriculture, investigation, expertise, urbanization, service, quality life, and societal transformation are all related, either directly or indirectly, to the rise of the industrial sectors. If the industrial divisions have increased production and development, the aforementioned areas will develop. Industrial entities uncontrollably devour natural resources and emit a significant amount of pollutants into the atmosphere [1]. Making the best use of the resources that are already available and reducing industry capacity waste are alternative ways to increase productivity and quality while lowering unit costs. Commercial and industrial enterprises are contributing to environmental degradation by failing to implement effective waste disposal practices. Increased carbon footprint and other related pollutants have consequently caused serious health problems. Due to growing ozone and carcinogens in the air, the chemical and meteorological changes brought on by CO_2 will result in a rise in mortality rates. A method for determining and enhancing an organization's sustainability is called Green Lean Six Sigma (GLS) [2]. This strategy merely yields short-term gains with no lasting sustainable advantages. The integration of social factors in industrial processes is necessary to consider the need for a healthy work environment and optimum labor practices.

For the program's progressive implementation, top-ranked barriers like those relating to management and the environment should be addressed first. High-spec end product manufacturing is a problem that lean manufacturing cannot solve. Six Sigma emerges at this point, which lowers the defect rate and results in high specifications. It is a project-based methodology that lowers possibilities for faults by up to 3.4 M. A new sustainable development technique called GLS has emerged as a result of the Lean Six Sigma (LSS) methodology [3]. It lowers environmental emissions, wastes, and flaws, increasing organizational sustainability. The quantity of waste and faults that are eliminated rises as a result of the LSS's adoption of green technology. The literature contains only few examples of GLS as a novel methodology. It has been used to model enablers and barriers, improve sustainability, and the locomotive industry. Lean manufacturing, also known as green technology (GT), and Six Sigma are two methods that promote the fusion of three various methods under the GLS banner. Therefore, the combination of approaches creates a fresh strategy for reducing environmental impact, increasing resource efficiency, and enhancing long-term sustainability in the industry, promoting a resilient and responsible path for businesses and the planet [2].

The conditions that encourage the organization to use a new strategy are known as enablers. The facilitators of GLS have been extensively studied by researchers, whether they are Green Lean principles alone or Lean Six Sigma concepts. There is not a lot of information in the literature about how to identify and model the interactions of GLS enablers. Due to its novelty and cultural disparity, GLS only found a relatively limited number of applications in the manufacturing industry. Barriers are defined as administrative and practical difficulties that prevent an organization from achieving its goals in a timely manner. To get a competitive edge over rivals, any business organization needs to quickly identify important obstacles or hurdles. Based on the perspectives of experts, 20 sustainable Lean Six Sigma (LSS) enablers can be categorized into various groups, offering a comprehensive framework for enhancing operational efficiency, minimizing waste, and fostering a culture of continuous improvement [4]. According to the literature, LSS implementation has a positive impact on an organization's financial and ecological performance. The incorporation and application of green technology with LSS are not without difficulties though. The efficient implementation of sustainability-focused initiatives is hampered by a lack of funding for clean technologies projects, a weak organizational support structure, lack of resources, the absence of tools and practices, and uncertain gains. When the suggested model was put to the test in the marine sector, it showed improvement in both the economic and environmental aspects of sustainability, as well as the conventional emphasis of productivity [5].

It has been established that management innovation serves as a bridge for the introduction of efficient GLS practices for the industry's sustainable growth. Garza-Reyes [6] presented the Green LSS business strategy, which combines the green lean and Six Sigma methodologies. The generated model was put to the test during the installation of a pile cap. A structured framework for combining green technology with lean and Six Sigma was established. Based on the employee responses, they ranked the enablers and performance metrics in order of importance. The provided insights added to the theoretical body of knowledge that supports the creation of commercial strategies and environmental considerations [6–11]. According to a research, GLS and sustainability present both obstacles and opportunities. One of the main obstacles to the successful implementation of the sustainable GLS program during the product development process is a lack of managerial commitment. A systematic model is used to address the issue by using Six Sigma. They discovered that the framework's successful implementation caused the level of defects to drop from 16,000 ppm to 6,000 ppm. In the construction industry, a methodical approach to removing various obstacles to the GLS program implementation was created [5].

12.2 Challenges to the Manufacturing Sector

Any country's economy grows mostly through its manufacturing sector. Rapid industrialization-related economic expansion promotes capital formation, urbanization, greater use of natural resources, reduction of poverty, and reduction of unemployment. High-capacity waste creates non-optimal resource usage, which is a significant difficulty for the manufacturing sector. Strict government regulations on environmental emissions, consumer demands for sustainable products, and associated societal issues provide difficulties [6]. Large sums of money are being spent by industrial groups to develop environmentally friendly manufacturing and consumption practices. To manufacture the highest-quality products, a variety of theories and methods, including Lean, Green, Six Sigma, etc., have emerged. However, a singular strategy is unable to comprehensively handle all of the problems connected to sustainability. Green technology aims to lessen harmful environmental effects such eutrophication, acidification, photochemical oxidation, and global warming. Despite addressing environmental challenges, GT is unable to eliminate process variance. The Six Sigma method is required to create a process with minimal variation.

It is essential to combine all three methods into the single special technique known as GLS, which can tackle all three issues at once. The GLS approach to sustainable development lowers waste, defects, and environmental emissions while increasing production and profitability. Through the methodical implementation of related tools and reuse, recycle, and reduce at various phases of the product, it addresses the contemporary difficulties facing the sector [12].

12.3 GT and Manufacturing Development Procedure

Lean's positive public spillover is green technology, since it eliminates waste and lessens negative environmental effects. It includes methods and tools for improving biologically inactive layers of arid land, such as green supply chains, eco-friendly design, reverse logistics, eco-friendly buildings, and landfills [13].

12.3.1 Identification of the Present State

Green manufacturing system development is a multistep procedure. Green stream mapping (GSM), impact analysis, a series of surveys with a specific structure, and other methods may be utilized at this stage (Figure 12.1).

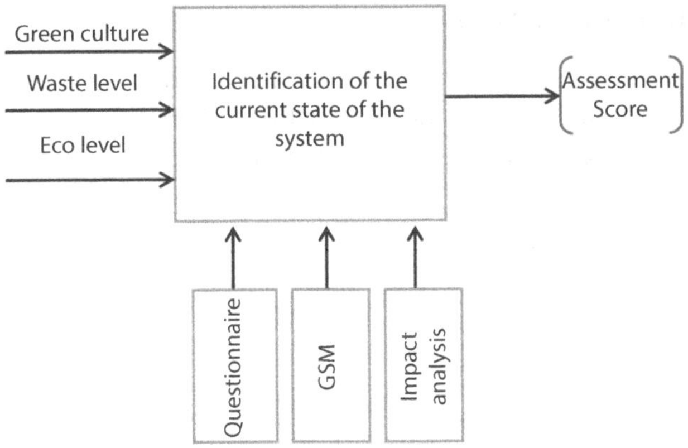

Figure 12.1 Identification current system state.

The level of the system per time of the assessment is computed based on an assessment score.

12.3.2 Planning

Improvement measures are implemented after a production plan and assessment score (Figure 12.2). Life cycle analysis, optimization of the many factors, and consumption analysis are all completed. An improvement plan needs to have a strategic impact on each aspect of the organization's sustainability [14].

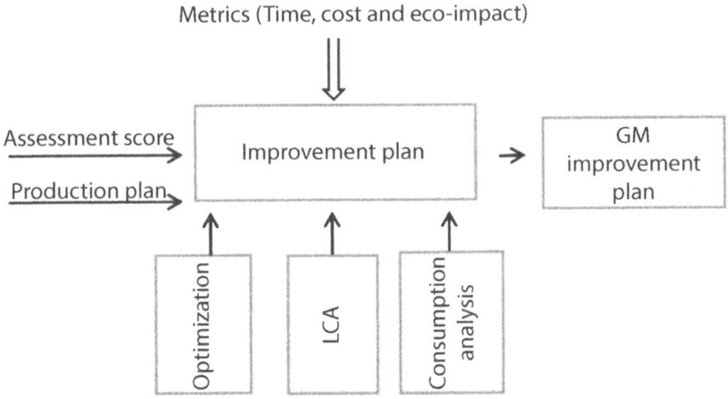

Figure 12.2 Implementation plan.

12.3.3 Implementation

The green manufacturing system was created as an improvement scheme for greater material, energy, and production efficiency (Figure 12.3). The numerous measures are designed to determine the system under consideration's sustainability content. These include the amount of recyclable materials, the presence of prohibited substances, the intensity of greenhouse gases, etc.

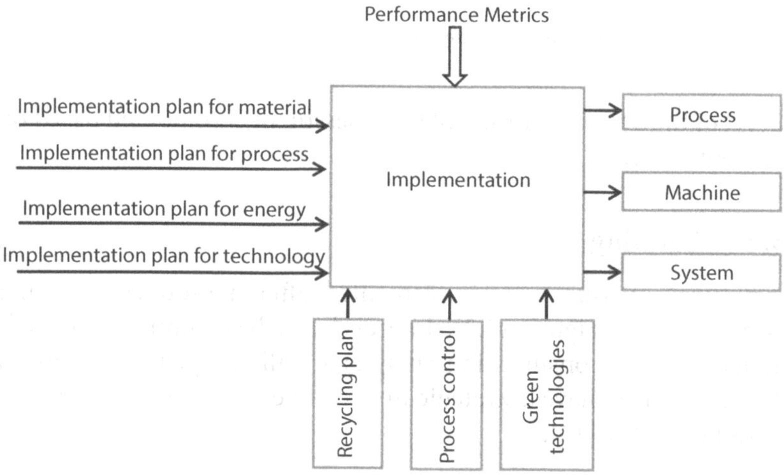

Figure 12.3 Implementation plan.

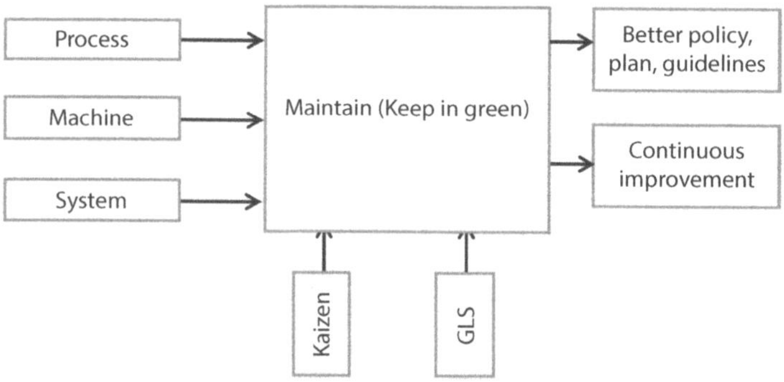

Figure 12.4 Implementation and sustainability.

12.3.4 Sustainability

It is periodically monitored how well the system is improving the financial capacity, production, and environmental emissions in order to assess the effectiveness of the selected strategy (Figure 12.4). GT is able to apply practices like reduce, reuse, and recycle to lessen its impact on the environment, but it cannot eliminate process variance. To create a process with reduced variation, a Six Sigma technique is required.

12.4 Green Lean Manufacturing Terminologies

12.4.1 Lean Manufacturing

Lean manufacturing is a method for reducing waste that increases the value of the product by reducing waste to the absolute minimum. The Toyota Production System developed by Japanese engineers Taiichi Ohno and Shigeo Shingo is credited with giving rise to the contemporary idea of lean manufacturing [4]. By aligning the flow with consumer demand and pursuing perfection via ongoing improvement, it reduces non-value-added activities and eliminates waste. Transport, catalog, motion waiting, production, processing, and faults are some of the wastes connected to lean manufacturing. GLS Kaizen introduces improvements to existing rules, strategies, and guidelines, enabling a systematic and iterative approach to eliminate these wastes, optimize processes, and cultivate a culture of constant enhancement within the organization. Elimination of these System of Process Machines Lean manufacturing techniques successfully used leads to continuous improvement. The majority of studies on lean elements in the literature have been found to be single-aspect studies, with just a very small number of studies focusing on several lean element aspects. However, the association has to focus on all components, such as value stream mapping, cellular manufacturing, u-line system, etc., for the lean implementation to be comprehensive.

12.4.2 Green Lean Interactions

The focus of industries has shifted from conventional factors like quality and customer happiness to sustainability due to growing environmental concerns. Sustainability includes balancing resources to satisfy public demand as well as upholding the three Ps (people, planet, and profit). Lean is a notion that many sectors employ to cut waste, and green is a concept that some businesses use to cut energy use and highlight trash-recycling methods. The Green Lean (GL) methodology places an emphasis on minimizing resource usage through waste reduction and minimizing adverse environmental effects through the elimination of hazardous gases. Customers like to buy products that are of the highest quality, cost-effective, and most importantly, delivered on schedule [6].

12.4.3 Restrictions of the Green Lean Approach

This strategy has some drawbacks, such as the fact does not employ statistical methods. The GL approach primarily entails techniques like green building, reverse logistics, 5S, poka-yoke, and design for the environment. The integrated GL method does not address product variety from a variable point of view, but it does have an impact on customers' perceptions in attributive ways.

12.4.4 Six Sigma

A corporate improvement strategy called Six Sigma finds and eliminates the root causes of errors, process waste, etc. Six Sigma was invented by Motorola in 1987, and after achieving success at General Electric, interest in it surged. In a nutshell, this methodology begins with the consumers by comprehending their needs and then makes following adjustments to the procedure. The core tenet of Six Sigma is that if flaws in a process can be estimated, a plan can be made in advance to fix them. It is a highly structured method used to limit process variability so that the failure rate is less than 3.4 per million.

12.4.5 Define-Measure-Analyze-Improve-Control (DMAIC) Methodology

For any project, DMAIC cycle is the process that is utilized to successfully implement the Six Sigma strategy. Figure 12.5 depicts the DMAIC cycle's steps, which involve the following actions [15].

<u>Define:</u> One of the key components of a successful Six Sigma implementation is choosing the appropriate task or process for improvement. Any corporate organization must make the appropriate project selections to guarantee that they are meeting the needs of their Six Sigma consumer base.
<u>Measure:</u> The measurement phase's main objectives are to evaluate the effectiveness of the current process and identify any potential input variables that might have an effect on the outcome.
<u>Analyze:</u> The relationship between the input and output factors is described during the analysis phase. Improve: Improvement actions are chosen and put into action based on the findings of the analysis phase. Typically,

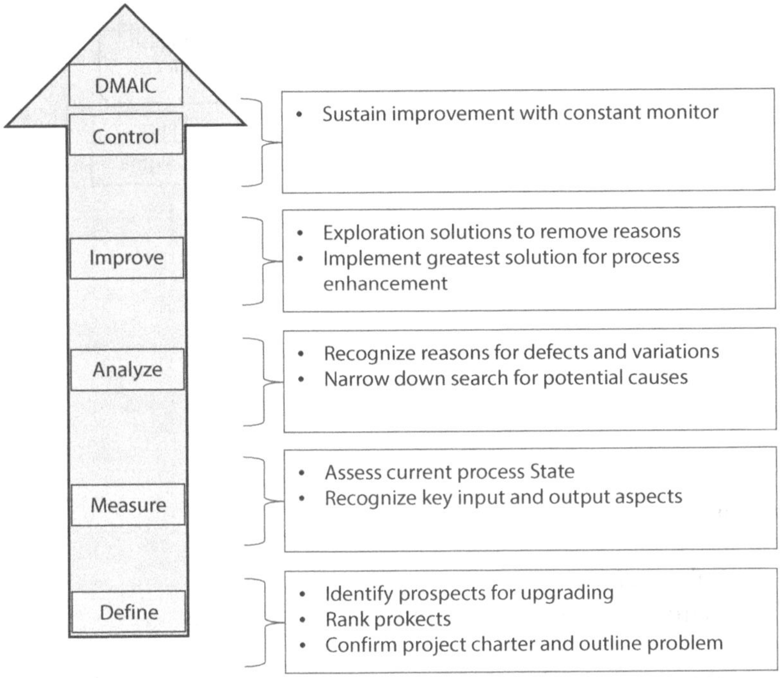

Figure 12.5 DMAIC methodology.

optimization is required to maximize the effectiveness of the process. Control: It is crucial to make sure that progress is sustainable after the solution has been put into place and evaluated.

12.4.6 Green Lean Six Sigma

GLS are three distinct strategies created in the future to satisfy client demand. The manufacturing industries have been compelled to switch from conventional to sustainable processes due to severe rivalry, shorter product life, and international laws to reduce emissions [13].

- Find opportunities to upgrade Rank projects
- Verify the project charter and describe the issue
- Identify current procedures and state
- Recognize important characteristics of input and output
- Recognize the causes of variances and flaws
- Concentrate your search for causes
- Investigation of remedies to get rid of causes
- Implement the best process improvement idea

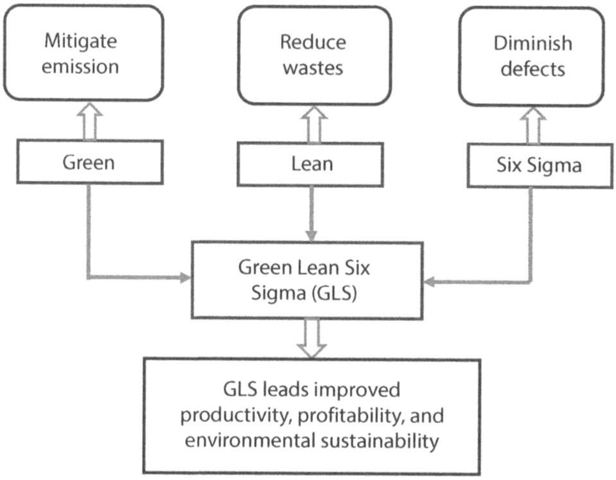

Figure 12.6 Green Lean Six Sigma.

- Enhancement
- Maintain progress by keeping an eye on things

The primary goal of DMAIC is to boost corporate productivity by methodically eliminating numerous non-value-adding processes. Utilizing an integrated GLS approach lowers waste, defects, and environmental emissions while increasing organizational efficiency and profitability. GLS is depicted in a model in Figure 12.6. The integrated GLS has been condensed to make organizations more sustainable and waste-free. The following phase involves identifying the primary contributing factor to the issue and potential areas for change utilizing methods like life cycle assessment (LCA), cause and effect diagrams, etc.

12.4.7 Capacity and Capacity Waste

12.4.7.1 Concept of Capacity

Capacity is the ability to generate an output without increasing input variables [14]. The capacity of the plant affects every piece of infrastructure and piece of machinery required to produce the final product. The capacity restriction is the weakest link in the manufacturing process. When capacity is considered, many manual or mechanical operations in a factory may not be used to their full potential. The capacity can be raised by spreading equipment throughout the plant's subprocesses. Due to its significant degree of ambiguity, capacity is challenging to evaluate and manage.

12.4.7.2 *Capacity Utilization: Concept and Significance*

Capacity utilization is the relationship between output and potential output. It is the proportion of the plant's observed output to its intended output. The capacity utilization index is used to rate a plant's capacity. Because it is used to rate the plant's capacity, capacity utilization plays a crucial part in the success of a corporation. Planning to prevent resource wasting is much simpler if resource wastage is assessed.

12.4.7.3 *Estimation of Capacity Waste*

The concept of capacity is explained using a framework that is easy for everyone to comprehend. The literature presents numerous theories and myths on ability perception, assessment, and management. In this framework, the terms related to capacity are presented.

<u>Theoretical Installed Capacity:</u> Operational time is determined by collaborators or established using proper methods.
<u>Theoretical Rated Capacity:</u> Operational time is downrated due to subpar methodology, flawed work measurement, and low labor productivity.
<u>Planned Capacity:</u> Labor is employed for fewer shifts, resulting in less capacity than rated capacity.
<u>Real Capacity:</u> Labor is employed to work fewer shifts per year, resulting in lower capacity than rated capacity.

12.5 Real-Time Problem Formulation and Research Approach

Earth's temperature has significantly increased as a result of rising emissions of greenhouse gases and other pollutants. Among the strategies used by corporations to cut waste and raise production quality is GLS. However, each method has disadvantages that can be avoided by using a third strategy, such as organic or noncorrosive production. Industrial managers are hesitant to use this strategy within their firms despite the development of GLS because they lack the preparation and are afraid of failure. Organizations must implement readiness measures, also known as relevant operational tools and procedures, in order to deploy a new strategy. GLS lacks integration measures, implementation framework, preparation measures, and literature evaluation. The systematic framework and related toolkit, which offer a step-by-step road map for the realization of the GLS program, are also necessary for the execution of GLS.

The issue is evaluating GLS effectiveness in reducing capacity waste, lean waste, faults, and the environmental impact. Enablers and impediments for the successful implementation of GLS were identified in order to tackle the issue. The LCA framework was then created to access environmental effect. The goal of the current research problem is to examine the manufacturing challenge through the prism of sophisticated laboratory research. According to the literature, poor project selection has led to the failure of 40% of Six Sigma programs. First, using a literature review, GLS enablers have been identified for all industries, and they have been further confirmed using professional judgment. In this study, interpretative structural modeling had been employed to determine the contextual link among enablers. Therefore, it is crucial to identify and rank the enablers that drive the adoption of GLS in order to avoid GLS failure. Developing a manufacturing framework for Life Cycle Assessment (LCA) using the GRA-best worst method (BWM) to pinpoint and assess LCA challenges, while establishing the integration of Green Lean Six Sigma (GLS) and formulating a comprehensive framework. Implementing the GLS framework within an industrial context and applying it to a practical scenario. The practical implications of the manufacturing environment for practitioners.

BWM that yields more reliable findings is employed in this study to prioritize GLS enablers. Identification of the adoption obstacles for GLS is essential if the program is to run well. First, GLS obstacles were discovered in the literature, and then, using principal component analysis, they were further grouped into logical groups. GLS enablers were prioritized in this study using BWM, which yields more reliable results. It is crucial to pinpoint adoption hurdles for GLS if you want the program to run smoothly. Prior to further modeling into logical categories utilizing principal component analysis, GLS barriers were initially discovered through literature. The metrics of environmental measures must be evaluated as soon as the GLS hurdles have been recognized. A system made up of various unit processes can be given a complete environmental examination using the LCA method. Using LCA, wastes from a system are quantified and investigated, and the main causes of these non-value-added activities are further determined. It is crucial to evaluate the metrics of environmental measures when the GLS barriers have been recognized. The LCA tool offers a thorough environmental study of a system made up of various unit processes. The LCA-based quantification and analysis of wastes from a system also helped to identify the main causes of these non-value-added activities.

12.5.1 Integration Measure and Model of GLS

An encompassing strategy seeks to optimize processes, finances, operations, and emissions. The GLS strategy unifies the three methodologies' finest features under one heading. One new opportunity for industrial businesses to increase sustainability is the convergence of GLS. Research organizations that have used the GLS have outperformed those that have used individual techniques. The integration model's major objective is to outline the crucial information that industrial organizations need in order to enhance sustainable performance. The limitations that prevent organizational efforts to enhance sustainability dynamics pose difficulties for the GLS integration [16, 17]. The GLS frameworks and related tools are called supporting mechanism integrated GLS.

12.5.2 Green Lean Six Sigma Framework

Industrial organizations frequently utilize the Lean and Six Sigma approaches to minimize waste and process variation. However, as a result of stricter government regulations and growing environmental consciousness, businesses must change their operating models to be more environmentally friendly. This need for the green technology to be included into Lean Six Sigma results in the creation of a revolutionary strategy. There is no GLS framework that can be used by manufacturing and service firms in the literature. The GLS shown in the current study is based on DMAIC and is applicable to all corporate enterprises. It was carried out using the next five steps (Figure 12.7).

12.5.2.1 Project Identification

The General Standards Laboratory employs a project-based method for handling waste, flaws, and environmental concerns. According to the research, poor project selection is to blame for the failure of 40% of Six Sigma programs. By addressing each department or section separately, they carried out project by project in a progressive manner. It is crucial to pick a project that offers the greatest potential for sustainable advancement. An extensive analysis of the industry's main segments is done for this. The comprehensive analysis covers environmental emission levels, wastes, and flaws for various market categories. Sustainable customer expectations are translated into technical and environmental features using critical to quality method, which is an essential tool for project selection. So, a suitable project is chosen based on the present needs of the clients as well as business and environmental considerations [18, 19].

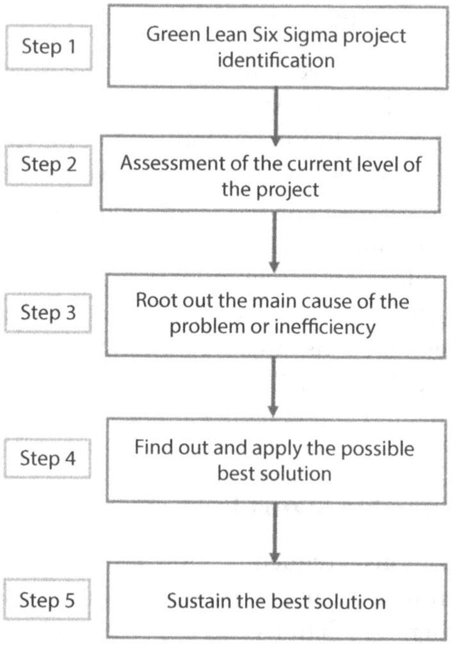

Figure 12.7 GLS framework.

12.5.2.2 Assessment of the Project

The estimation of the system's or project's present level is the focus of the first stage of the GLS framework. The project's standard deviation and sigma level are assessed based on the facts and data that have been gathered. The effectiveness of the chosen GLS project is evaluated by GLS indices. A useful method to evaluate the present state of many connected wastes is value stream mapping (VSM). The individual stage cycle time and material usage are estimated by VSM [20].

12.5.2.3 Root and Cause Discussion

The GLS framework identifies major factors that contribute to high-level wastes, emissions, and defects and compares process cycle efficiency to industry-leading benchmarks. Tools such as Failure Modes and Effects Analysis, LCA are used to identify potential causes of faults [6].

12.5.2.4 *Finding Solutions*

Solutions are evaluated using design of experiments, LCA, business-related, regulatory, and other criteria. The practitioner should be willing to modify or combine solutions to choose the best one [14]. The current VSM is updated to reflect future changes.

12.5.2.5 *Sustain the Best Solution*

VSM and LCA are used to measure waste and emission reduction. A pilot project for an eco-friendly good or service is deemed successful if it is maintained for an extended period of time. The chosen solution is maintained if the performance parameters that have been reevaluated are superior to those from the measuring step. GLS is fully implemented in the sector, which improves sustainability and boosts repute on a global scale.

12.5.3 Green Lean Six Sigma Tools

Industrial firms frequently employ GL methodology, but relatively little efforts have been made to bring the GLS into reality. Organizations are hesitant to implement GLS because of the approach's incipient progress and a culture that is resistant to change. The enablers that help a new strategy be implemented successfully have been considered by the facilitators. An organization's readiness to adopt a new strategy is determined by the enabler [4, 19].

The critical failure factors that operate as barriers prevent an organization from achieving its objectives. These particular management and technical difficulties prevent the companies from attaining their goals. Improved sustainability is achieved through managerial dedication, knowledge of GLS tools, and data assimilation. Numerous studies have indicated that one of the major obstacles to the GLS's execution is a lack of top management support.

The radar graphic has been used to assess the GLS tools. SIPOC charts and environmental VSM were discovered to be the most frequently used tools. In addition, reverse logistics, cause-and-effect diagrams, and process capability were shown to be the other most often utilized tools. Most industrial enterprises rely on lean techniques to address environmental challenges. Lean tools are used to achieve both lean and green goals [21]. GLS's importance keeps growing because of its favorable effects on sustainability, VOC, and productivity. This goodwill ultimately improves the company's reputation on a worldwide scale. In order to contribute to a healthier environment, the majority of firms aim to incorporate green technologies into their current processes. Lack of an appropriate route map is one of the organizations' major problems.

The authors offer a conceptual integration of GLS to aid professionals in adopting the culture, as well as a thorough DMAIC-based framework for implementation in a particular department or area of the business. In the initial stage of a General Waste Management System framework, lean and green practices are used to choose possible projects, assess current circumstances, and look for potential solutions.

12.6 Green Lean Six Sigma Barriers

The obstacles or roadblocks that prohibit a company from implementing a novel strategy are known as barriers. The first topic covered in this chapter is GLS manufacturing barriers. Using principal component analysis (PCA), the barriers have been found in the literature and logically grouped by subject matter experts. IF-DEMATEL was used to do additional analysis on the grouped barriers, and other MCDM approaches were used to confirm the findings. Finally, practitioners were given advice on mitigation measures and their implications. GLS is a sustainable development strategy that improves process performance by lowering wastes, variances, and emissions that harm the environment [22, 23]. GLS encourages the 3Rs (reduce, reuse, and recycle) and improves material efficiency while streamlining the procedure. When obstacles or roadblocks are removed from a system or process, a program can be carried out without incident. Despite the fact that the study is on success factors or facilitators, hurdles to the construction industry do remain. However, there are no studies on exploration or creation of the contextual relationships among GLS obstacles to comprehend the nature of hurdles. PCA was used to further classify the 18 barriers to GLS adoption into six logical groups after being discovered through a literature review and expert answers. The interaction among the categorized GLS adoption hurdles must be demonstrated in order to determine how each barrier affects the others. Organizations can only concentrate on the most important GLS hurdles, so those were prioritized using the IF-DEMATEL and BWM techniques. The BWM was used to further validate the ranks of the barriers to make sure that each group had an equal chance of succeeding.

12.6.1 Research Approaches for Barriers

12.6.1.1 Identification and Clustering of Barriers

The Cronbach's alpha test showed the reliability of the questionnaire with an alpha coefficient of 0.83. The manufacturing industry's mid- and high-level management practitioners as well as LSS staff members received the questionnaire [13, 24].

The Shapiro–Wilk test value p-0.05 indicates that the data are regularly distributed. An important method for determining homogeneity and comparing means across various groupings of data is an analysis of variance (ANOVA). The ANOVA test was run with the null hypothesis [19]. ER obstacles are more important for performance improvement than CI barriers. This suggests that there are not many differences between these two sets of barriers in terms of their removal priorities. The barriers were divided into a reasonable number of groups using PCA for the exploratory factor analysis. The dataset's factorability and sample sufficiency were evaluated. Bartlett's test of sphericity scored highly, with a chi-square of 301 and a p-value of 0.000, while the KMO test result was determined to be 0.052. The GLS barriers were divided into various groups with the same properties using the eigenvalue and percentage of variance approaches of PCA. The GLS uses a range of tools and techniques to help businesses become more environmentally sustainable. GLS is a project-based methodology, and the key to its success is choosing projects that consider all three facets of sustainability. The project selection calls for experienced GLS experts with knowledge of various environmental factors.

With regard to incorrect area identification, knowledgeable individuals with experience in GLS strategies and technologies, and the nature of environmental sustainability, the industrial organization implementing GLS faces challenges. The most crucial component for the implementation of any new strategy within an organization is top management support. Lack of knowledge of VOCs and inefficient transportation are barriers that fall under the category of knowledge base barriers. To utilize this sustainable development strategy to its maximum capacity, it is crucial that employees receive training in the many GLS techniques and toolkit. Lack of training results in improper tool use and GL area selection, which ultimately raises the possibility of the project failing. Organizational members are made aware of various GLS and regulatory frameworks through training in GLS issues. The primary drivers of GLS success are the alignment of organizational goals with GLS, holding everyone accountable for the organization's sustainability, and financial support. Everyone is accountable for implementing sustainability measures due to the forward-looking attitude of the organization. The category of organization hurdles is made up of a lack of alignment between organizational goals and GLS objectives, cultural fragmentation, and financial limitations. The 3R (reduce, reuse, and recycle) opportunities are created by the continuous improvement mindset within the firm.

Environmental barriers: Environmental barriers are those obstacles that prevent organizations from improving their sustainability or environmental performance. This category of obstacles consists of five obstacles: ignorance of different GLS tactics, ignorance of the environment, selection of the incorrect GLS tool, incorrect identification of lean and green areas, and a lack of competent GLS staff. The success of the GLS program depends on having a solid understanding of green and lean metrics as well as environmental concerns related to the process. Environmental barriers make up this category of obstacles, which in the last 3 years has contributed 10.853% of the overall variance between projects [25].
Management-related barriers: Management barriers are those that are caused by the management's inability to support and function. This collection of obstacles consists of weak organizational culture, a lack of management support, and reluctance to change. The organization must demonstrate management commitment, technological adaptation, and a forward-thinking culture in order to incorporate sustainability-related elements [22].
Knowledge base barriers: One of the most frequent challenges faced by manufacturers and service providers is a lack of knowledge of the various forms of VOC. Another is an inefficient transportation infrastructure.
Training-related barriers: Some of the most typical explanations for GLS bottlenecks include a lack of training, standardization and scheduling practices, and an insufficient regulatory framework [13].
Organizational related barriers: Organizational hurdles, which include those caused by cultural differences, financial limitations, and synergistic discrepancies between continuous improvement techniques and strategic objectives, account for 9% of the total variance explained. Organizational hurdles include those that pertain to creating a culture of shared learning, raising funds for the adoption of clean technology, and ingraining a green culture.
Continuous improvement barriers: According to the Chartered Institute of Procurement and Supply, the bundled barrier represents 10.442% of the overall variation and is stacked with obstacles to adopting sustainable practices and thinking about continuous improvement.

12.6.1.2 *Classification and Prioritization*

Using IF-DEMATEL, this phase involved classifying the accumulated barriers into cause-and-effect barriers. To guarantee the reliability of the findings, the sensitivity analysis was also carried out. Additionally, BWM and then GRA were used primarily to validate the results. The IF-DEMATEL execution process involves the following steps [19, 22–25].

Step 1: IF-DEMATEL compares grouped barriers using a set of language variables and IFS, with five language scales ranging from "no influence" to "very strong influence."
Step 2: The weights of the DMs are determined.
Step 3: To create an aggregated IF decision matrix during a group decision-making process, each decision-(DMs) maker's individual opinions must be combined into a collective opinion. The operator after him offered a step-by-step analysis of the DMs' viewpoints to address this.
Step 4: Defuzzification transforms fuzzy output into a precise value, giving language variables and assigned numbers their qualitative worth based on DMs' perceptions.
Step 5: The matrix of normalized direct relationships has been found.
Step 6: Create a matrix of all relationships.
Step 7: In order to create the (D+R) and (D-R) vectors, the whole relationship matrix's rows and columns have been added together.

12.6.2 Practical and Theoretical Implications

GLS measures help practitioners identify bottlenecks and potential solutions for the increased sustainability dynamics of the industry. This study provides insights to develop a comprehensive GLS framework that will reduce emissions through reduction of wastes, defects, and rework. Policymakers can adopt new policies through the replacement of traditional operational dynamics with GL measures. Manufacturing organizations need to understand and analyze the barriers in the implementation of GLS, which have been categorized in cause and effect through IF-DEMATEL. Top-ranked barriers like management and environmental centered barriers should be tackled first, and optimization concepts should be applied to find the best solution for environmental related problems [8, 26].

12.7 Conclusion

Six Sigma, green technology, and lean manufacturing have all been covered. The main objective is to increase organizational efficiency by methodically removing jobs that do not provide value. Different tools, widely used procedures, and links between all three can be merged under the auspices of GLS. The study can be accessed by managers to understand how various GLS enablers are associated and connected, which will ultimately result in the GLS program's successful execution. Training, knowledge base, and continuous improvement challenges are all directly impacted by the cause barriers. It also provides a complete methodology for assessing and comprehending environmental impacts. To assist managers in implementing this sustainable strategy in the industrial sector, GLS tools have been developed.

References

1. Singh, C., Singh, D., Khamba, J.S., Understanding the key performance parameters of green lean performance in manufacturing industries. *Mater. Today Proc.*, 46, 111–115, 2021.
2. Nallusamy, S., Dinagaraj, G.B., Balakannan, K., Satheesh, S., Sustainable green lean manufacturing practices in small scale industries-A case study. *Int. J. Appl. Eng. Res.*, 10, 62, 143–146, 2015.
3. Bergmiller, G. and McCright, P., Parallel models for lean and green operations, in: *Industrial Engineering Research Conference*, pp. 1138–1143, 2009.
4. Garza-Reyes, J.A., Green lean and the need for six sigma. *Int. J. Lean Six Sigma*, 6, 3, 226–248, 2015.
5. Gholami, H., Jamil, N., Mat Saman, M.Z., Streimikiene, D., Sharif, S., Zakuan, N., The application of green lean Six Sigma. *Bus. Strateg. Environ.*, 30, 4, 1913–1931, 2021.
6. Kaswan, M.S. and Rathi, R., Green lean six sigma for sustainable development: Integration and framework. *Environ. Impact Assess. Rev.*, 83, 106396, 2020.
7. Boopathi, S. and Sivakumar, K., Optimal parameter prediction of oxygen-mist near-dry wire-cut EDM. *Int. J. Manuf. Technol. Manage.*, 30, 3–4, 164–178, 2016.
8. Boopathi, S., An extensive review on sustainable developments of dry and near-dry electrical discharge machining processes. *J. Manuf. Sci. Eng.*, 144, 5, 1–37, 2022.
9. Boopathi, S., Cryogenically treated and untreated stainless steel grade 317 in sustainable wire electrical discharge machining process: A comparative study. *Environ. Sci. Pollut. Res.*, 1–10, 2022. (in Press) https://doi.org/10.1007/s11356-022-22843-x.
10. Boopathi, S. and Sivakumar, K., Experimental investigation and parameter optimization of near-dry wire-cut electrical discharge machining using multi-objective evolutionary algorithm. *Int. J. Adv. Manuf. Technol.*, 67, 9–12, 2639–2655, 2013.
11. Sampath, B., Sureshkumar, M., Yuvaraj, T., Velmurugan, D., Experimental investigations on eco-friendly helium-mist near-dry wire-cut edm of m2-hss material. *Mater. Res. Proc.*, 19, 175–180, 2021.
12. Kumar, S., Kumar, N., Haleem, A., Conceptualisation of sustainable green lean six sigma: an empirical analysis. *Int. J. Bus. Excell.*, 8, 2, 210–250, 2015.

13. Hussain, K., He, Z., Ahmad, N., Iqbal, M. *et al.*, Green, lean, six sigma barriers at a glance: A case from the construction sector of Pakistan. *Build. Environ.*, 161, 106225, 2019.
14. Caiado, R., Nascimento, D., Quelhas, O., Tortorella, G., Rangel, L., Towards sustainability through green, lean and six sigma integration at service industry: Review and framework. *Technol. Econ. Dev. Econ.*, 24, 4, 1659–1678, 2018.
15. Monday, L.M., Define, measure, analyze, improve, control (DMAIC) methodology as a roadmap in quality improvement. *Glob. J. Qual. Saf. Healthc.*, 5, 2, 44–46, 2022.
16. Sagnak, M. and Kazancoglu, Y., Integration of green lean approach with six sigma: An application for flue gas emissions. *J. Clean. Prod.*, 127, 112–118, 2016.
17. Kaswan, M.S., Rathi, R., Khanduja, D., Integration of green lean six sigma: A novel approach for sustainable development. *Int. J. Six Sigma Compet. Advant.*, 12, 4, 389–405, 2020.
18. Cherrafi, A., Elfezazi, S., Govindan, K., Garza-Reyes, J.A., Benhida, K., Mokhlis, A., A framework for the integration of Green and Lean Six Sigma for superior sustainability performance. *Int. J. Prod. Res.*, 55, 15, 4481–4515, 2017.
19. Hariyani, D., Mishra, S., Sharma, M.K., A descriptive statistical analysis of barriers to the adoption of integrated sustainable-green-lean-six sigma-agile manufacturing system (ISGLSAMS) in Indian manufacturing industries. *Benchmarking*, 2022. (in Press) https://doi.org/10.1108/BIJ-11-2021-0701.
20. Touriki, F.E., Benkhati, I., Kamble, S.S., Belhadi, A., El fezazi, S., An integrated smart, green, resilient, and lean manufacturing framework: A literature review and future research directions. *J. Clean. Prod.*, 319, 128691–128701, 2021.
21. Šomplák, R., Pavlas, M., Kropáč, J., Putna, O., Procházka, V., Logistic model-based tool for policy-making towards sustainable waste management. *Clean Technol. Environ. Policy*, 16, 7, 1275–1286, 2014.
22. Xia, D., Zhang, M., Yu, Q., Tu, Y., Developing a framework to identify barriers of Green technology adoption for enterprises. *Resour. Conserv. Recycl.*, 143, 99–110, 2019.
23. Kaswan, M.S., Rathi, R., Reyes, J.A.G., Antony, J., Exploration and investigation of green lean six sigma adoption barriers for manufacturing sustainability. *IEEE Trans. Eng. Manage.*, 2021.
24. Cherrafi, A., Elfezazi, S., Garza-Reyes, J.A., Benhida, K., Mokhlis, A., Barriers in Green Lean implementation: A combined systematic literature review and interpretive structural modelling approach. *Prod. Plan. Control*, 28, 10, 829–842, 2017.
25. Sinha, P. and Yadav, M.V., Identification of performance improvement green lean six sigma barriers: A critical literature review. *Int. J. Sci. Technol. Manage.*, 8, 2, 1–10, 2021.

26. Boopathi, S., An investigation on gas emission concentration and relative emission rate of the near-dry wire-cut electrical discharge machining process. *Environ. Sci. Pollut. Res.*, 29, 86237–86246, 2022.

13

Design for Sustainable Methods in Additive Manufacturing

Akesh B. Kakarla* **and Ing Kong**

School of Computing, Engineering and Mathematical Sciences, La Trobe University, Bendigo, VIC, Australia

Abstract

During the Industry 4.0, additive manufacturing (AM) has evolved into a rapidly developing technology in the fields of research and industry. AM is a technique used in manufacturing the products of complex three-dimensional geometry. The main advantages of AM technologies are design freedom, high customization, waste minimization, and flexibility. Moreover, AM shows enormous potential in reducing the consumption of natural resources in product development and enables more eco-friendly manufacturing processes. Therefore, AM is considered more sustainable than traditional techniques. In this chapter, the role of AM in the sustainable manufacturing process is demonstrated based on product development. Moreover, the implications of sustainability in the AM process and the life cycle assessment of product development using AM are analyzed.

Keywords: Industry 4.0, eco-friendly, life cycle assessment, sustainable

13.1 Introduction

Natural resources, such as water and energy, are the main components for life on Earth. However, humans have been consuming resources rapidly, causing a significant impact on the environment. According to United Nations environmental reports, approximately 140 billion tons of natural resources will be utilized by humans by 2050 at the current rate of consumption [1, 2]. The trend of exploitation of natural resources is leading

**Corresponding author*: a.kakarla@latrobe.edu.au

S. Thirumalai Kumaran and Tae Jo Ko (eds.) Sustainable Machining and Green Manufacturing, (271–288)

to the deterioration of biodiversity. Furthermore, resources such as earth metals, minerals, oils, gases, and ores will be depleted and most likely disappear, leaving future generations with limited or no natural resources. Therefore, there has been a surge in awareness of environmental issues through sustainable practices to protect the ecosystem from destructive human actions [3–6]. Sustainability development is a rapidly emerging process that helps reduce the overutilization of the Earth's resources (Figure 13.1) [7]. Sustainability principles are based on recycling, reusing, and reducing (the three Rs). In this concept, the manufacturing process's sustainability is significant to manage climate change and the depletion of natural resources. Mihelcic *et al.* [8] defined sustainability in the manufacturing process as safeguarding the quality of life on Earth without diminishing the resources. Furthermore, they quantified that future industrial prospects are accessible with constructive possession of humankind's social, financial, and environmental endeavors. The primary objective of sustainability in the manufacturing process is to save natural resources [5, 9]. Sustainability comprises the product from cradle to reincarnation, the supply chain management and the adoption of product life cycle to reduce the impact on the ecosystem in the most efficient way [9]. In recent times, industries of all types have been involved in promoting the "save the environment" concept. In this advanced manufacturing era, researchers and industries are focused on implementing sustainable manufacturing

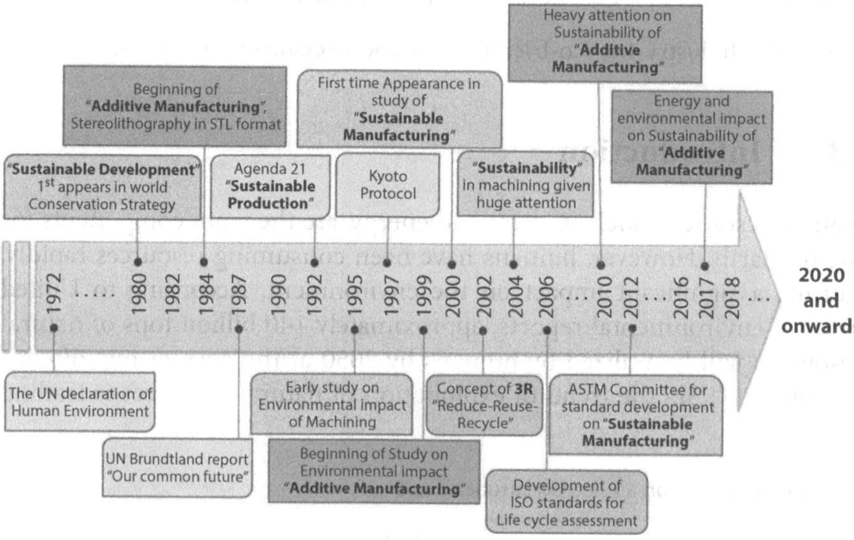

Figure 13.1 Strategic development of sustainability in the manufacturing industry [14].

processes to achieve comprehensive progressive plans to reduce pollution and the usage of natural resources along the product life cycle. Several studies have been conducted on the manufacturing process of product development to understand the environmental impacts [10–12]. The impacts of the process were considered based on energy consumption, using raw materials and lubricants. Subsequently, research has been conducted to minimize consumption in manufacturing techniques and to reduce ecological impact. From this perspective, additive manufacturing (AM) has rapidly evolved with plentiful benefits in product development, particularly satisfying the sustainable manufacturing aspect [13].

AM is also called rapid manufacturing or three-dimensional (3D) printing. The technique is recognized as modern technology in Industry 4.0 to implement as a sustainable manufacturing process [13, 15, 16]. The evolution of AM is illustrated in Figure 13.2. The operating principle of AM is based on depositing the materials layer by layer to develop the final product using computer-aided design. AM was introduced in the 1980s and substantially impacted the manufacturing process with its significant benefits, such as a high degree of freedom, low cost, and high efficiency [16–18]. The process is suitable for product development with advanced materials such as polymers [19], composites [20], cement and concrete [21], clays [22], meta-materials [23, 24], and tissue and organs [25]. The technique reduces the waste materials to the maximum extent [26, 27]. In addition, the energy utilization in the manufacturing of a product is significantly decreased, and relatively few manufacturing steps in producing the complex geometry reduce the total costs [7]. Moreover, the approach

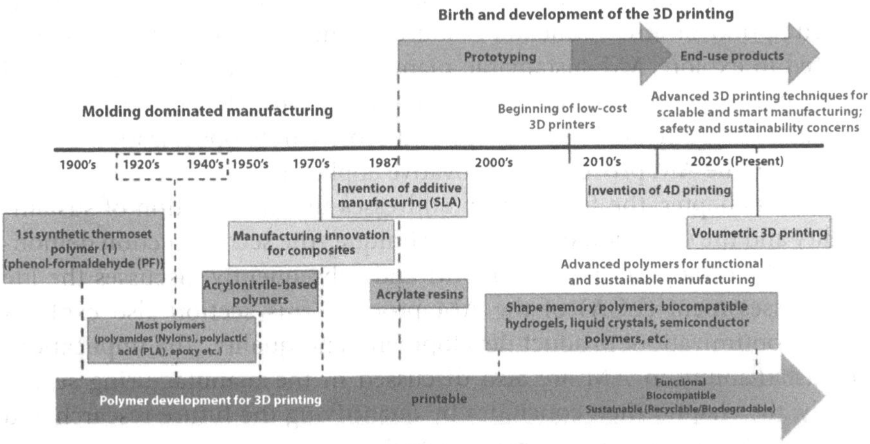

Figure 13.2 Advancement of AM in the manufacturing industry [29].

has a minimal carbon footprint in the product design and development stages compared to conventional industrial techniques [6, 28]. Hence, 3D printing is considered a more sustainable manufacturing process technique owing to all of these features.

During its early stages, AM was mainly used for prototyping products in automobile [30], aerospace [30], medical [31], and dental applications [32, 33]. The AM technique is promising for producing prototypes with the desired materials and for the rapid fabrication of products. In addition, the process offers a design freedom for engineers to develop a product that can be produced in sustainable approaches. There are various methods involved in the AM techniques such as stereolithography, fused filament extrusion, vat photopolymerization, direct energy deposition, and many others. The techniques are considered based on the material (liquid/melt/soften) to manufacture the product [34]. Stereolithography is based on the photopolymerization of the liquid materials such as resins that can be solidified to produce the final product when exposed to ultraviolet radiation. Laser-based technique uses a laser to melt powder materials, while fused deposition modeling extrudes melted material through nozzles to create parts. It uses thermoplastic materials that feed into the machine in the form of filaments. Basically, all of these AM techniques are based on similar principles wherein the end product is manufactured in layer-by-layer accumulation of the material [34, 35].

In the last few years, there is a significant demand for AM that was valued at approximately $20 billion in 2021 and is predicted to reach $20 million by 2030 [36–38]. Moreover, it is anticipated that AM can be implemented to develop complex products for mass production [39]. However, there are concerns about the ecological impacts of AM, particularly regarding the consumption of environmental resources. Thus, studies have been conducted to explore AM sustainable manufacturing, and it has been found as the most favorable and efficient technology [6, 40, 41]. Furthermore, the research has determined the energy and eco-friendly optimization opportunities of the AM process for innovative application.

In this chapter, the first section addresses the introduction of sustainability and the evolution of the AM technique. The second section demonstrates the ecological impacts of AM and subsequently discusses the life cycle assessment (LCA) of the AM process. This section also explores design optimization, product development, and quality. The implications of sustainability in AM are also discussed in the manufacturing sector. Lastly, the chapter then concludes by quantifying the future research and effects of AM on sustainable manufacturing.

13.2 Ecological Impacts of Additive Manufacturing

Over the past few years, there has been a growing importance in sustainable product development due to the climate change resulting from rapid utilization of fossil fuels [42]. Therefore, assessing the sustainability of products produced using manufacturing systems has become vital for industries. The assessment methods considered in each step of product development can lead to a conclusion for granted sustainability. The assessment addresses the energy consumption, raw materials, product redesign, repair, recyclability, waste management, and life of products. AM methods must demonstrate their ecological benefits by considering all main elements.

13.2.1 Materials

AM deposits the materials layer by layer to develop 3D products. A wide variety of materials, such as metals [43], ceramics [44], polymers [45], and composites [46] are used to produce 3D models through AM technologies. The materials are mainly in the form of wires or molten liquids or powders of polymers [46]. In general, materials developed from polymer by-products can often be recycled [47]. However, the properties of the materials can be altered due to additives, cross-contamination, and a mixture of several other products. Moreover, polymers generate toxic gases due to melting during AM extrusion, which causes pollution. Therefore, materials with standardized scales of toxicity, flammability, and reactivity should be reviewed and chosen for producing parts with AM.

Polymeric materials are the most widely used and vital in the AM technique [48]. They include thermoplastics, elastomers, thermosets, polymer mixtures, hydrogels, polymer nanocomposites, and functional polymers [49]. The materials are employed based on the product requirements to satisfy the final product's physical properties. The polymeric material is extruded in a molten or semi-molten setup using a mobile nozzle or print head to produce a 3D model. Polylactic acid [50] and acrylonitrile butadiene styrene (ABS) are commonly used polymers in extrusion-based AM techniques. However, other materials that can be semi-liquefied at a suitable temperature are also employed for AM techniques. Nevertheless, in terms of sustainability and environmental aspects, polylactic acid is considered less toxic and biodegradable than acrylonitrile butadiene styrene. Considering the recent advances in AM techniques, several sustainable materials, such as cellulose-based polymeric and recycled plastic, have been developed to be utilized in AM procedures to decrease environmental impacts.

Distinctive metallic materials such as alloys of aluminum, titanium, cobalt, copper, nickel, iron, and stainless steel are most generally used in the AM technique [51, 52]. The metallic materials in the AM technique are used in several aerospace components because of the excellent material properties. Most of these alloy materials are produced using gas atomization in powder or in wire form based on the AM process [51]. In addition, ceramics such as alumina, zirconium, hydroxyapatite, bioglass, and calcium phosphate are used in the AM techniques for producing biomedical components [53]. The materials are produced using varied ceramic synthesis approaches with controlled particle size distributions. However, not all metals or ceramics can be recycled due to the acuteness with ecological conditions such as temperature, oxygen, and humidity. Furthermore, the recycled powders after processing may have varied properties than pure metals or ceramics. Therefore, alternative approaches are required in product design and processing through addition of additives for product quality requirements.

13.2.2 Energy Consumption

Power consumption is a vital element of evaluating the sustainability in the manufacturing process. Despite its many benefits, AM has yet to be confirmed as an eco-friendly manufacturing technique due to its high energy consumption from melting polymeric materials in the heating process or lasering or curing resins. According to Verhoef *et al.* [36], energy consumption by AM is approximately 5%–27% based on the scenarios. The analysis revealed that half of the energy consumption was related to raw materials and the other half to transportation and construction. Additionally, the studies reported that electricity was used as significant energy, but AM still utilized considerably less energy than robotics and other traditional machining processes. A comprehensive review of energy consumption in AM was demonstrated by Liu *et al.* [54]. The study revealed that most energy consumption could be minimized by improving capacity utilization and part design optimization. Similarly, Yang *et al.* [55] reported on stereolithography-based AM machine energy consumption. The investigation evaluated the influences of diverse factors and their relations to overall energy utilization during the machining process. The reports revealed that maximum energy consumption was associated with the build volume of the part. Therefore, to minimize energy consumption and achieve good product quality, surface optimization of the product design was necessary. Many other studies [54, 56] have evaluated AM's energy consumption for sustainability-related performance. The investigations revealed that part orientation, layer thickness, support design, and time spent producing the part were contributors of the high energy consumption. Energy consumption should be minimized by optimizing the product's design parameters. Nonetheless, AM methods' energy consumption has not been fully explored by the current literature, given the varied parameters and exploration of contribution to total energy consumption.

13.3 Life Cycle Analysis

LCA is an internationally accepted way to assess the ecological consequences of processes, activities, and products. The LCA of AM at distinct stages is shown in Figure 13.3. In terms of product life cycle, the assessments on long-term influences of AM regarding the environment are still at an early stage [57]. Ford and Despeisse [35] carried out an LCA of AM from the viewpoint of feed stocks, product development, and product recyclability. The studies revealed that AM inherently supported sustainable production. Similarly, Gebler *et al.* [6] conducted a quantitative analysis on consumption of energy and carbon dioxide emissions through AM technologies by 2025 from a global perspective. The assessment concluded that emissions and energy consumption exhibited potential sustainability over the entire life cycle of produced products. Faludi *et al.* [58] analyzed AM's environmental impact by comparing the AM technique with traditional computer numerical control machining. The outcomes reveal that waste generation through computer numerical control technique had a major

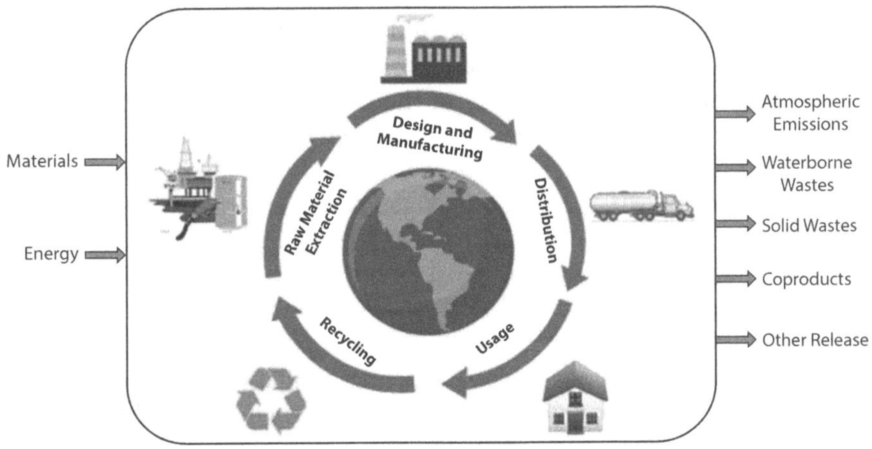

Figure 13.3 Illustration of life cycle assessment of the AM process [60].

impact on the ecological system, while AM techniques had minimal waste generation and were more sustainable. Similarly, Kafara *et al.* [59] assessed the conventional versus the AM in terms of mould core development for carbon fiber–reinforced polymers. The studies revealed that AM machining has a lower ecological impact than the conventional manufacturing process, achieved through reducing the usage of resources. However, studies conducted thus far on the LCA of AM are still in the preliminary stages and do not contain all relevant information to support decisions based on maximum sustainability. Therefore, an AM LCA of materials, toxicity, safety, and social impact, in addition to energy consumption, should be conducted.

13.4 Implications of Sustainable Development in AM

The effects of sustainability in AM are grouped into four stages. They are product design and process of product, material, manufacturing of product, and closing the loop [35, 61].

13.4.1 Design and Process of Product

In general, AM allows for designing complex components with a greater degree of freedom in terms of shape and size, with the addition of minimizing the assembling parts and materials. Designers and product development engineers play a significant role in determining a future of produced goods in the manufacturing domain. The unsustainable activity's designers' view can encourage a shift toward sustainability. Initially, the effort of integrating sustainability should be initiated in the design layout to consider the ecological consequences of the entire product. For instance, prototyping of the design is costly in the traditional manufacturing process from the perspective of material waste, energy, and time, but AM can produce with reduced material, energy, and time. Furthermore, AM can produce products with more complex geometry due to its high degree of freedom. Designing products with simple assemblies and configuration assists in the remanufacture and reuse of the parts, and such products are simple to repair. Despeisse and Ford [61] demonstrated the adoption of sustainability in the AM process by enhancing resource efficiency in the manufacturing sector. Based on case studies, the analysis revealed that AM can develop new business possibilities of reuse, repair, refurbishment, and remanufacturing. Furthermore, the industries are evolving to discover that AM can be extended to product life cycle and closing loopholes.

13.4.2 Product Redesign

The advancement of AM allows the integration of sustainability in the technological process and redesign of products. The redesign process is considered through product optimization approaches. Optimization can be achieved through a series of redesigns based on topology, speeding up the workflow, product quality and durability, and customer willingness to purchase the product again. Dalpadulo *et al.* [62] studied the redesigning approach of assertive devices to exploit AM potentials for producing devices. The results demonstrate that devices can be produced using AM after systematic redesigning of a part. The redesign mainly includes material replacement and a technology shift from the traditional manufacturing process, leading to faster workflow with increasing performance.

Many industries, such as aerospace [63], automotive [64], and medical [65], have had their manufacturing processes replaced by AM and have engaged in the redesigning of product structures to involve sustainability in product development. For instance, aeroplane seat buckles are redesigned and produced through AM to minimize the weight. Furthermore, heat sinks have been redesigned and produced using AM to enhance airflow and thermal efficiently. The utilization of the AM technique considerably reduced the scale of production and displayed maximum performance. Moreover, the production of aero engines involved significant waste during the production stage. In general, the proportion of material input to final product ranges from 4:1 to as high as 20:1 based on component. The project EU FP7 MERLIN, involving European aerospace organizations (Rolls-Royce, Turbomeca, and Fraunhofer ILT), addressed the downward impact on the environment through the implementation of AM when developing components [35]. The results show that the utilization of the AM technique (laser material deposition) saved approximately 60% of material and 30% of time and consumed the same level of energy after product optimization. In another project, General Electric demonstrated the noteworthy capabilities of AM through redesigning the LEAP engine that launched in 2016 [61]. General Electric manufactured fuel nozzles using AM that are used as components in engines. The produced nozzle is five times stronger, with efficient fuel flow to boost the combustion capability. Furthermore, the redesign product reduced the weight by 25% and simplified and combined the 20 varied components into a single component [61].

13.4.3 Process Redesign

AM includes a wide range of subtechnologies and single-manufacture equipment that can become more energy and resource efficient. The integration of AM technology has a substantial impact on the production process in the manufacture system workflow in the business model and supporting sustainability. Therefore, optimization in product design leads to improvements in the production process of products. For instance, Salcomp, a company that manufactures power supplies for mobile phones and other electronic devices, intended to reduce the cooling period during the injection molding of products. Salcomp engineers managed to redesign the vent structures of the molds using the AM technology and produced a mold to dissipate heat rapidly [66]. The production process decreased from 14s to 8s, with improved quality and enhanced rate of production. Similarly, the construction sector [67–69] utilizes a high amount of material, energy, and water so it requires a process redesign. A 12-m-long 3D-printed pedestrian bridge (MX3D) [67] was constructed in Amsterdam. This bridge demonstrates the reduction in materials, logistics, and energy, which drastically decreases the cost and environmental impact, achieved with the AM technology. Nevertheless, the construction industry is highly conservative and slow to modify due to concerns over the life cycle of the infrastructure built by the AM technology [67].

13.4.4 Raw Materials

AM technology not only influences the product design and process but also influences the material used as inputs to produce the sustainable products. The technologies utilized a wide range of substances (liquids, filaments, pastes, and powders) produced from natural resources as inputs without comprising the product quality [40]. The nature of the material is dependent on the working principle of AM technology. In terms of sustainability, the processing of raw materials can be reconsidered based on the product's need to minimize environmental impacts [6, 7, 61]. For instance, for decades, most physical components in the automobile and aerospace sectors were produced using metal. However, after the integration of the AM technique, components have been produced using polymers with a similar strength to metal components. The production of components using polymers significantly reduced energy and material consumption. Furthermore, the components can be recycled to reduce the ecological impacts caused by waste landfill and waste generation [35, 40]. For instance, the fused deposition modeling (FDM) method uses polymeric materials in filament form to produce parts, and material waste generated through the process can be recycled and reused as filament in the FDM technology [70]. Hart *et al.* [71] demonstrated the recycling of waste polymer generated from meal bags into a 3D printable filament. The results show that physical properties were considerably similar to native meal bag material [71]. Similarly, Turku *et al.* [72] researched mechanical and physical properties of ABS, polystyrene, and polyvinylchloride plastic waste filaments produced using a Filabot extruder. The analysis determined that the mechanical and thermal properties of the recycled polymers were considerably fewer than those of the virgin polymers [72]. The feasibility of using the polymer waste or recycling the waste generated from the product development has improved the sustainability of AM technologies. However, a more detailed study should be conducted to understand the product quality and the viability for mass production to reduce the use of natural resources.

13.4.5 Transformation of By-Product Into Product

In general, after the product life ends or damaged or the product does not meet the purchaser requirements, it will either be repaired or end up in a landfill. Repairing a product depends on the cost and ease of its restoration. Nevertheless, to repair and produce a replacement component that is not on demand through traditional manufacturing is expensive [35]. In addition, producing the original manufacturer parts and preserving the inventory encompasses demand uncertainty and cost of maintenance. However, AM technology can benefit the production of spare components with cost-effective design files that can be easily shared or stored on the cloud [35, 73, 74]. In terms of sustainability, AM is a considered as an efficient manufacturing method with minimal waste generation than conventional machining process. According to Chen *et al.* [75], the evolution of digital manufacturing, especially AM technologies, reveals a promising future for the manufacturing industry, with major influences in production paradigms from the batch size of one. Furthermore, the study reveals that AM enables the reduction of material utilization and fabricate the products based on consumer needs. For instance, the production of aeroplane engine components using a traditional four-axis milling process involves a high amount of raw material [35, 76]. The blisk was manufactured using AM techniques to reduce material waste, and studies have shown that approximately 60% of the material can be saved [76]. AM was also integrated in the production design of the repair of the damaged blisk to produce components on demand that extend operational life.

13.4.6 Closed-Loop Manufacturing

Closed-loop manufacturing (CLM) systems strive for sustainability through enhancing environmental and economic goals [40]. The negative ecological impacts, such as waste, energy consumption, and logistics, can be reduced through a CLM process. The AM process demonstrates potential in the direction of CLM standards. Mohammed *et al.* [77] demonstrated a CLM process with ABS filaments using AM. The results show that ABS can be 100% recycled after two recycling phases. Moreover, the mechanical and thermal characteristics of the recycled ABS increased, unlike those of the virgin filament, so the ABS can be used for various industrial applications to produce products [77]. The data imply that AM can be a closed-loop and low-carbon footprint manufacturing process that is sustainable over multiple recycling phases.

13.5 Conclusions

AM, as an emerging technique in the manufacturing sector, demonstrates the significant scope for sustainability in product development. Unlike traditional techniques, AM integration in product development reduces time, material consumption, and waste generation. Moreover, the waste materials can be recycled to develop contemporary designs that significantly assist in CLM processes. Based on the implications of sustainability and the LCA of AM, it is evident that AM has the advantage to transform the manufacturing system toward more sustainable approaches. However, further studies are needed to analyze the recycled materials, energy consumption, and challenges related to incorporating AM in mass production to increase sustainability.

References

1. *Decoupling Natural Resource Use and Environmental Impacts from Economic Growth (Summary)*, United Nations Environmental Program, (n.d.). https://www.unep.org/resources/report/decoupling-natural-resource-use-and-environmental-impactseconomicgrowth-summary#:~:text=By2050%2Chuman ity could devour,rate of natural resource consumption.
2. *Global Resource Consumption to Triple by 2050: UN*, PHYS ORG, 2011.
3. Bernstein, H., Industrialization, development, and dependence, in: *Introduction to the Sociology of "Developing Societies"*, pp. 218–235, Macmillan Education UK, London, 1982.
4. Hyvarinen, J., The 2005 world summit: UN reform, security, environment and development. *Rev. Eur. Community Int. Environ. Law*, 15, 1, 1–10, 2006.
5. Mehrpouya, M., Vosooghnia, A., Dehghanghadikolaei, A., Fotovvati, B., The benefits of additive manufacturing for sustainable design and production, in:

Sustainable Manufacturing, pp. 29–59, Elsevier, 2021, https://doi.org/10.1016/B978-0-12-818115-7.00009-2.
6. Gebler, M., Schoot Uiterkamp, A.J.M., Visser, C., A global sustainability perspective on 3D printing technologies. *Energy Policy*, 74, 158–167, 2014.
7. Peng, T., Kellens, K., Tang, R., Chen, C., Chen, G., Sustainability of additive manufacturing: An overview on its energy demand and environmental impact. *Addit. Manuf.*, 21, 694–704, 2018.
8. Mihelcic, J.R., Crittenden, J.C., Small, M.J., Shonnard, D.R., Hokanson, D.R., Zhang, Q., Chen, H., Sorby, S.A., James, V.U., Sutherland, J.W., Schnoor, J.L., Sustainability science and engineering: The emergence of a new metadiscipline. *Environ. Sci. Technol.*, 37, 23, 5314–5324, 2003.
9. Cai, W., Lai, K., Liu, C., Wei, F., Ma, M., Jia, S., Jiang, Z., Lv, L., Promoting sustainability of manufacturing industry through the lean energy-saving and emission-reduction strategy. *Sci. Total Environ.*, 665, 23–32, 2019.
10. Yip, W.S., Zhou, H., To, S., A critical analysis on the triple bottom line of sustainable manufacturing: Key findings and implications. *Environ. Sci. Pollut. Res.*, 30, 41388–41404, 2023, https://doi.org/10.1007/s11356-022-25122-x.
11. Haapala, K.R., Zhao, F., Camelio, J., Sutherland, J.W., Skerlos, S.J., Dornfeld, D.A., Jawahir, I.S., Clarens, A.F., Rickli, J.L., A review of engineering research in sustainable manufacturing. *J. Manuf. Sci. Eng.*, 135, 4, 041013, 2013, https://doi.org/10.1115/1.4024040.
12. Sabzi, M., Mousavi Anijdan, S.H., Shamsodin, M., Farzam, M., Hojjati-Najafabadi, A., Feng, P., Park, N., Lee, U., A review on sustainable manufacturing of ceramic-based thin films by chemical vapor deposition (CVD): Reactions kinetics and the deposition mechanisms. *Coatings*, 13, 1, 188, 2023.
13. Ahmadi, M., Rahmatabadi, D., Karimi, A., Koohpayeh, M.H.A., Hashemi, R., The role of additive manufacturing in the age of sustainable manufacturing 4.0, in: *Sustainable Manufacturing in Industry 4.0*, pp. 57–78, Springer Nature Singapore, Singapore, 2023.
14. Majeed, A., Zhang, Y., Ren, S., Lv, J., Peng, T., Waqar, S., Yin, E., A big data-driven framework for sustainable and smart additive manufacturing. *Robot. Comput. Integr. Manuf.*, 67, 102026, 2021.
15. Go, J. and Hart, A.J., A framework for teaching the fundamentals of additive manufacturing and enabling rapid innovation. *Addit. Manuf.*, 10, 76–87, 2016.
16. Guo, N. and Leu, M.C., Additive manufacturing: Technology, applications and research needs. *Front. Mech. Eng.*, 8, 3, 215–243, 2013.
17. Huang, Y., Leu, M.C., Mazumder, J., Donmez, A., Additive manufacturing: Current state, future potential, gaps and needs, and recommendations. *J. Manuf. Sci. Eng.*, 137, 1, 014001, 2015, https://doi.org/10.1115/1.4028725.
18. Piedra-Cascón, W., Krishnamurthy, V.R., Att, W., Revilla-León, M., 3D printing parameters, supporting structures, slicing, and post-processing

procedures of vat-polymerization additive manufacturing technologies: A narrative review. *J. Dent.*, 109, 103630, 2021.
19. Ligon, S.C., Liska, R., Stampfl, J., Gurr, M., Mülhaupt, R., Polymers for 3D printing and customized additive manufacturing. *Chem. Rev.*, 117, 15, 10212–10290, 2017.
20. Chapiro, M., Current achievements and future outlook for composites in 3D printing. *Reinf. Plast.*, 60, 6, 372–375, 2016.
21. Jo, J.H., Jo, B.W., Cho, W., Kim, J.-H., Development of a 3D printer for concrete structures: Laboratory testing of cementitious materials. *Int. J. Concr. Struct. Mater.*, 14, 1, 13, 2020.
22. Abdallah, Y.K. and Estévez, A.T., 3D-printed biodigital clay bricks. *Biomimetics*, 6, 4, 59, 2021.
23. Bertoldi, K., Vitelli, V., Christensen, J., van Hecke, M., Flexible mechanical metamaterials. *Nat. Rev. Mater.*, 2, 11, 17066, 2017.
24. Sadeqi, A., Rezaei Nejad, H., Owyeung, R.E., Sonkusale, S., Three dimensional printing of metamaterial embedded geometrical optics (MEGO). *Microsyst. Nanoeng.*, 5, 1, 16, 2019.
25. Dey, M. and Ozbolat, I.T., 3D bioprinting of cells, tissues and organs. *Sci. Rep.*, 10, 1, 14023, 2020.
26. Mehrpouya, M., Azizi, A., Janbaz, S., Gisario, A., Investigation on the functionality of thermoresponsive origami structures. *Adv. Eng. Mater.*, 22, 8, 2000296, 2020.
27. Janbaz, S., Weinans, H., Zadpoor, A.A., Geometry-based control of instability patterns in cellular soft matter. *RSC Adv.*, 6, 24, 20431–20436, 2016.
28. Fruggiero, F., Lambiase, A., Bonito, R., Fera, M., The load of sustainability for Additive Manufacturing processes. *Procedia Manuf.*, 41, 375–382, 2019.
29. Park, S., Shou, W., Makatura, L., Matusik, W., Fu, K., 3D printing of polymer composites: Materials, processes, and applications. *Matter*, 5, 1, 43–76, 2022.
30. Mohanavel, V., Ashraff Ali, K.S., Ranganathan, K., Allen Jeffrey, J., Ravikumar, M.M., Rajkumar, S., The roles and applications of additive manufacturing in the aerospace and automobile sector. *Mater. Today Proc.*, 47, 405–409, 2021.
31. Culmone, C., Smit, G., Breedveld, P., Additive manufacturing of medical instruments: A state-of-the-art review. *Addit. Manuf.*, 27, 461–473, 2019.
32. Chen, R.K., Jin, Y., Wensman, J., Shih, A., Additive manufacturing of custom orthoses and prostheses—A review. *Addit. Manuf.*, 12, 77–89, 2016.
33. Bhargav, A., Sanjairaj, V., Rosa, V., Feng, L.W., Fuh, J.Y.H., Applications of additive manufacturing in dentistry: A review. *J. Biomed. Mater. Res. Part B Appl. Biomater.*, 106, 5, 2058–2064, 2018.
34. Bikas, H., Stavropoulos, P., Chryssolouris, G., Additive manufacturing methods and modelling approaches: A critical review. *Int. J. Adv. Manuf. Technol.*, 83, 1–4, 389–405, 2016.
35. Ford, S. and Despeisse, M., Additive manufacturing and sustainability: An exploratory study of the advantages and challenges. *J. Clean. Prod.*, 137, 1573–1587, 2016.

36. Verhoef, L.A., Budde, B.W., Chockalingam, C., García Nodar, B., van Wijk, A.J.M., The effect of additive manufacturing on global energy demand: An assessment using a bottom-up approach. *Energy Policy*, 112, 349–360, 2018.
37. Pal, A.K., Mohanty, A.K., Misra, M., Additive manufacturing technology of polymeric materials for customized products: Recent developments and future prospective. *RSC Adv.*, 11, 58, 36398–36438, 2021.
38. Attaran, M., The rise of 3-D printing: The advantages of additive manufacturing over traditional manufacturing. *Bus. Horiz.*, 60, 5, 677–688, 2017.
39. Bi, Z., Wang, G., Thompson, J., Ruiz, D., Rosswurm, J., Roof, S., Guandique, C., System framework of adopting additive manufacturing in mass production line. *Enterp. Inf. Syst.*, 16, 4, 606–629, 2022.
40. Colorado, H.A., Velásquez, E.I.G., Monteiro, S.N., Sustainability of additive manufacturing: The circular economy of materials and environmental perspectives. *J. Mater. Res. Technol.*, 9, 4, 8221–8234, 2020.
41. Godina, R., Ribeiro, I., Matos, F., T. Ferreira, B., Carvalho, H., Peças, P., Impact assessment of additive manufacturing on sustainable business models in industry 4.0 context. *Sustainability*, 12, 17, 7066, 2020.
42. Romanello, M., Di Napoli, C., Drummond, P., Green, C., Kennard, H., Lampard, P., Scamman, D., Arnell, N., Ayeb-Karlsson, S., Ford, L.B., Belesova, K., Bowen, K., Cai, W., Callaghan, M., Campbell-Lendrum, D., Chambers, J., van Daalen, K.R., Dalin, C., Dasandi, N., Dasgupta, S., Davies, M., Dominguez-Salas, P., Dubrow, R., Ebi, K.L., Eckelman, M., Ekins, P., Escobar, L.E., Georgeson, L., Graham, H., Gunther, S.H., Hamilton, I., Hang, Y., Hänninen, R., Hartinger, S., He, K., Hess, J.J., Hsu, S.-C., Jankin, S., Jamart, L., Jay, O., Kelman, I., Kiesewetter, G., Kinney, P., Kjellstrom, T., Kniveton, D., Lee, J.K.W., Lemke, B., Liu, Y., Liu, Z., Lott, M., Batista, M.L., Lowe, R., MacGuire, F., Sewe, M.O., Martinez-Urtaza, J., Maslin, M., McAllister, L., McGushin, A., McMichael, C., Mi, Z., Milner, J., Minor, K., Minx, J.C., Mohajeri, N., Moradi-Lakeh, M., Morrissey, K., Munzert, S., Murray, K.A., Neville, T., Nilsson, M., Obradovich, N., O'Hare, M.B., Oreszczyn, T., Otto, M., Owfi, F., Pearman, O., Rabbaniha, M., Robinson, E.J.Z., Rocklöv, J., Salas, R.N., Semenza, J.C., Sherman, J.D., Shi, L., Shumake-Guillemot, J., Silbert, G., Sofiev, M., Springmann, M., Stowell, J., Tabatabaei, M., Taylor, J., Triñanes, J., Wagner, F., Wilkinson, P., Winning, M., Yglesias-González, M., Zhang, S., Gong, P., Montgomery, H., Costello, A., The 2022 report of the Lancet Countdown on health and climate change: Health at the mercy of fossil fuels. *Lancet*, 400, 10363, 1619–1654, 2022.
43. Herzog, D., Seyda, V., Wycisk, E., Emmelmann, C., Additive manufacturing of metals. *Acta Mater.*, 117, 371–392, 2016.
44. Lakhdar, Y., Tuck, C., Binner, J., Terry, A., Goodridge, R., Additive manufacturing of advanced ceramic materials. *Prog. Mater. Sci.*, 116, 100736, 2021.
45. González-Henríquez, C.M., Sarabia-Vallejos, M.A., Rodriguez-Hernandez, J., Polymers for additive manufacturing and 4D-printing: Materials, methodologies, and biomedical applications. *Prog. Polym. Sci.*, 94, 57–116, 2019.

46. Wang, Y., Zhou, Y., Lin, L., Corker, J., Fan, M., Overview of 3D additive manufacturing (AM) and corresponding AM composites. *Compos. Part A Appl. Sci. Manuf.*, 139, 106114, 2020.
47. Pakkanen, J., Manfredi, D., Minetola, P., Iuliano, L., About the use of recycled or biodegradable filaments for sustainability of 3D printing, In: Campana, G., Howlett, R., Setchi, R., Cimatti, B. (Eds.), *Sustainable Design and Manufacturing, 2017, SDM 2017. Smart Innovation, Systems and Technologies,* vol. 68, pp. 776–785, Springer, Cham. https://doi.org/10.1007/978-3-319-57078-5_73.
48. Cruz Sanchez, F.A., Boudaoud, H., Camargo, M., Pearce, J.M., Plastic recycling in additive manufacturing: A systematic literature review and opportunities for the circular economy. *J. Clean. Prod.*, 264, 121602, 2020.
49. Stansbury, J.W. and Idacavage, M.J., 3D printing with polymers: Challenges among expanding options and opportunities. *Dent. Mater.*, 32, 1, 54–64, 2016.
50. Beltrán, F.R., Arrieta, M.P., Moreno, E., Gaspar, G., Muneta, L.M., Carrasco-Gallego, R., Yáñez, S., Hidalgo-Carvajal, D., de la Orden, M.U., Martínez Urreaga, J., Evaluation of the technical viability of distributed mechanical recycling of PLA 3D printing wastes. *Polym. (Basel),* 13, 8, 1247, 2021.
51. Blakey-Milner, B., Gradl, P., Snedden, G., Brooks, M., Pitot, J., Lopez, E., Leary, M., Berto, F., du Plessis, A., Metal additive manufacturing in aerospace: A review. *Mater. Des.*, 209, 110008, 2021.
52. Ahn, D.-G., Direct metal additive manufacturing processes and their sustainable applications for green technology: A review. *Int. J. Precis. Eng. Manuf. Technol.*, 3, 4, 381–395, 2016.
53. Galante, R., Figueiredo-Pina, C.G., Serro, A.P., Additive manufacturing of ceramics for dental applications: A review. *Dent. Mater.*, 35, 6, 825–846, 2019.
54. Liu, Z.Y., Li, C., Fang, X.Y., Guo, Y.B., Energy consumption in additive manufacturing of metal parts. *Procedia Manuf.*, 26, 834–845, 2018.
55. Yang, Y., Li, L., Pan, Y., Sun, Z., Energy consumption modeling of stereolithography-based additive manufacturing toward environmental sustainability. *J. Ind. Ecol.*, 21, S1, S168–S178, 2017, https://doi.org/10.1111/jiec.12589.
56. Hopkins, N., Jiang, L., Brooks, H., Energy consumption of common desktop additive manufacturing technologies. *Clean. Eng. Technol.*, 2, 100068, 2021.
57. Drizo, A. and Pegna, J., Environmental impacts of rapid prototyping: An overview of research to date. *Rapid Prototyp. J.*, 12, 2, 64–71, 2006.
58. Faludi, J., Bayley, C., Bhogal, S., Iribarne, M., Comparing environmental impacts of additive manufacturing vs traditional machining via life-cycle assessment. *Rapid Prototyp. J.*, 21, 1, 14–33, 2015.
59. Kafara, M., Süchting, M., Kemnitzer, J., Westermann, H.-H., Steinhilper, R., Comparative life cycle assessment of conventional and additive

manufacturing in mold core making for CFRP production. *Procedia Manuf.*, 8, 223–230, 2017.
60. Gao, C., Wolff, S., Wang, S., Eco-friendly additive manufacturing of metals: Energy efficiency and life cycle analysis. *J. Manuf. Syst.*, 60, 459–472, 2021.
61. H., Kiritsis, D., von Cieminski, G. (Eds.), *Advances in Production Management Systems: Innovative Production Management Towards Sustainable Growth. APMS 2015. IFIP Advances in Information and Communication Technology*, vol. 460, Springer, Cham, https://doi.org/10.1007/978-3-319-22759-7_15
62. Dalpadulo, E., Gherardini, F., Pini, F., Leali, F., Integration of topology optimisation and design variants selection for additive manufacturing-based systematic product redesign. *Appl. Sci.*, 10, 21, 7841, 2020.
63. Najmon, J.C., Raeisi, S., Tovar, A., Review of additive manufacturing technologies and applications in the aerospace industry, in: *Additive Manufacturing for the Aerospace Industry*, pp. 7–31, Elsevier, 2019, https://doi.org/10.1016/B978-0-12-814062-8.00002-9.
64. Srivastava, M. and Rathee, S., Additive manufacturing: Recent trends, applications and future outlooks. *Prog. Addit. Manuf.*, 7, 2, 261–287, 2022.
65. Salmi, M., Additive manufacturing processes in medical applications. *Mater. (Basel)*, 14, 1, 191, 2021.
66. Gunbay, S., Hopkins, M., Hu, G., Gately, N.M., Devine, D.M., Hayes, C., Additive manufacturing of tooling for use in mass production processes, in: *Polymer-Based Additive Manufacturing*, pp. 73–96, Springer International Publishing, Cham, 2019.
67. Buchanan, C. and Gardner, L., Metal 3D printing in construction: A review of methods, research, applications, opportunities and challenges. *Eng. Struct.*, 180, 332–348, 2019.
68. Kyvelou, P., Buchanan, C., Gardner, L., Numerical simulation and evaluation of the world's first metal additively manufactured bridge. *Structures*, 42, 405–416, 2022.
69. Gardner, L., Kyvelou, P., Herbert, G., Buchanan, C., Testing and initial verification of the world's first metal 3D printed bridge. *J. Constr. Steel Res.*, 172, 106233, 2020.
70. Rett, J.P., Traore, Y.L., Ho, E.A., Sustainable materials for fused deposition modeling 3D printing applications. *Adv. Eng. Mater.*, 23, 7, 2001472, 2021.
71. Hart, K.R., Frketic, J.B., Brown, J.R., Recycling meal-ready-to-eat (MRE) pouches into polymer filament for material extrusion additive manufacturing. *Addit. Manuf.*, 21, 536–543, 2018.
72. Turku, I., Kasala, S., Kärki, T., Characterization of polystyrene wastes as potential extruded feedstock filament for 3D printing. *Recycling*, 3, 4, 57, 2018.
73. Valilai, O.F. and Houshmand, M., Depicting additive manufacturing from a global perspective; Using Cloud manufacturing paradigm for integration and collaboration. *Proc. Inst. Mech. Eng. Part B J. Eng. Manuf.*, 229, 12, 2216–2237, 2015.

74. Rudolph, J.-P. and Emmelmann, C., A Cloud-based platform for automated order processing in additive manufacturing. *Procedia CIRP*, 63, 412–417, 2017.
75. Chen, D., Heyer, S., Ibbotson, S., Salonitis, K., Steingrímsson, J.G., Thiede, S., Direct digital manufacturing: Definition, evolution, and sustainability implications. *J. Clean. Prod.*, 107, 615–625, 2015.
76. Mathew, J., Hauser, C., Stoll, P., Kenel, C., Polyzos, D., Havermann, D., MacPherson, W.N., Hand, D.P., Leinenbach, C., Spierings, A., Koenig-Urban, K., Maier, R.R.J., Integrating fiber fabry-perot cavity sensor into 3-D printed metal components for extreme high-temperature monitoring applications. *IEEE Sens. J.*, 17, 13, 4107–4114, 2017.
77. Mohammed, M.I., Wilson, D., Gomez-Kervin, E., Tang, B., Wang, J., Investigation of closed-loop manufacturing with acrylonitrile butadiene styrene over multiple generations using additive manufacturing. *ACS Sustain. Chem. Eng.*, 7, 16, 13955–13969, 2019.

14

Optimization of Fused Deposition Modeling Control Parameters Using Hybrid Taguchi and TOPSIS Method

B. Singaravel[1*], T. Niranjan[2], M. Vasu Babu[3] and K. Nagarjuna[1]

[1]*Department of Mechanical Engineering, Vignan Institute of Technology and Science, Deshmukhi, Hyderabad, Telangana, India*
[2]*Department of Mechanical Engineering, Mahatma Gandhi Institute of Technology, Hyderabad, Telangana, India*
[3]*Department of Electronic and Instrumentation Engineering, Vignan Institute of Technology and Science, Deshmukhi, Hyderabad, Telangana, India*

Abstract
Additive manufacturing technologies are used to make prototypes rapidly. The three-dimensional (3D) printing process is one of the widely preferred methods that uses solid-based wire as raw materials. Raw materials play a significant role to form accurate and quality products. In this work, an investigation is carried out on fused deposition modeling (FDM) to analyze the effect of different raw materials and control parameters on quality, time taken for printing, hardness, and strength. Three different raw materials such as nylon, acrylonitrile butadiene styrene (ABS), and polyethylene terephthalate glycol (PETG) are considered. Control parameters considered in the study are layer thickness and extrusion speed. Performance is evaluated using the following: surface roughness, time taken for printing, tensile strength, and hardness of the printed specimen. Experiments are executed with the help of Taguchi L_9 orthogonal array. The hybrid Taguchi and Technique for Order of Preference by Similarity to Ideal Solution (TOPSIS) technique is employed for the prediction of optimum control factors. The result of the investigation indicated that control parameters significantly influenced the selected response.

Keywords: FDM, raw materials, control parameters, optimization

**Corresponding author*: singnitt@gmail.com

S. Thirumalai Kumaran and Tae Jo Ko (eds.) Sustainable Machining and Green Manufacturing, (289–304) © 2024 Scrivener Publishing LLC

14.1 Introduction

Additive manufacturing technologies are widely used for time and cost reduction related to manufacturing that uses the layer-by-layer manufacturing concept. Fused deposition modeling (FDM) is a widely preferred method due to its significant advantages in terms of minimum cost for production, easy handling of machine, and better strength of the prepared sample. In the execution of the FDM, initially, a CAD model is created and sliced into Standard Triangular Language (STL) format. STL files are used to divide into the number of layers as per machine parameters. In the FDM, wire-type polymer-based solid material is used. Raw materials are wound in coil type and sent to the extrusion nozzle tip. During extrusion, temperature in the nozzle tip is used to melt the filament. The molten filament stage is printed on three movable axis-controlled platforms. The first layer is generated based on solidification using molten material from the extrusion process. Then, the movable platform is moved for the second layer formation. It will be repeated up to the entire thickness of the selected CAD geometry. In any investigation, an important step is to design and plan the experiments to be executed. In the present investigation, the Taguchi orthogonal array is used to execute the experiments. The quality of the printed specimen can be analyzed using tensile strength. Hardness after three-dimensional (3D) printing of samples must be provided for wear resistance before further application. Investigation related to time taken for printing is used to realize the resource consumption. Surface roughness is an indication of functional quality [1–5].

Optimization of control parameters in machining leads to better efficiency. The Taguchi method is a standard optimization tool that is used to optimize control parameters and enhance quality. The important steps involved are the selection of a suitable orthogonal design, experiment execution, and signal-to-noise (S/N) ratio calculation [6–8]. This technique is based on a single objective optimization technique but in real-time applications is based on multiple objectives. Hence, researchers make use of the gray concept to solve multiple objective optimization problems. Nowadays, the multi-criteria decision-making (MCDM) method is attractive to researchers due to its easy execution and not requiring any software and assumption parameters. It is a statistical concept used for solving a number of alternatives from a set of alternatives. The Technique for Order of Preference by Similarity to Ideal Solution (TOPSIS) is a type of MCDM tool that involves the basic concept of smallest distance of positive solution and largest distance of negative solution. The extraction of both Taguchi and MCDM advantages is used for solving multi-objective optimization problems [9–12].

14.2 Literature Review

Durgashyam *et al.* [13] investigated the FDM process using polyethylene terephthalate glycol (PETG) material. Optimization steps were executed using Taguchi-based optimization to analyze the experimental results. Layer thickness, infill density, and extrusion feed were considered as process parameters. Responses such as tensile strength and bending strength were performed on 3D printed samples. The result of the experimentation observed that layer thickness significantly influenced the performance of the samples. Vishwas and Basavaraj [14] printed with acrylonitrile butadiene styrene (ABS) material using additive manufacturing methods. The mechanical performance study was performed. Input parameters such as layer thickness, orientation angle, and shell thickness were considered. The result concluded that process parameters were significantly influenced the machining performance. Ramesh and Panneerselvam [15] used nylon material for the 3D printing application using the FDM process. Machining parameters thickness of layer, infiltration density, and printing speed were used. The mechanical performance was investigated, and the material was observed to have better strength with an optimum level of properties. Vishwas *et al.* [16] used two different materials (ABS and nylon) for 3D printing application. The Taguchi steps were executed for result analysis. Tensile strength was investigated, and the results indicated that process parameters significantly affected the mechanical performance. Vikneswaran *et al.* [17] used three different materials (PETG, Polylactic acid (PLA), and ABS) for 3D printing application and conducted a mechanical strength study. The result revealed a high value of tensile strength with PETG material and a high value of stiffness with ABS. Raja and Rajan [18] used the TOPSIS method to select the appropriate FDM machine. Fuzzy concept was combined with TOPSIS to select the best FDM machine. Nine different models with features were considered. The result revealed that TOPSIS was useful for the selection and purchase of the FDM machine by considering the appropriate weight. Chinchanikar *et al.* [19] used hybrid genetic algorithm and MCDM techniques to optimize the FDM control parameters. Mechanical performance and surface quality were investigated with the control factor variation. The result concluded that the hybrid technique utilized predicted the optimum control parameters. Vinodh [20] used the gray-based Taguchi and MCDM concept for the selection of an automobile component. Relative importance was considered and calculated using the analytical hierarchy process. The result revealed that 27% had similar results with other the optimization techniques. Vinodh and Shinde [21] used the MCDM approach for the selection of optimum control factors. The effect of control parameters on surface roughness and build time was investigated. Singh *et al.* [22] conducted machining on polymer-based composite and performed an optimization study using the hybrid Taguchi and TOPSIS method. The best alternative was predicted using a high value of relative closeness. Balasubramaniyan and Selvaraj [23] used an integrated optimization tool for the estimation of optimum control factors using Taguchi and TOPSIS. Dimensional accuracy parameters were considered after the turning pro-cess. The result pointed out that the integrated optimization tool used solved multi-criteria problems in an efficient manner. Chohan *et al.* [24] used the FDM process for printing 3D objects, and control factors were optimized using the Taguchi-based TOPSIS approach. The advantages of Taguchi, TOPSIS, and analysis of variance were utilized to perform the optimization study. The result pointed out that the combined approach was useful for the effective estimation of optimum control parameters. Rinanto *et al.* [25] used a hybrid optimization technique for the prediction of optimum control factors. Minimization of energy consumption and enhancement for tensile strength after 3D printing specimens were investigated. The results of the optimization study of minimized energy consumption and enhanced strength were predicted accurately.

From the literature, the importance of raw materials in FDM-based 3D printing application is observed. At the same time, performance needs to be studied by varying the control parameters of the FDM machine. Optimization of control parameters leads to better efficiency. The Taguchi method is a unique optimization method, but it will be preferred to solve single-objective optimization problems. To solve simultaneous optimization problems, the Taguchi method needs to be combined with tools like the gray concept. In this work, MCDM-based TOPSIS is combined with Taguchi method and executed for optimization of control parameters. Hence, in the present investigation, the influence of control parameters on response parameters is investigated and optimized using a hybrid Taguchi-based optimization technique.

14.3 Experimental Setup

In this work, three raw materials were used, such as nylon, ABS, and PETG. Nylon possesses better durability and strength. Due to its significant properties, it is widely preferred to be used in spring and brackets. ABS has good mechanical properties, dimension stability, and physical properties. PETG has minimum shrinkage and good adhesion strength. As mentioned earlier, the FDM-type 3D printer is used for printing the specimen. Figure 14.1a represents the FDM machine that is used for this work. CREO software is used for slicing the 3D CAD modeling of the actual image to be printed. Figure 14.1b shows the interface of CREO and CURO software. Control parameters considered are the type of raw material, thickness of the layer to be produced, and extrusion speed. Table 14.1 represents the control parameters selected in this investigation. Figure 14.2 indicates the specimen printed using FDM. Experiments are performed with the help of the L_9 Taguchi experimental plan. The universal testing machine (UTM) is used to carry out the tensile test on the 3D printed specimen. In this regard, ATM standard of D638 is used to perform the tensile test. Figure 14.3 represents the specimen after performing the tensile test. The hardness of the 3D printed specimen is determined using the Rockwell hardness tester. The surface roughness and layer thickness of the printed samples are considered. The Talysurf surface roughness tester was used to measure the roughness value. Table 14.2 shows the experimental results based on the orthogonal array.

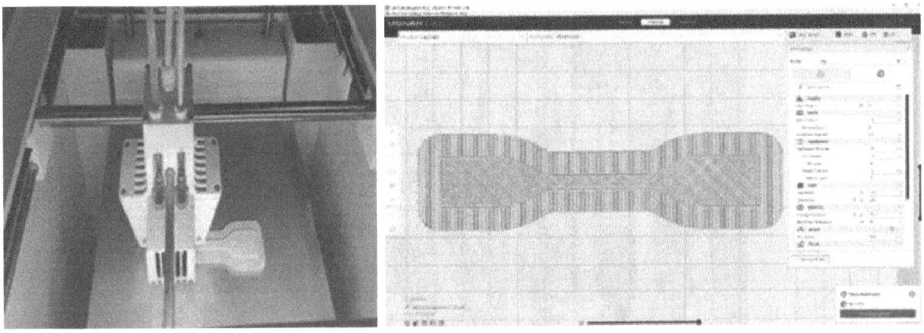

(a) (b)

Figure 14.1 (a) FDM machine, (b) CURO software interface.

Table 14.1 Control parameters and its level.

	Level 1	Level 2	Level 3
Type of raw materials	PETG	Nylon	ABS
Layer thickness in mm	0.1	0.2	0.3
Extrusion speed in mm/sec	50	60	70

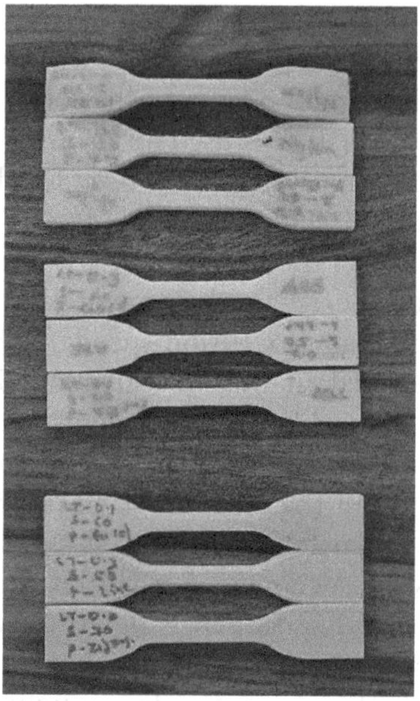

Figure 14.2 Specimens printed using the FDM.

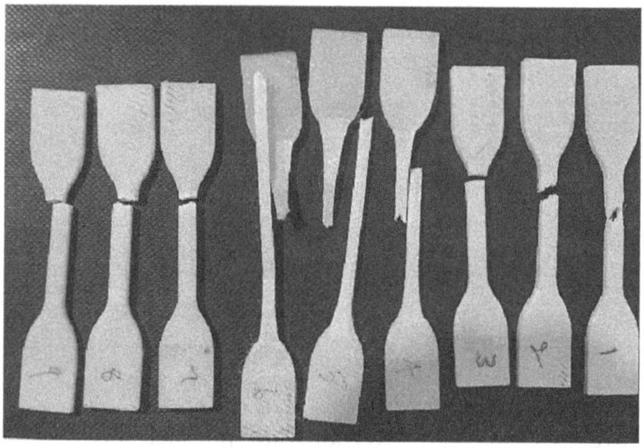

Figure 14.3 Samples after the tensile test.

Table 14.2 Experimental results.

S. no.	Raw materials	Layer thickness (mm)	Extrusion speed (mm/sec)	Surface roughness (µm)	Hardness in C scale	Printing time (min)	Tensile strength (MPa)
1	PETG	0.1	50	7.88	78	42	36.11
2	PETG	0.2	60	8.71	67	36	24.13
3	PETG	0.3	70	9.42	59	28	20.58
4	Nylon	0.1	60	5.62	92	38	57.28
5	Nylon	0.2	70	6.45	88	33	38.64
6	Nylon	0.3	50	7.32	81	25	31.97
7	ABS	0.1	70	6.14	86	36	37.44
8	ABS	0.2	60	7.21	77	27	22.18
9	ABS	0.3	50	8.32	68	22	37.19

14.4 Methodology

Industrial problems are multi-objective and need to be solved simultaneously. Various multi-objective techniques are available, namely, Taguchi-based, metaheuristic-based, and MCDM-based optimization techniques. MCDM-based methods are widely preferred decision-making tools to estimate a better combination of control parameters. It is effectively used for solving multi-objective problems with different alternatives. Different MCDM methods are presented in which TOPSIS is the most popular tool. It is based on the estimation of positive and negative ideal solutions. The main objective of TOPSIS is to maximize the beneficial criteria and minimize the non-beneficial criteria. The final outcome of the TOPSIS should be toward a positive ideal solution and farthest from the negative ideal solution [26–29]. The following are the procedures needed to be executed.

Step 1: Objective description

In this work, type outputs such as tensile strength and hardness are maximized. Type outputs such as surface roughness and printing time are minimized.

Step 2: Calculation of the signal-to-noise ratio

In the Taguchi procedure, the orthogonal array was selected and the experiments based on the orthogonal array were performed. Then, the S/N ratio will be calculated and used to understand the optimum effect of input and output. The optimum combination leads to better input and output.

The maximized type objective function needs to use the-larger-the-better characteristics

$$\frac{S}{N} = -10 * Log_{10}\left(\frac{1}{n}\sum\nolimits_{i-1}^{n}\frac{1}{y^2}\right) \tag{14.1}$$

The minimized type objective function needs to use the-larger-the-better characteristics

$$\frac{S}{N} = -10 * Log_{10}\left(\frac{1}{n}\sum\nolimits_{i-1}^{n}1y^2\right) \tag{14.2}$$

Table 14.3 represents the calculated value of the S/N ratio.

Table 14.3 Decision, normalized, and weighted matrix.

S. no.	Decision matrix				Normalized matrix				Weighted matrix			
1	-17.9	37.8	-32.46	31.15	0.654	0.628	1	0.549	0.163	0.157	0.25	0.137
2	-18.8	36.5	-31.12	27.65	0.848	0.286	0.761	0.155	0.212	0.071	0.190	0.038
3	-19.4	35.4	-28.94	26.26	1	0	0.372	0	0.25	0	0.093	0
4	-14.9	39.2	-31.59	35.16	0	1	0.845	1	0	0.25	0.211	0.25
5	-16.1	38.8	-30.37	31.74	0.266	0.899	0.627	0.615	0.06	0.224	0.156	0.153
6	-17.2	38.1	-27.95	30.09	0.511	0.713	0.197	0.430	0.127	0.178	0.049	0.107
7	-15.7	38.6	-31.12	31.46	0.171	0.848	0.761	0.584	0.042	0.212	0.190	0.146
8	-17.1	37.7	-28.62	26.91	0.482	0.599	0.316	0.073	0.120	0.149	0.079	0.018
9	-18.4	36.6	-26.84	31.40	0.759	0.319	0	0.5780	0.189	0.079	0	0.144

Step 3: Decision matrix construction
The decision matrix is formed with the help of the measured output and its S/N ratio value. Table 14.3 indicates the calculated decison matrix.

Step 4: Normalization of the different scores into an equivalent score
The output considered in the work uses different units of measure, which is required to bring an equivalent score. Normalization is the process of converting different units into a single equivalent unit. The following equation is used for normalization purposes.

If the output value (y_{ij}) is to be maximized, then it has the-larger-the-best concept. The actual score can be normalized (x_{ij}) as follows:

$$x_{ij} = \frac{y_{ij} - \min y_{ij}, i = 1,2,3\ldots n}{\max y_{ij}, i = 1,2,3\ldots n - \min y_{ij}, i = 1,2,3\ldots n} \tag{14.3}$$

If the output value (y_{ij}) is to be minimized, then it has the-smaller-the-best concept. The original score can be normalized (x_{ij}) as follows:

$$x_{ij} = \frac{\min y_{ij}, i = 1,2,3\ldots n - y_{ij}}{\max y_{ij}, i = 1,2,3\ldots n - \min y_{ij}, i = 1,2,3\ldots n} \tag{14.4}$$

where n represents the number of experiments. In this investigation, Equation 14.3 is used for maximizing the output data and Equation 14.4 is used for minimizing the output data involved in finding the normalized score of the experimental values. Table 14.3 indicates the normalized value estimated.

Step 5: Relative importance matrix
Responses have different impacts on cumulative objective. Hence, the relative importance of each objective is playing an important role. Each response is multiplied by its relative imperative also known as weight factor.

$$W_{ij} = N_{ij} \times W_j \tag{14.5}$$

where N_{ij} indicates the value of the normalized matrix and W_j indicates the relative importance of each response. In this work, all responses are assumed to have equal importance.

Step 6: Calculation of the positive ideal solution (PID⁺) and the negative ideal solution (NID⁻)
The PID^+ and the NID^- are determined using Equations 14.6 and 14.7.

$$A^+ = \{(\max W_{ij} \mid j\epsilon J), (\min W_{ij} \mid j\epsilon J')\} \tag{14.6}$$

$$A^- = \{(\min W_{ij} \mid j\epsilon J),(\max W_{ij} \mid j\epsilon J')\} \tag{14.7}$$

where J = 1,2,3,....,n
J is associated with the maximization criteria J' = 1,2,3,....,n
where J' is associated with the minimization criteria.

Step 7: Separation measure estimation
The separation of the positive ideal solution is estimated as follows:

$$S_i^+ = \sqrt{\sum\left(W_{ij} - A_j^+\right)^2} \tag{14.8}$$

j = 1, where i = 1, 2,..., m
The separation of the negative ideal solution is estimated as follows:

$$S_i^- = \sqrt{\sum\left(W_{ij} - A_j^-\right)^2} \tag{14.9}$$

j = 1, where i = 1, 2,..., m

Step 8: Relative closeness prediction (C_i) to the positive ideal solution

$$C_i = S_i^- / (S_i^+ + S_i^-) \tag{14.10}$$

The high value of C_i indicates an optimum combination of the alternatives.

Step 9: Rank the optimum alternatives using the relative closeness value
Table 14.4 indicates the value of separation of the positive and the negative ideal solutions, relative closeness value, and their rank.

Table 14.4 Separation measures of the positive and negative ideal solutions and relative closeness value.

S. no.	PID	NID	RC
1	0.2080	0.2258	0.5205
2	0.0128	0.1078	0.8931
3	0.1567	0.1567	0.5
4	0.2113	0.4347	0.6729
5	0.1022	0.3414	0.7695
6	0.0367	0.3138	0.8952
7	0.0914	0.3358	0.7860
8	0.1321	0.2621	0.6649
9	0.0856	0.3055	0.7809

14.5 Results and Discussion

In the present investigation, the FDM-based 3D printing process is used for printing 3D specimens for tensile test. As stated earlier, three different raw materials such as PET, nylon, and ABS were used. The FDM machine control parameters thickness of layer to be printed and extrusion speed are considered. The Taguchi-based experimental design is used for execution of the 3D printing. Specimen quality is estimated using tensile strength, time taken for printing, hardness after 3D printing, and surface roughness in the layer formation side. The hybrid Taguchi-based TOPSIS technique is executed for the optimization of the process parameters.

The hybrid Taguchi and TOPSIS technique is utilized to select and rank higher preferred output-based alternatives from available alternatives. In this technique, the S/N ratio followed by the relative closeness played a significant role. A high value of relative closeness is an indication of a higher preferred output. Generally, TOPSIS is used to predict the shortest higher preferred output combination and farthest from the low output combination. As an indication of the high value of relative closeness, the alternatives are arranged as follows: 8-2-9-6-5-1-3-7-4. The optimum FDM control parameters are predicted based on the maximum measure of relative closeness, and the values are 0.3 mm layer thickness and 50 mm/

sec extrusion speed with nylon as raw material. The main effect plot based on relative closeness indicated the optimum combination of output such as tensile strength (31.97 MPa), hardness (81 RHN in C scale), printing time (25 min), and surface roughness (7.32 μm). The result of optimization is based on simultaneous consideration of output. Nylon as raw material provides better bonding strength and stiffness after printing the specimen. Another advantage of nylon is it holds the filament shape accurately when compared to other raw materials due to its higher melting point. Control parameters such as layer thickness and extrusion speed also significantly influenced the process performance. A high value of tensile strength is noticed with the minimum value of layer thickness. Results of experiments 1, 4, and 7 indicated the high value of tensile strength. This is due to good bonding in nature and axial load capacity using a low value of layer thickness of the printed specimen. Spreading of raw materials followed by solidification is influenced by extrusion speed. A low value of extrusion speed is used to obtain better tensile strength. A higher value of extrusion speed may lead to poor quality due to material deposition defect and voids. The hardness of the printed specimen is greatly influenced by layer thickness. A high value of hardness is obtained with a low value of layer thickness. The reason might be that a greater number of layers were stacked and provide better resistance to penetration. Extrusion speed significantly influences the quality of the printed specimen. An extrusion speed of 50 mm/sec resulted in better strength and hardness. This is due to a better bonding property of extruded material at 50 mm/sec. The low value of strength and hardness with a higher extrusion speed was due to inadequate contact time for the spread and filling of the material. The time taken for effective printing provides for proper filament melting, extrusion, spreading, and solidification. Control parameters played a significant role in effective printing with minimum time taken. The minimum value of layer thickness and extrusion speed provided better strength and hardness. The result of surface roughness is low at the minimum value of layer thickness and extrusion speed. Increases in layer thickness led to a higher surface roughness. It is understood that surface roughness is directly proportional to layer thickness.

It is realized that the proposed hybrid technique based on TOPSIS is easy to comprehend and has simple steps for execution. When compared to other MCDM methods and Taguchi methods (MOORA, VIKOR, GRA), it involves a simple ratio analysis and minimum numbers of steps. It will be helpful for decision-makers without software skills. Another notable advantage is not involving any assumption-based constant (for example, v in VIKOR and ξ in GRA methods). The above advantages of the hybrid Taguchi and TOPSIS technique are highly preferable.

14.6 Conclusions

- The FDM is used to prepare the tensile specimen using three different materials as per the ASTM standard. Three different materials are printed, and their performance in terms of tensile strength, hardness, time, and surface roughness is evaluated.
- The hybrid Taguchi and TOPSIS approach is used for the prediction of optimum control factors to enhance the FDM process. Optimization was achieved with control parameters such as a layer thickness of 0.3 mm and an extrusion speed of 50 mm/sec with nylon as raw material.
- The result of the experiment revealed that the nylon material gave better mechanical properties due to its good adhesion properties, bonding in nature, and more axial load capacity.
- Control parameters significantly influenced the performance. Layer thickness led to high tensile strength and hardness. Extrusion speed led to spreading of raw material and solidification.
- The hybrid technique of Taguchi and TOPSIS is a statistical method that is easy to execute, eliminating the requirement of software and assumption of any constant. Also, it is simple to use and a reliable technique.

References

1. Rashed, K., Kafi, A., Simons, R., Bateman, S., Fused filament fabrication of nylon 6/66 copolymer: Parametric study comparing full factorial and Taguchi design of experiments. *Rapid Prototyp. J.*, 28, 1111–1128, 2022.
2. Navaneetha, K., Prasad, B.A., Bhargavi, M., Singaravel, B., Modeling of two link manipulator used for 3D printing. *IOP Conf. Ser. Mater. Sci. Eng.*, 1057, 012023, 2021.
3. Kumar, R., Sharma, H., Saran, C., Tripathy, T.S., Sangwan, K.S., Herrmann, C., A comparative study on the life cycle assessment of a 3D printed product with PLA, ABS & PETG materials. *Procedia CIRP*, 107, 15–20, 2022.

4. Dey, A. and Yodo, N., A systematic survey of FDM process parameter optimization and their influence on part characteristics. *J. Manuf. Mater. Process.*, 3, 64, 2019.
5. Prakash, K.S., Nancharaih, T., Rao, V.S., Additive manufacturing techniques in manufacturing-an overview. *Mater. Today Proc.*, 5, 3873–3882, 2018.
6. Singaravel, B. and Selvaraj, T., Application of Taguchi method for optimization of parameters in turning operation. *J. Manuf. Sci. Eng.*, 16, 183–18, 2016.
7. Singaravel, B. and Selvaraj, T., Application of desirability function analysis and utility concept for selection of optimum cutting parameters in turning operation. *J. Adv. Manuf. Syst.*, 15, 1–11, 2016.
8. Divya, C., Raju, L.S., Singaravel, B., Niranjan, T., Performance investigation of micro hole textured cutting inserts on power consumption and its measuring methodology in turning process. *Alex. Eng. J.*, 61, 3125–3130, 2022.
9. Singaravel, B., Chakradhar, B., Rajan, D.S., Kumar, A.K., Optimization of friction stir welding process parameters using MCDM method. *Mater. Today Proc.*, 76, 597–601, 2022.
10. Devaraj, S., Ramakrishna, M., Singaravel, B., Application of TOPSIS method for prediction of optimum design parameters of micro hole textured cutting inserts in turning of Al-MMC. *Int. J. Veh. Struct. Syst.*, 13, 471–476, 2021.
11. Niranjan, T., Singaravel, B., Raju, S.S., Optimization of hole quality parameters using TOPSIS method in drilling of GFRP composite. *Mater. Today Proc.*, 62, 2109–2114, 2022.
12. Divya, C., Raju, L.S., Singaravel, B., Application of MCDM methods for process parameter optimization in turning process—A review, in: *Recent Trends in Mechanical Engineering*, pp. 199–207, 2021.
13. Durgashyam, K., Reddy, M.I., Balakrishna, A., Satyanarayana, K., Experimental investigation on mechanical properties of PETG material processed by fused deposition modeling method. *Mater. Today Proc.*, 18, 2052–2059, 2019.
14. Vishwas, M. and Basavaraj, C.K., Studies on optimizing process parameters of fused deposition modelling technology for ABS. *Mater. Today Proc.*, 4, 10994–11003, 2017.
15. Ramesh, M. and Panneerselvam, K., Mechanical investigation and optimization of parameter selection for nylon material processed by FDM. *Mater. Today Proc.*, 46, 9303–9307, 2021.
16. Vishwas, M., Basavaraj, C.K., Vinyas, M., Experimental investigation using taguchi method to optimize process parameters of fused deposition modeling for ABS and nylon materials. *Mater. Today Proc.*, 5, 7106–7114, 2018.
17. Vikneswaran, S.K., Nagarajan, P., Dinesh, S.K., Kumar, K.S., Megalingam, A., Investigation of the tensile behaviour of polylactic acid, acrylonitrile butadiene styrene, and polyethylene terephthalate Glycol materials. *Mater. Today Proc.*, 66, 1093–1098, 2022.
18. Raja, S. and Rajan, A.J., A decision-making model for selection of the suitable FDM machine using fuzzy TOPSIS. *Math. Probl. Eng.*, 2022, 1–15, 2022.

19. Chinchanikar, S., Shinde, S., Shaikh, A., Gaikwad, V., Ambhore, N.H., Multi-objective optimization of FDM using hybrid genetic algorithm-based multi-criteria decision-making (MCDM) techniques. *J. Inst. Eng. (India) D*, 1–15, 2023.
20. Vinodh, S., Parametric optimization of fused deposition modelling process using grey based Taguchi and TOPSIS methods for an automotive component. *Rapid Prototyp. J.*, 27, 155–175, 2020.
21. Vinodh, S. and Shinde, P., Parametric optimization of 3D printing process using MCDM method. *Precision Product-Process Design and Optimization: AIMTDR*, pp. 141–159, 2018.
22. Singh, A., Datta, S., Mahapatra, S.S., Application of TOPSIS in the Taguchi method for optimal machining parameter selection. *J. Manufacturing Sci. & Production*, 11, 1–3, 49–60, 2011.
23. Balasubramaniyan, S. and Selvaraj, T., Application of integrated Taguchi and TOPSIS method for optimization of process parameters for dimensional accuracy in turning of EN25 steel. *J. Chin. Inst. Eng.*, 40, 4, 267–274, 2017.
24. Chohan, J.S., Kumar, R., Singh, T.B., Singh, S., Sharma, S., Singh, J., Mia, M., Pimenov, D.Y., Chattopadhyaya, S., Dwivedi, S.P., Kapłonek, W., Taguchi S/N and TOPSIS based optimization of fused deposition modelling and vapor finishing process for manufacturing of ABS plastic parts. *Materials*, 13, 5176, 2020.
25. Rinanto, A., Pujiyanto, E., Nugroho, A., No, J.M., Karangasem, L., Prasetyo, H., Optimization of tensile strength and energy consumption on FDM rapid prototyping process using taguchi and PCR-TOPSIS methods. *4th International Conference on Science and Technology*, 2018.
26. Singaravel, B. and Selvaraj, T., Optimization of machining parameters in turning operation using combined TOPSIS and AHP method. *Teh. Vjesn.*, 22, 6, 1475–1480, 2015.
27. Singaravel, B., Shankar, D.P., Prasanna, L., Application of MCDM method for the selection of optimum process parameters in turning process. *Mater. Today Proc.*, 5, 5, 13464–13471, 2018.
28. Divya, C., Raju, L.S., Singaravel, B., A review of TOPSIS method for multi criteria optimization in manufacturing environment. *Intelligent Techniques and Applications in Science and Technology: Proceedings of the First International Conference on Innovations in Modern Science and Technology*, vol. 1, pp. 719–727, 2020.
29. Rao, K.M., Kumar, D.V., Shekar, K.C., Singaravel, B., Optimization of EDM process parameters using TOPSIS for machining AISI D2 steel material. *Mater. Today Proc.*, 46, 701–706, 2021.

15

Sustainable Machining of Monel 400 Using Cryogenic Treated Tool

S. Balakrishnan[1]*, K. Senthilkumar[2] and S. Thirumalai Kumaran[2]

[1]*Department of Mechanical Engineering, Francis Xavier Engineering College, Tirunelveli, Tamil Nadu, India*
[2]*Department of Mechanical Engineering, PSG Institute of Technology and Applied Research, Coimbatore, Tamil Nadu, India*

Abstract
Wide arrays of high-temperature materials, including nickel-based alloys, are used to manufacture components of aeroengines, eyeglass frames, and musical instruments. Monel 400 is an important nickel-based alloy and a combination of nickel, copper, iron, and manganese. Monel 400 is the hardest material, stronger than steel and a corrosion-resistant material. It has wide engineering applications. Despite this, the study of machining Monel 400 remains difficult due to its unique properties (renowned for its toughness, maintained over a considerable range of temperature). Similar to this, the manufacturing environment is very concerned about the ongoing creation of carbide milling cutters and their coating technology. For a milling cutter to create a good surface finish and longer tool life at a lower cost, cryogenic coating is crucial. In this research, the use of uncoated and cryogenically coated carbide tools to perform end milling on Monel 400 plates with various process parameters has been examined. Factors like depth of cut, cutting speed, and feed rate as well as the best machining conditions for surface roughness (Ra) were determined. The studies were designed and executed using an L9 orthogonal array.

Keywords: Monel 400, EDM milling, cryogenic tool, Taguchi

**Corresponding author*: krishbhel@gmail.com

S. Thirumalai Kumaran and Tae Jo Ko (eds.) Sustainable Machining and Green Manufacturing, (305–316) © 2024 Scrivener Publishing LLC

15.1 Introduction

Monel is an alloy made up of two or more elements, one of which is a metal, and the end product has metallic qualities. This is a nickel-based material with copper, iron, and manganese compounds mixed in [1]. This alloy is indeed harder and stronger than steel, and it is widely recognized for its corrosion resistance and rust resistance [2]. Monel is primarily used in chemical and marine units [3]. This material is also used in the manufacturing of hydrocarbon processing equipment, valves, taps, shafts, fittings, fasteners, heat exchangers, and eyeglass frames. This provides good resistance to brackish or seawater. This exhibits a high level of acid resistance when de-aerated, especially to hydrochloric and hydrofluoric acids. At room temperature, the alloy is slightly magnetic. However, Monel is generally more expensive than stainless steel [4]. Generally, the milling operation is very suitable for machining these alloys. In general, milling is an excellent method for machining these alloys. Most commonly, milling is thought of as a material removal technique in which unwanted material can be removed from a part for various purposes. Milling is commonly used to generate level surfaces. The components of a milling operation are a workpiece, a fixture, and a cutter. End mills are used for a variety of operations to shape and drill holes in a workpiece, including milling, profiling, contouring, slotting, counter boring, drilling, and reaming. On the face and body edge of these are cutting teeth. This has numerous cutting options for a wide range of materials. During milling, the workpiece is attached to a fixture, which is mounted on the platform of the milling machine. Milling machines also have cutters, which are cutting tools with sharp teeth that rotate at high speeds. By feeding the workpiece into the spinning cutter, material is chopped away in the form of little chips. The high tolerances and surface finishes that milling can provide make it a perfect method for adding precision features to an existing item. It is widely known that Monel 400 is the strongest metal. This is extremely difficult to cut and has an irregular surface finish. In order to overcome these issues, Selvakumar *et al.* [5] employed the L18 orthogonal array to investigate die corner accuracy using wire-electrical discharge machining and found that pulse parameter variation had an impact on corner accuracy. Finally, Taguchi additive model predictions were used to construct a sample technology table for a certain machining situation. Evans [6] looked into the nickel content of a nickel-copper-zinc alloy. Nickel is a silvery white metal that is more difficult to work with than iron. Sulfide and lateritic are the two forms that may be profitably mined. It is rarely recycled on its own, but it is frequently combined with other metals. Melting the metal mixture and producing a remelt ingot is a frequent recycling procedure. Nickel and its alloys have numerous applications. The study predicts that nickel and its alloys will be essential for the development of renewable energy. Jet planes, spaceships, and missiles all use super alloys in their construction. Nickel-based batteries are utilized in hybrid automobiles. It is also used as a catalyst and in coin production. Furthermore, the impact of ECM of Monel 400 has been studied by Kalaimathi *et al.* [7] using the parameters of applied voltage, interelectrode, and electrolyte. The central composite design based on RSM is utilized as an experimental strategy. Based on the parameter interactions, the effect and optimum parameters are discovered. Kumar *et al.* [8] researched into the four wire electrical discharge machining (WEDM) input parameters: pulse off time, pulse on time, discharge current, and servo voltage for machining rate and surface roughness. Machining rate was calculated using a quadratic model, and surface roughness was calculated using a two-factor interaction. Despite the rough cut's large discharge energy, a single trim cut can dramatically improve the surface finish. In the WEDM operation, Rajyalakshmi [9] assessed the essential process parameters of wire feed on material removal rate, peak current, pulse off time, surface roughness, and pulse on time. The settings were examined using a multiresponse particle swarm optimization approach. The impact of significant variables is also examined using the analysis of variance. As a result, it has been confirmed that the ideal set of machining parameters can improve machining condition. Giridharan *et al.* [10] investigated the use of cubic boron nitride (CBN) coatings in end mill cutters to increase the surface finish and tool life at a low cost. The effects of many variables on surface roughness and material removal rate (MRR) can vary, including speed, feed, and cut depth. Taguchi method is used to optimize the process. Surface roughness is predicted by the regression, which has a very excellent fit. In addition, Sajgalik *et al.* [11] evaluated the use of a ceramic milling tool to improve surface quality cutting conditions. Parida and Maity [12] focused on the three nickel-based alloys for chip manufacturing using flame heating in the hot turning process. Cutting force, tool wear, surface roughness, tool life, and chip tool contact have all been measured at room and hot temperatures, with significant decreases in cutting force, tool wear, chatter generation, tool life, and chip tool contact as a result. Inconel 718 and Monel 400 were employed by Khan *et al.* [13] as two separate potential materials for the electrical discharge machining procedure. To calculate the experimental results of both materials, machining time is held constant while electrical parameters are changed to create plasma spark for the diffusion process. The investigation confirmed that the drilling is completely round, and there is no deviation. Verma *et al.* [14] explored whether there is a significant harm to the environment and the workers' health when cutting fluids are used in the machining process. Minimum quantity lubrication (MQL) influences the surface integrity of the machined workpiece and minimizes the amount of cutting fluid and machining forces needed during an operation. According to the investigation, the workpiece turned with MQL had a lower hardness crust than a workpiece turned without MQL based on the microhardness profile in the subsurface region. This would result in a longer fatigue life for the hard-turned component.

According to the literature, high spindle speed can reduce medium surface roughness to low surface roughness, which is a concern when cutting Monel 400. In this study, the Monel 400 workpiece is cut with a cryogenic treated carbide end mill cutter tool to reduce residual stress, increase wear resistance. And optimum parameters are evaluate for milling machining such as depth of cut, feed rate and spindle speed and their surface roughness.

15.2 Materials and Methods

15.2.1 Fabrication of Workpieces

The nickel-copper alloy Monel alloy 400 (Cu = 28%–34%; Fe = 2.5% max; Mn = 2% max; Si = 0.5% max; C = 0.3% max; S = 0.024%; Ni = remaining) exhibits remarkable corrosion resistance in a variety of hostile environments, from sub-zero to 800°F. It is a single-phase solid-solution alloy. It is one among the few alloys that can handle fluorine, hydrofluoric acid, and hydrogen fluoride. In all concentrations up to the boiling point, hydrofluoric acid has displayed excellent resilience. Under reducing circumstances, alloy 400 is resistant to a variety of sulfuric and hydrochloric acids. There are numerous different varieties of Monel accessible commercially, including alloy 401, alloy 400, Monel 404, alloy K500, and alloy R405. Super alloy Monel is another name for Monel 400. Hexagonal, round, tube, pipe, plate, strip, sheet, and wire are a few of the often-encountered shapes for this alloy. Copper and nickel concentrations in Monel are the same as those in nickel ore produced in Ontario. This alloy has a high level of corrosion resistance. It is made up of two crystalline solids that combine to form a single solid. These materials are known for their acid and alkali resistance, as well as their excellent mechanical strength and ductility. Monel 400 retains its strength

Table 15.1 Properties of Monel 400.

Density	8.8 g/cm^3
Tensile strength	517–620 Mpa
Yield strength	172–345 Mpa
Modulus of elasticity	179 Gpa
Poisson's ratio	0.32

even at sub-zero temperatures; hence, it is employed in these applications. The property of the selected material is described in Table 15.1.

15.2.2 Taguchi Experimental Design

The Taguchi approach, which is frequently used in production, allows for parameter optimization in machining via turning, milling, and boring. EDM wire cut EDM welding, grinding, and other processes are performed with the fewest number of specimens possible, resulting in the lowest total cost and time [15]. Some of the methodologies employed are the Taguchi method, mathematical programming, and response surface methodology. Taguchi's approaches are mostly employed for research purposes [16]. Surface finish is the study's output parameter, with depth of cut, feed, and velocity serving as its three input parameters. Three levels of variation are applied to the three parameters. Taguchi's 9 trial experiments with three parameters at three levels are used to conduct the research. Variable input parameter values are described in Table 15.2.

15.2.3 CNC Milling Operation

CNC milling process involves computer-controlled controls and rotating multipoint cutting tools used to remove material from a workpiece so that

Table 15.2 Input parameter values.

Cutting speed (rpm)	Feed rate (mm/min)	Depth of cut (mm)
60	25	0.4
80	50	0.6
100	75	0.8

a custom-designed part can be produced [17]. Adaptability is a strong suit of milling machines. Usually, it is used to generate flat surfaces, although it can also make uneven ones. It is also capable of drilling, boring, cutting gears, and creating slots. The type of milling machine that is most frequently seen in student shops is a vertical spindle milling machine with a swiveling head. Face milling creates flat surfaces and allows equipment to work to the exact length needed. The feed can be horizontal or vertical in face milling. A standard milling machine will not be able to deliver an exact value in a timely manner; however, CNC milling will. Figure 15.1 presents the photograph of CNC milling machine. Table 15.3 represents specifications of the selected milling machine.

Figure 15.1 CNC milling machine.

Table 15.3 Specifications of the CNC milling machine.

Components	**Specifications**
Travel X Y and Z axis	700 mm*320 mm*250 mm
T slot table size	1,000 mm*250 mm
Rapid feed rate	3,500 mm/min
Spindle speed	12,000–24,000 rpm
Machine weight	250 kg
Drive type	75 V DC stepper motor
Coolant	Fog buster sprayer system and run flood coolant

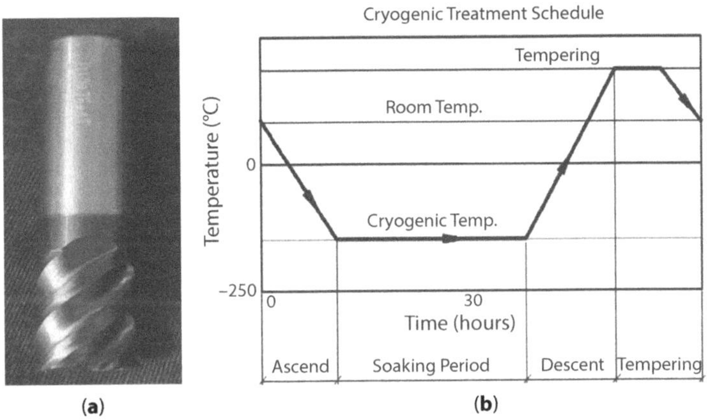

Figure 15.2 (a) Carbide end mill cutter; (b) cryogenic treatment schedule.

15.2.4 Cutter Selection

High-speed steel (HSS) is less rigid than solid carbide. It may be utilized at high speeds on cast iron, nonferrous metals, plastics, and other hard-to-machine materials and is heat resistant. Carbide end mills are more robust than HSS and can be operated 2–3 times quicker. They can also be used with high-pressure lubricants including water, water-soluble oil, and high-flashpoint oil. Monel 400 is best cut with a carbide end mill cutter. Figure 15.2 (a) depicts the cutter for a carbide end mill. Cryogenic solution is used to cure carbide end mill cutters at temperatures below -190°C. The residual stress in a cryogenic treated tool is removed, and wear resistance and composites are improved. The tool should be immersed in the solution for at least 48 hours. The capacity of cryogenic treatment is used to boost corrosion resistance. The cryogenic treatment schedule is presented in Figure 15.2 (b).

15.3 Results and Discussion

The design experiments that made use of orthogonal arrays led to the use of CNC milling. The surface finish determines the smoothness of a surface. If the surface generated is rough, it indicates that the surface finish will have an impact on the part's quality. The "Ra" value should have been kept to a minimum in order to get a satisfactory surface finish. A Perthometer is used to determine the roughness of the surface. Figure 15.3 depicts the variation of values while doing experiments.

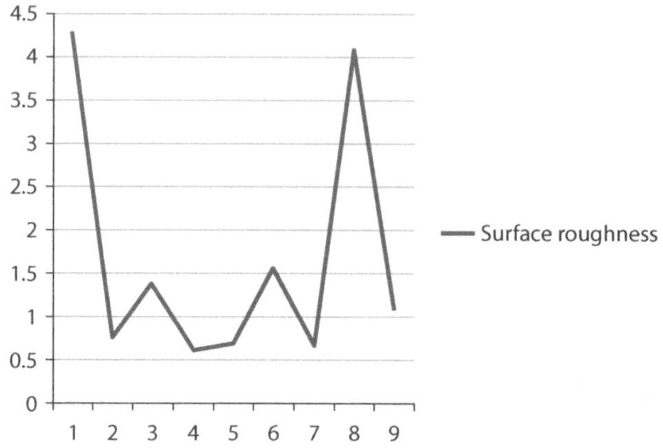

Figure 15.3 Surface roughness of the conducted experiments.

A static design can be described using a combination model. It is worth mentioning that in a combination model, the overall effect of all components equals the sum of individual factor impacts. In a static design, cross product terms involving at least two factors are not taken into consideration

Table 15.4 Responses of S/N ratio and means.

Responses for S/N ratio			
Level	**Cutting speed**	**Feed rate**	**Depth of cut**
1	-4.295	-1.530	-9.511
2	1.238	-2.221	1.881
3	-3.250	-2.557	1.323
Delta	5.533	1.027	11.392
Responses for means			
1	2.1057	1.8157	3.2780
2	0.9520	1.8513	0.8317
3	1.9640	1.3547	0.9120
Delta	1.1537	0.4967	2.4463
Rank	2	3	1

because it is assumed that individual factor effects are linear. The forecasted signal-to-noise (S/N) ratio and responses for means for smaller is better is presented in Table 15.4. The main effect of surface roughness with respect to S/N ratio and means is plotted as Figure 15.4.

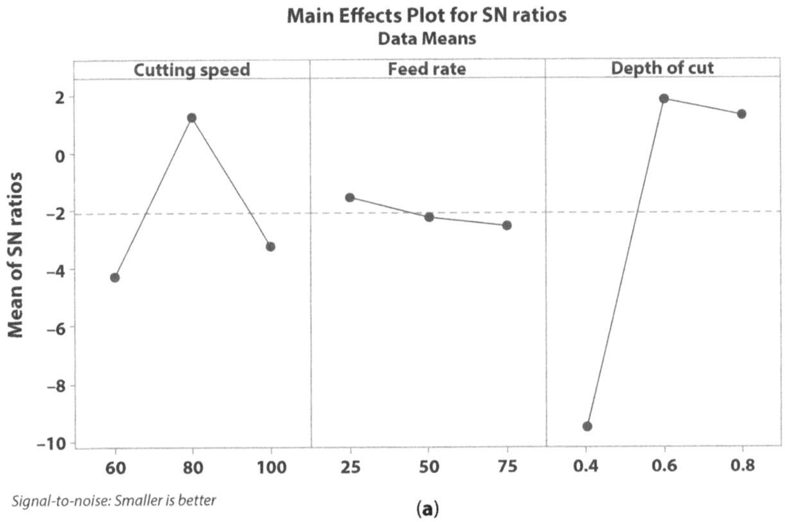

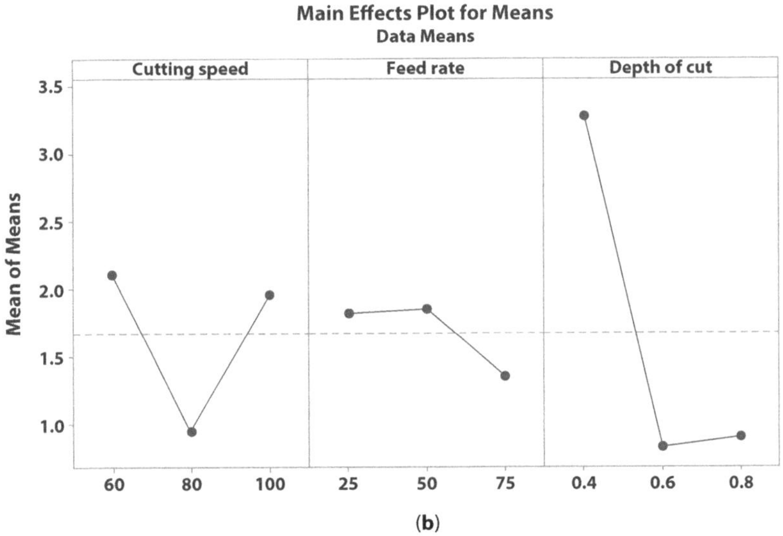

Figure 15.4 Effects of surface roughness (a) S/N ratio plot; (b) means plot.

After conducting the trials, it had been discovered that the depth of cut is the most important parameter in milling operations for achieving good surface roughness on Monel 400 plates. In surface roughness prediction, smaller is better. Monel 400 steel plate has an excellent machinability and produces a fair surface finish. The best settings for milling Monel 400 plates to achieve a good surface finish have been found to be spindle speed of 900 rpm, feed rate of 25 mm/min, and depth of cut of 0.6 mm. During the milling operation of Monel 400 steel plates, an excellent surface finish and minimal wear are attained. All parameters interact and are reliant on the

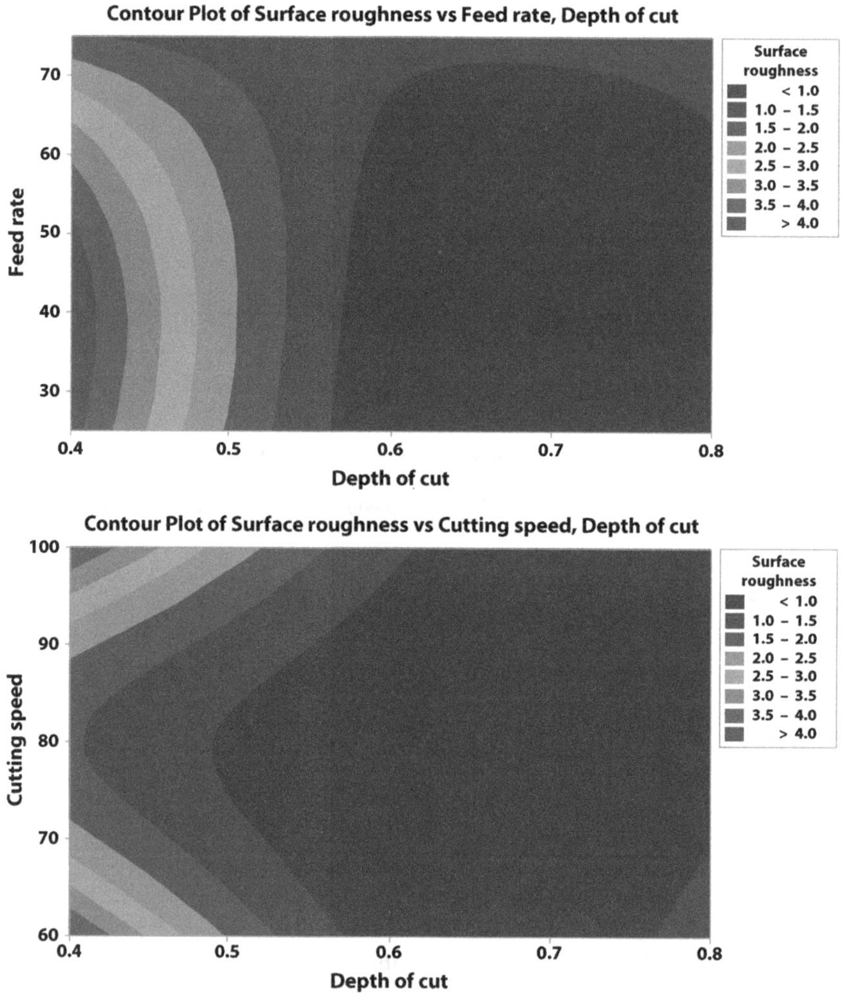

Figure 15.5 Surface roughness is represented in a contour plot.

milling operation during the milling process. Figure 15.5 depicts a contour plot representation and three-dimensional relationship between milling and surface roughness factors in the Monel 400 milling process, indicating three levels of spindle speed and the first level of cut depth.

15.4 Conclusion

The surface roughness of Monel 400 alloy during CNC milling had been examined in this study. The cryogenic treated carbide tool has been utilized for machining. The Taguchi technique is utilized to divide the control parameters into machine-controlled and uncontrollable elements, and the impact on the responses has been explored. Surface roughness has been most affected by the wire feed rate and depth of cut. On increasing surface roughness accuracy, the impact of input parameter interaction has been studied. The machining combinations were then classified based on the criteria and examined using the contour plot technique. Finally, it had been revealed that a depth of cut of 0.6 mm, a feed rate of 25 mm/min, and a spindle speed of 900 rpm are the optimum parameters for milling Monel 400 plates to get a good surface finish.

References

1. Bäumel, A., Drodten, P., Heitz, E. *et al.*, Nickel-copper alloys, in: *Corrosion Handbook*, Wiley-VCH Verlag GmbH & Co. KGaA, Weinheim, Germany, 2008.
2. Armstrong, R.W. and Walley, S.M., High strain rate properties of metals and alloys. *Int. Mater. Rev.*, 53, 105–128, 2008.
3. Rubow, K.L., Porous metal for aerospace. *Adv. Mater. Process*, 167, 26–28, 2009.
4. Dutta, R.S., Corrosion aspects of Ni–Cr–Fe based and Ni–Cu based steam generator tube materials. *J. Nucl. Mater.*, 393, 343–349, 2009.
5. Selvakumar, G., Sarkar, S., Mitra, S., Experimental investigation on die corner accuracy for wire electrical discharge machining of Monel 400 alloy. *Proc. Inst. Mech. Eng. B J. Eng. Manuf.*, 226, 1694–1704, 2012.
6. Evans, T.E., Nickel and nickel alloys, in: *Corrosion*, pp. 4:116–4:156, Elsevier, Amsterdam, Netherlands, 1994.
7. Kalaimathi, M., Venkatachalam, G., Makhijani, N.P. *et al.*, Investigations on machining of monel 400 alloys using electrochemical machining with sodium nitrate as electrolyte. *Appl. Mech. Mater.*, 592–594, 467–472, 2014.

8. Kumar, V., Kumar, V., Jangra, K.K., An experimental analysis and optimization of machining rate and surface characteristics in WEDM of Monel-400 using RSM and desirability approach. *J. Ind. Eng. Int.*, 11, 297–307, 2015.
9. Rajyalakshmi, G., Modeling and multi-objective optimization of WEDM of commercially monel super alloy considering multiple users preferences. *J. Pharm. Sci. Res.*, 8, 8, 902–908, 2016.
10. Giridharan, R., Kumar, P., Vijayakumar, P. *et al.*, Experimental investigation and design optimization of end milling process parameters on monel 400 by taguchi method. *Int. J. Mech. Eng. Technol.* [Internet], 8, 113–122, 2017.
11. Sajgalik, M., Czan, A., Drbul, M. *et al.*, Identification of technological parameters when machining Ni-alloys by monolithic ceramic milling tool. *Proc. Manuf.*, 14, 51–57, 2017.
12. Parida, A.K. and Maity, K., Comparison the machinability of inconel 718, inconel 625 and monel 400 in hot turning operation. *Eng. Sc. Technol. Int. J.*, 21, 364–370, 2018.
13. Khan, M.A., Gokul, A.K., Dharan, M.P.B. *et al.*, Machinability of nickel based alloys using electrical discharge machining process. *IOP Conf. Ser. Mater. Sci. Eng.*, 346, 012044, 2018.
14. Verma, J.K., Bartarya, G., Bhaskar, J., Effect of minimum quantity lubrication on tool wear and surface integrity during hard turning of EN31 steel. In: Shunmugam, M.S. and Kanthababu, M. (Eds.), *Advances in Forming, Machining and Automation. AIMTDR 2018*. Lecture Notes on Multidisciplinary Industrial Engineering. Springer, Singapore. https://doi.org/10.1007/978-981-32-9417-2_16
15. Lee, H.T. and Yur, J.P., Characteristic analysis of EDMed surfaces using the Taguchi approach. *Mater. Manuf. Process.*, 15, 781–806, 2000.
16. Basavarajappa, S. and Chandramohan, G., Wear studies on metal matrix composites: A Taguchi approach. *J. Mater. Sci. Technol.*, 21, 845–850, 2005.
17. Afkhamifar, A., Antonelli, D., Chiabert, P., Variational analysis for CNC milling process. *Proc. CIRP*, 43, 118–123, 2016.

Index

Printed in the USA/Agawam, MA
September 16, 2026

910661.091